Quantum Theory of Polymers

NATO ADVANCED STUDY INSTITUTES SERIES

Proceedings of the Advanced Study Institute Programme, which aims at the dissemination of advanced knowledge and the formation of contacts among scientists from different countries

The series is published by an international board of publishers in conjunction with NATO Scientific Affairs Division

A Life Sciences B Physics	Plenum Publishing Corporation London and New York
C Mathematical and Physical Sciences	D. Reidel Publishing Company Dordrecht and Boston
D Behavioral and Social Sciences	Sijthoff International Publishing Company Leiden
E Applied Sciences	Noordhoff International Publishing Leiden

Series C – Mathematical and Physical Sciences

Volume 39 – Quantum Theory of Polymers

Quantum Theory of Polymers

Proceedings of the NATO Advanced Study Institute on Electronic Structure and Properties of Polymers held at Namur, Belgium, 31 August – 14 September, 1977

edited by

JEAN-MARIE ANDRÉ
Laboratoire de Chimie Théorique Appliquée
Facultés Universitaires de Namur, Namur, Belgium

JOSEPH DELHALLE
Laboratoire de Chimie Théorique Appliquée
Facultés Universitaires de Namur, Namur, Belgium

JÁNOS LADIK
Lehrstuhl für Theoretische Chemie
Universität Erlangen-Nürnberg, Erlangen, B.R.D.

D. Reidel Publishing Company

Dordrecht:Holland / Boston:U.S.A.

Published in cooperation with NATO Scientific Affairs Division

Library of Congress Cataloging in Publication Data

Nato Advanced Study Institute on Electronic Structure and Properties of Polymers, Facultés universitaires de Namur, 1977.
Quantum theory of polymers.

(NATO advanced study institutes series: Series C, Mathematical and physical sciences; v. 39)
Bibliography: p.
Includes index.
1. Polymers and polymerization — Congresses. 2. Quantum chemistry — Congresses. I. André, Jean-Marie, 1944- II. Delhalle, Joseph, 1945- III. Ladik, János. IV. North Atlantic Treaty Organization. V. Title. VI. Series.
QD380.N36 1977 547'.84 77-26725
ISBN-13: 978-94-009-9814-8 e-ISBN-13: 978-94-009-9812-4
DOI: 10.1007/978-94-009-9812-4

Published by D. Reidel Publishing Company
P.O. Box 17, Dordrecht, Holland

Sold and distributed in the U.S.A., Canada, and Mexico
by D. Reidel Publishing Company, Inc.
Lincoln Building, 160 Old Derby Street, Hingham, Mass. 02043, U.S.A.

Softcover reprint of the hardcover 1st edition 1978

TABLE OF CONTENTS

PREFACE

The NATO Advanced Study Institute on "Electronic Structure and Properties of Polymers" was held at the Facultés Universitaires de Namur (F.U.N.) from August 31 till September 14, 1977.

We wish to express our deepest gratitude to the Scientific Affairs Division of NATO, the main sponsor of this Institute, and to the Facultés Universitaires Notre Dame de la Paix and their Board who gave us generous financial help as well as accommodation for the School. Our sincere thanks to Dr Tilo Kester from the NATO Scientific Affairs Division and Prof. Roger Troisfontaines, Rector and President of the Facultés Notre Dame de la Paix.

This volume contains the main lectures of the Institute. It is our great pleasure to thank all the lecturers for their most excellent and interesting lectures and for the clarity of their manuscripts.

During the School the participants and lecturers felt that though there has been considerable progress in recent years in the methods applicable to the quantum theoretical treatment of polymers, not very many calculations of their properties have been performed. This is the reason that the title of this volume has been changed to "Quantum Theory of Polymers".

The School started with a review by André and Delhalle of the SCF LCAO crystal orbital (CO) theory in its *ab initio* and semiempirical forms with applications to simple polymers by calculating, besides the band structures, the density of states and some other physical properties. Veillard discussed in detail the techniques of integrations which play a key role in *ab initio* MO and CO calculations. Verbist gave a review of the theory and methods of X-ray photoelectron spectroscopy and its applications to the electronic structure of polymers which can provide a good test for the quality of theoretical band structure calculations.

The next step was to discuss the problem of the correlation energy of solids and the applicability of the methods developed to polymers.

March treated in a very elegant way the problem of correlation in extended systems taking the ground state energy as a functional of the density and also applying different model Hamiltonians. Collins discussed the problem of excited states in solids and polymers using Green's function technique, treating the long range correlation effects on the basis of the electronic polaron model, and also the short range exciton correction. Čížek gave a detailed account of the coupled cluster expansion method (using diagrammatic techniques) which seems to be very promising for the treatment of the short range correlation in the ground state of polymers. Harris discussed the Fourier representation of the total energy of solids and different aspects of the correlation problem in the ground state of extended systems. Brandow combined modern (linked cluster) perturbation theory with a diagrammatic technique developed both for closed- and open-shell systems expressions for the correlation energy, and derived very useful effective π-electron Hamiltonians in the polymer theory.

A further topic of the School was the theory of disordered systems and its application to surface states and chemisorption. Martino derived the coherent potential approximation (CPA) with the aid of Green's function formalism and outlined the possibilities of its applications to polymers. McCubbin presented some other methods for the treatment of structurally disordered systems taking as an example the case of the rotational disorder of polyethylene. Del Re, using simple models, discussed the end effects and the effect of chemisorption (also in a time-dependent form) in the case of a linear chain.

Schuster reviewed the different methods for the treatment of hydrogen bonds and intermolecular interactions between single molecules and their applicability to extended systems (clusters). Ladik in his first series reported *ab initio* calculations on some periodic DNA and protein models and the correction of the resulting bands for excitonic and long range correlation effects. In his second series, calculations on $(SN)_x$ and on the TCNQ-TTF system were reported and the applicability of the CPA method for aperiodic polymers was discussed.

Csavinszky reviewed in his first series the theory of the transport properties of semiconductors based on Boltzmann's transport equation, and in his second series discussed the quantum theory of the mechanical properties of semiconductors. In both cases the applicability of the theory to polymers was pointed out. Finally Suhai in his lecture extended the transport theory of polymers for the case of narrow energy bands and presented different applications to biopolymers and highly conducting polymers.

The lectures of the Institute were kept on a rather high level and they actually covered nearly all the important aspects of the quantum mechanical treatment of extended systems and the most representative applications to polymers.

The secretarial burden fell on Mrs Marie-Claude André, Miss Patricia Lonnoy and Miss Sabine Patzak who worked expertly and smilingly throughout the Institute until the final preparation of these proceedings. We wish to thank them most heartedly. Mr G. Kleiner, Dr J. Fripiat and Mr D. Vercauteren did a wonderful job in arranging all the practical details. Special thanks are due to our hostesses: Claudine Derouane and Chantal Legrand who took such good care of everyone during the Institute.

Finally we gratefully acknowledge the help of all the other people - members of the Chemistry Department, students, etc. - who took an active and useful part in arranging the many practical details and the splendid excursions during the Institute.

Jean-Marie André
Joseph Delhalle
János Ladik

Erlangen, Namur, September 1977

LIST OF THE PARTICIPANTS

1. J. LEJEUNE (Belgium)
2. I.P. BATRA (India)
3. P. BULLIVANT (Great Britain)
4. D.P. VERCAUTEREN (Belgium)
5. J.M. ANDRE (Belgium)
6. F. LOIZILLON (France)
7. P. OTTO (West Germany)
8. M.L. ALMEIDA (Portugal)
9. M.C. ANDRE (Belgium)
10. J. LADIK (West Germany)
11. Y. SCHLESINGER (Israël)
12. J.E. ROOTS (Sweden)
13. P. WINKLER (West Germany)
14. H. BOUDEVSKA (Bulgaria)
15. J.A.G. DELHALLE (Belgium)
16. G. ORLANDI (Italy)
17. C. LAMOTTE (Belgium)
18. R.C. AHUJA (The Netherlands)
19. J.D. ANDRADE (U.S.A.)
20. M. ROHDE (West Germany)
21. D.L. VUKOVIC (Yugoslavia)
22. O. B.NAGY (Belgium)
23. R.D. SINGH (West Germany)
24. F. BIONDI (Italy)
25. N.H. MARCH (Great Britain)
26. I. BOZOVIC (Yugoslavia)
27. T.C. COLLINS (U.S.A.)
28. P. SCHUSTER (Austria)
29. Mr. KASPAR (West Germany)
30. M.P.S. COLLINS (Canada)
31. F. MARTINO (U.S.A.)
32. J.L. HOUBEN (Italy)
33. T. MAH (Great Britain)
34. F. HARRIS (U.S.A.)
35. P. CSAVINSZKY (U.S.A.)
36. R.A. WHITESIDE (U.S.A.)
37. Mrs. GRIVSKY (U.S.A.)
38. M. SEEL (West Germany)
39. P. LONNOY (Belgium)
40. J. MULLER (Sweden)
41. G. DEL RE (Italy)
42. D.R. BECK (Greece)
43. E.M. GRIVSKY (U.S.A.)
44. C. DEROUANE (Belgium)
45. S.F. ABDULNUR (West Germany)
46. J. RIESS (Switzerland)
47. C.E. THOMPSON (U.S.A.)
48. G. KAPSOMENOS (Greece)
49. A. KARPFEN (Austria)
50. J. KOUTECKY (West Germany)
51. C. DEMANET (Belgium)
52. Mr. CSAVINSZKY (Hungary)
53. A. VEILLARD (France)
54. V. BONACIC-KOUTECKY(W.Germany)
55. C. LEGRAND (Belgium)
56. B. CSAVINSZKY (U.S.A.)
57. J.L.E. BREDAS (Belgium)
58. C. FURTADO (Portugal)
59. W.L. Mc CUBBIN (Iran)
60. J.J. VERBIST (Belgium)

LCAO METHODS FOR BAND STRUCTURE CALCULATIONS OF POLYMERS

Jean-Marie ANDRE and Joseph DELHALLE

Laboratoire de Chimie Théorique Appliquée
Facultés Universitaires N.D. de la Paix
61, rue de Bruxelles, B-5000 - Namur (Belgium)

I. INTRODUCTION

The electronic structure of polymers has become in the last years a subject of great interest. Polymers as constituents of plastics are important from the point of view of structural materials; biopolymers play a fundamental role in life process and molecular crystals built up from polymers like $(SN)_x$ or the TTF-TCNQ system exhibit special physical characteristics. It is now hoped that more emphasis will be given to the synthesis of those polymers which present specific features.

Since 1974, the quantum theory of polymers became basically important with the discovery of superconductivity in $(SN)_x$ chains and by the great number of papers on the experimental investigation of the TTF-TCNQ system. It is our opinion that a detailed knowledge on the band structure of polymers must be obtained for a proper understanding of conduction and mechanical properties as well as of the various collective states.

The aim of this paper is to present the general LCAO theory, actually used, for studying the electronic properties of regular macromolecules. Since bonds in a polymer are not different from usual bonds in organic molecules, the methods already proved successfull in the investigation of the electronic structure of molecules can be used for describing the electronic properties of polymers. In this sense, the one-electron energy bond model has been and is still one of the mosed used approaches to the description of the electronic states in solids and stereoregular polymers.

J.-M. André et al. (eds.), Quantum Theory of Polymers, 1-22.

In using a molecular approach, electronic states of large chains can be visualized by drawing the orbital energies (filled and empty) associated with chains of increasing lengths. In a series of homogeneous molecules, the number of energy levels increases with the molecular size and correspondingly the distance between these energy levels vanishes. Such a situation is illustrated in Figure I for the series : methane, ethane, propane, butane, pentane, ... polyethylene.

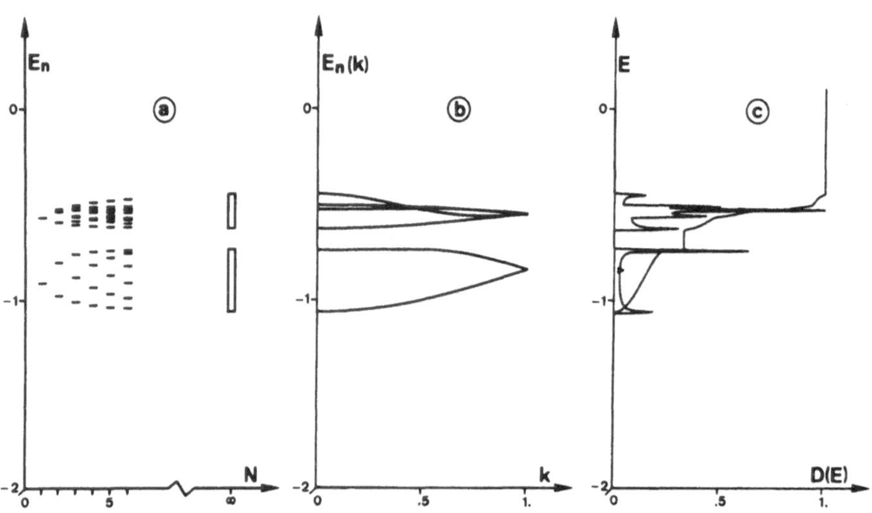

Figure I : a) distribution of the valence levels going from methane to polyethylene (extended Hückel), in a.u.
b) band structure of polyethylene (extended Hückel-CO)
c) density of electronic states associated to the band structure shown in I-b.

For the infinite regular chain, another description can be given. Due to the one-dimensional periodicity, the monoelectronic levels (orbital energies in the molecular description) may be represented as a multivalued function of a reciprocal wave number defined in the inverse space dimension. The set of all those branches (energy bands) plotted with respect to the reciprocal wave number (k-point) in a well defined region of the reciprocal space (first Brillouin zone) is the band structure of the polymers. In the usual terminology, we note the analogy between the occupied levels and the valence bands, the unoccupied levels and the conduction band. As an example, the occupied energy scale of polyethylene is made up from two bands of mainly $1s_C$ character, two bands describing the localized C-C bonds and four bands representing the C-H energy terms.

Besides the representation of the molecular levels into energy bands, the right-hand side of Figure I shows a third representation which is the so-called density of electronic states of the polymer. The density function indicates the variation of the number of allowed energy levels with the energy within the band, i.e. the number of allowed electron states per unit energy range.

The paper is divided in two parts; part one gives the summary of principles to be used for systems with periodical properties, indicates the general L.C.A.O. periodical forms and summarizes how to obtain band structures from the knowledge of basic atomic orbitals. A brief review of the main procedures and approximations used is also given. Part two introduces basic quantities associated with the band structure.

II. THE LCAO METHOD FOR POLYMERS

II.1. *General formalism* [1-3]

The aim of the present part is to deduce the LCAO equations for the case of infinite periodic one-dimensional polymers. This is an interesting case of a one-dimensional periodicity although the wave-functions are truly three-dimensional.

The one-electron theory (which means the band theory in solid state physics) defines a set of polymeric orbitals to represent the wave-functions of a single electron in a periodic potential of the nuclei and the other electrons. The optimal set of polymer orbitals for a given atomic basis set is constructed in the usual way by solving a set of Hartree-Fock equations. The SCF monoelectronic operator has an explicit form :

$$h(i) = -\frac{1}{2}\nabla^2(i) - \sum_h^N \sum_\alpha^A \frac{Z_\alpha}{|r_i - r_\alpha - ha|} + \sum_{n'} \sum_{k'} \{2\, J_{k'n'}(i) - K_{k'n'}(i)\} \qquad (1)$$

with the coulomb and exchange terms being detailed as

$$J_{k'n'}(i)\ \phi_n(k,r_i) = \int \phi^*_{n'}(k',r_j)\ \frac{1}{r_{ij}}\ \phi_{n'}(k',r_j)\,dv_j\ \phi_n(k,r_i) \qquad (2)$$

$$K_{k'n'}(i)\phi_n(k,r_i) = \int \phi^*_{n'}(k',r_j) \frac{1}{r_{ij}} \phi_n(k,r_j)dv_j \; \phi_n{}'(k',r_i) \tag{3}$$

The terms involved in equation (1) express respectively the kinetic energy operator, the attraction of a single electron with all nuclei, α, centered in all cells, h, the averaged electrostatic potential of all electrons and the averaged exchange interaction. N is the total number of interacting cells.

From Bloch's theorem [4], we know that a polymeric orbital has the LCAO form :

$$\phi_n(k,r) = N^{-\frac{1}{2}} \sum_j^N \sum_p^\omega e^{ikja} C_{np}(k)\chi_p(r-r_\alpha-ja) \tag{4}$$

In this expression, ω refers to the number of atomic functions (basis set) in one unit cell and a is the length of the cell. By forming the expectation value of the monoelectronic Hartree-Fock operator and by using the variational procedure for the LCAO coefficients $C_{np}(k)$, we obtain for the whole polymer the following system of equations :

$$\sum_p C_{np}(k) \{\sum_j e^{ikja} h^j_{pq} - E_n(k) \sum_j e^{ikja} S^j_{pq}\} = 0 \tag{5}$$

The compatibility condition of the former system of equations :

$$|\sum_j e^{ikja} h^j_{pq} - E(k) \sum_j e^{ikja} S^j_{pq}| = 0 \tag{6}$$

will produce in the reduced scheme the band structure, $\{E_n(k)\}$, as a multivalued function of k. We note that h^j_{pq} is a matrix element of the monoelectronic operator h(i) between the atomic orbital χ_p centered in the origin unit cell and the atomic orbital χ_q centered in cell j. S^j_{pq} has the same meaning for the unit operator. Both matrix elements decrease exponentially with the distance between the orbitals giving rise to a natural convergency of the summation over cells appearing in the secular systems (5) and dterminants (6).

In the Hartree-Fock self-consistent procedure, the matrix elements h^j_{pq} have the general form :

$$h^j_{pq} = T^j_{pq} - \sum_h \sum_\alpha Z_\alpha U^j_{pq}(h,\alpha) + \sum_k \sum_n \sum_h \sum_l \sum_r \sum_s C^*_{nr}(k)\, C_{ns}(k)$$

$$e^{-ika(h-l)} \{2(\begin{smallmatrix} oj \\ pq \end{smallmatrix}|\begin{smallmatrix} hl \\ rs \end{smallmatrix}) - (\begin{smallmatrix} oh \\ pr \end{smallmatrix}|\begin{smallmatrix} jl \\ qs \end{smallmatrix})\} \tag{7}$$

when defining

- the kinetic integrals

$$T^j_{pq} = -\frac{1}{2}\int \chi_p(r-r_\alpha)\ \nabla^2(r)\ \chi_q(r-r_\beta-ja)\ dv \tag{8}$$

$$= \langle \chi^o_p \mid T \mid \chi^j_q \rangle$$

- the nuclear attraction

$$U^j_{pq}(h,\alpha) = \int \chi_p(r-r_\beta)\ \frac{1}{|r-r_\alpha-ha|}\ \chi_q(r-r_\gamma-ja)dv \tag{9}$$

$$= \langle \chi^o_p \mid V^h \mid \chi^j_q \rangle$$

and the electron repulsion integrals

$$\left(\begin{matrix}oj\\pq\end{matrix}\middle|\begin{matrix}hl\\rs\end{matrix}\right) = \iint \chi_p(r_1-r_\alpha)\chi_q(r_1-r_\beta-ja)\ \frac{1}{|r_1-r_2|}\ \chi_r(r-r_\gamma-ha)$$

$$\chi_s(r-r_\delta-la)\ dv_1dv_2 \tag{10}$$

In all of these equations, α,β,γ and δ refer to atomic centers.

The iterative part of the computation involves elements of the so-called density matrix, R^j_{pq}, they are computed by integration over the first Brillouin zone of the polymer

$$R^j_{pq} = \sum_{kn} C^*_{np}(k)\ C_{nq}(k)\ e^{+ikja} = \frac{1}{L_{RC}}\int \sum_n C^*_{np}(k)C_{nq}(k)e^{ijka}\ dk \tag{11}$$

such types of integration extended over the first Brillouin zone are frequent in the quantum theory of polymers. They generally require numerical techniques; Simpson or Gaussian quadrature formulas are often used.

As already found in equation (1), Fock matrix elements consist of three physically different terms : a kinetic contribution, an electrostatic attraction and repulsion contribution and an exchange part.

- The kinetic term :

$$T^j_{pq} = \int \chi^o_p(r)\ \{-\frac{1}{2}\nabla^2\}\ \chi^j_q(r)\ dv \tag{12}$$

decreases exponentially with distance between orbital χ_p^o (centered in cell o) and orbital χ_q^j (centered in cell j). Recalling the probability density interpretation of the wave function in quantum theory, the orbital product :

$$P_{pq}^{j}(r) = \chi_p^o(r)\ \chi_q^j(r) \tag{13}$$

is the probability density of an electron shared by the orbitals $\chi_p^o(r)$ and $\chi_q^j(r)$. The kinetic integral can be interpreted as the mean kinetic energy of a distribution of charge given by $P_{pq}^j(r)$. The explicit dependence, however, is not straightforward because of the essential nature of the operator ∇^2. The kinetic energy integrals are usually small when the overlap involving the same orbitals is small thus producing the fast decrease with respect to the distance between orbital centres.

- The electrostatic attraction and repulsion terms :

$$- \sum_h \sum_\alpha Z_\alpha\ U_{pq}^{j}(h,\alpha) + \sum_h \sum_l \sum_r \sum_s D_{rs}^{l-h} \left(\begin{smallmatrix} oj \\ pq \end{smallmatrix}\middle|\begin{smallmatrix} hl \\ rs \end{smallmatrix}\right) \tag{14}$$

are simply the Coulomb energy of a negative charge distribution $P_{pq}^j(r)$ in the field of the positive framework of all the nuclei and of the negative electron distribution $P_{rs}^{l-h}(r)$. Due to cell neutrality, the total charge is zero so that the electrostatic contributions of cells far away from the center of the charge distribution $P_{pq}^j(r)$ goes to zero. In this sense, even if individual nuclear attraction and electron repulsion integrals decrease very slowly as a function of the distance (r^{-1} dependance), there is a strict cancellation at medium or large distances between the diverging nuclear attraction and electron repulsion terms since positive nuclear charges compensate the negative electron density of an equal number of electrons. This is easily illustrated by investigating the contributions of electrostatic interactions from cell h with the density $P_{pq}^j(r)$:

$$\int P_{pq}^{j}(r_1) \left(- \sum_\alpha \frac{Z_\alpha}{|r_1 - r_\alpha - ja|} + \sum_l \sum_r \sum_s R_{rs}^{l-h} \int \frac{P_{rs}^{l-h}(r_2)}{|r_1 - r_2|}\, dr_2 \right) dr_1 \tag{15}$$

If the center of the charge density $P_{pq}^j(r)$ is $\overline{C}_{pq}$ and the one of $P_{rs}^{l-h}(r)$ is $\overline{C}_{rs}$ and if $\overline{C}_{pq}$ and $\overline{C}_{rs}$ are sufficiently far apart then $r_{12} \simeq |\overline{C}_{pq} - \overline{C}_{rs}| \simeq r_{1\alpha} \simeq |\overline{C}_\alpha - \overline{C}_{pq}|$. As a consequence we reasonably approximate the former expression by

$$\simeq \frac{S_{pq}^{j}}{|\overline{C}_\alpha - \overline{C}_{pq}|} \left(- \sum_\alpha Z_\alpha + \sum_l \sum_r \sum_s R_{rs}^{l-h}\ S_{rs}^{l-h} \right) \tag{16}$$

This leads to an absolute cancellation since the two terms within the braces are, from left to right, the total nuclear charge and the total number of electrons (normalization condition). Proper handling of such cancellation properties leads to important savings of computing time [5-7] and correct evaluation of physical quantities.

- The exchange contribution :

$$-\sum_h \sum_l \sum_r \sum_s D_{rs}^{l-h} \left({}^{h}_{pr} \middle| {}^{lj}_{sq} \right) \tag{17}$$

corrects for the self-electron repulsion included in the Coulomb electrostatic potential and also includes the effect of the Pauli principle on the independent electron model. Clearly, if the centres of two functions χ_p and χ_q are widely separated the exponential form of the basis orbitals ensures that the product $\chi_p\chi_q$ will be small everywhere so that the leading terms in the exchange contributions to a matrix element are

$$\simeq -\sum_r \sum_s D_{rs}^{j} \left({}^{oo}_{pr} \middle| {}^{jj}_{sq} \right) \simeq - D_{pq}^{j} \left({}^{oo}_{pp} \middle| {}^{jj}_{qq} \right) \tag{18}$$

It is then to point out that exchange contributions can be important for those matrix elements between widely separated orbitals since the decrease of both two-center repulsion integrals $\left({}^{oo}_{pp} \middle| {}^{jj}_{qq} \right)$ and elements of density matrices is very slow (r^{-1}) with the distance and not an exponential one as for other contributions.

After completion of the integral calculations, energy terms and band structure are iteratively obtained from the previously defined matrices. The complex eigenvalue problem is solved for the k-points requested by the numerical integration. Diagonalizations are generally achieved with the very efficient and self-contained routine CBORIS [8] which uses successively the Cholesky decomposition, the Householder similarity transformation and the QL algorithm.

It is to be noted that the dimensions of the matrix equations to be solved are equal to the number of atomic orbitals, ω, per unit cell, the effect of the infinite lattice being included in the naturally converging but formally infinite sums.

The equations show that in a general way a polymer can be considered as a large molecule making the usual methods of quantum chemistry appropriate to investigate the electronic structure of polymers. As a consequence we are allowed to introduce the well-known approximation of molecular quantum mechanics into the formalism of polymeric orbitals. For instance, we can either restrict the number of electrons to be considered or approximate several electronic integrals or group of integrals. In this way, we will speak of "ab initio" or "non-empirical" techniques when considering all electrons and calculating all the necessary integrals. On the other hand, we will use semi-empirical methods if a reduced number of electrons (e.g. valence electrons) is taken into account and as a consequence the necessary integrals are approximated from experimental data.

II.2. *The parametric Hückel and extended Hückel methods for polymers* [9]

In the simple Hückel method, only π-electrons of purely conjugated molecules are taken into account and all interactions except for the nearest neighbours are neglected. This method is similar to the original formulation of the "Tight binding approximation" in solid states physics. This method obtained a considerable success by its simplicity and by the good validity of the involved approximations in the case of conjugated hydrocarbons. Due to the neglect of non-nearest interactions, only the first translation is retained in the secular system and determinant. As a consequence simple k-dependence of the π-energy bands are produced. Practical calculations are even easier since the atomic Hückel basis is assumed to be orthogonal; the overlap Bloch sums in those conditions reduce to delta functions. The method has been applied with success to polyene chains, to graphite and to simple conjugated polymers.

Around 1950, Mulliken, Wolfsberg and Helmolz suggested a very simple type of Hückel parametrization which is easily extended to σ-bonded systems. Hoffman took up this method and applied it to a large variety of saturated and conjugated molecules. This is by now well known as the extended Hückel theory (EHT). In this procedure, only valence orbitals are considered. One-center interactions are estimated from ionization potentials and electroaffinities. Two-center integrals are computed from one-center ones by the Mulliken's approximation and calibrated by a scale factor. Those integrals vary exponentially with distance so that Bloch sums are generally extended over ten or fifteen unit cells in secular systems. Due to the non-orthogonality, diagonalization procedures are more involved and sophisticated band structures can be generated. Since the procedure does not take into account the explicit electron-electron interactions, the method is non-iterative and directly produces a stable band structure. Modifications to introduce explicit dependence on electronic charges have been tempted giving rise to more complex iterative extended Hückel methods (IEHT). A large variety of polymers have been studied by the standard EHT techniques including polyethylene and its fluoro-derivatives, polydiacetylenes and substituted compounds, biopolymers,...

II.3. *The semi-empirical CNDO procedures for polymers* [9]

More sophisticated methods (as Pople's CNDO method) have no longer their justification on an entire evaluation of the matrix elements between basic orbitals but more precisely on a close analysis of energy terms. The matrix elements are splitted into their kinetic, attraction and repulsion parts and some integrals are either neglected or approximated by careful procedures. Fundamentally, the CNDO procedure attempts to reproduce by means of one-electron models the one-electron results to the many electron Hartree-Fock theory. Sometimes, the fit to experimental data results in an implicit inclusion of a part of the correlation effects although in a very indirect way.

II.4. *The ab initio LCAO method for polymers* [9]

The Hartree-Fock theory obtained its considerable success with the ab initio calculations on molecules of "chemical" sizes. In these calculations, all the electron are taken into account and all the necessary integrals are explicitly computed for a given basis of atomic orbitals. There is actually a considerable development of ab initio type calculations, specific examples will be given in forthcoming lectures[10-12].

At the present time, all the above cited methods are used to calculate energy band structure of polymers. Non-empirical methods are dependent on the fourth power of the number of orbitals considered while empirical methods are more dependent on the square power. If N is the number of cells for which all the overlap and Fock elements are greater than a given threshold (for instance 10^{-6}) and ω the size of the basis set in a unit cell, we have to calculate $N\omega(\omega+1)/2$ Fock integrals and also $N\omega(\omega+1)/2$ overlap terms The repulsion operator by its two-electron character gives rise to much more complex matrices. Those matrices (of number N^3) contain $\omega^2(\omega+1)/4$ integrals as first approximation. As a consequence, the time-consuming part of an "ab initio" program is of a general $N^3\omega^4/8$ time dependence [13]. Extended Hückel or CNDO band calculations have a much easier $N\omega^2$ dependence for the integral calculations. The time-consuming process is then the diagonalization and, as a consequence, those type of calculations are feasible on medium size computers in realistic computation times - since extended Hückel calculations avoid SCF iterative process - one might think that the computing time ought to be less than the time needed for the CNDO method. However, the former involves heavy matrix transformations in order to treat the nonorthogonal atomic and Bloch basis while the atomic and consequently the Bloch basis is assumed to be orthogonal in the CNDO-type methods. As a result, computing time are very similar for both methods.

III. BAND STRUCTURE CALCULATIONS AND ELECTRONIC PROPERTIES

As soon as the final eigenvalue equation (5) has been set up according to one of the various methods seen above, the problem of getting information to compare with experiment has to be considered. Indeed making guesses or interpretations on properties is the most fruitful and rewarding aspect of theoretical studies. Numerous experimental quantities can be correlated with terms deduced from electronic charge distribution and from band structure calculations. However band structure is not measured directly and one has to apply transformations to bring calculated data in a form readily comparable to experiment. This is typical of the electronic density of states distributions. Electronic charge distribution is also very attractive to consider when discussing chemical bonds; however, it is not easy to handle this three variables function. To circumvent the difficulty charge indices have been defined, they represent a more synthetic, though somewhat arbitrary, information.

It is the purpose of this part to introduce charge and one electron terms useful for the purpose of experimental interpretation. Additional and more involved quantities will be presented in the forthcoming lectures of this summer school.

III.1. *Electron Charge Indices Concept*

In molecular quantum chemistry the so-called Mulliken population analysis [14] proves to be valuable tool in discussing and comparing the results of electronic structure calculations. Though these populations assigned to the various regions of space are not uniquely defined (they depend on the choice of basis orbitals) they provide, when carefully handled, enlightenment on electron distribution changes. Besides the usual analysis on the participating atomic functions to molecular orbitals they can give estimates on atomic charges, overlap densities between nuclear centers and they are successfully correlated to ESCA chemical shifts through the Siegbahn model potential.

Similar quantities may also be deduced in the context of polymer energy band calculations. Hereafter we give some of those quantities; they are defined in the origin unit cell. In the followings n_o will refer to the number of doubly occupied bands, α and β to nuclear centers and L_{RC} to the length of the reciprocal cell.

a. P_{p_α} : electron charge in atomic function χ_p centered on atom

$$P_{p_\alpha} = \frac{1}{L_{RC}} \int \sum_{n=1}^{n_o} C^*_{np_\alpha}(k) \left[\sum_q C_{nq}(k)\, S^o_{p_\alpha q}(k)\right] dk \qquad (19)$$

b. P_α : electron charge on atom

$$P_\alpha = \sum_{P_\alpha} P_{p_\alpha} \qquad (20)$$

c. $P^n_{p_\alpha q_\beta}$: overlap population in band n between atomic functions χ_{p_α} and χ_{q_β}

$$P^n_{p_\alpha q_\beta} = \frac{2}{L_{RC}} \int C^*_{np_\alpha}(k)\, C_{nq_\beta}(k)\, S^o_{p_\alpha q_\beta}(k)\, dk \qquad (21)$$

d. $P^n_{\alpha\beta}$: overlap population in band n between atomic centers α and β

$$P^n_{\alpha\beta} = \sum_{p_\alpha} \sum_{q_\beta} P^n_{p_\alpha q_\beta} \qquad (22)$$

e. $P_{p_\alpha q_\beta}$: total overlap population between atomic functions χ_{p_α} and χ_{q_β}

$$P_{p_\alpha q_\beta} = \sum_{n=1}^{n_o} P^n_{p_\alpha q_\beta} \qquad (23)$$

f. $P_{\alpha\beta}$: total overlap population between atomic centers α and β

$$P_{\alpha\beta} = \sum_{p_\alpha} \sum_{q_\beta} P_{p_\alpha q_\beta} \qquad (24)$$

It is to be pointed out that polymer populations are averaged by integrating over the Brillouin zone; it presupposes energy bands, $E_n(k)$ and associated eigenvectors, $\{\underline{C}_n(k)\}$, have been correctly assigned an index n. This will be discussed in more detail when coming to electronic density of states problem. It is also possible to obtain corresponding quantities (a to f) for each wave number k, however this makes the analysis quite heavy to conduct.

III.2. Electron Charge Indices applications

To illustrate some of these quantities we give numbers on polyethylene calculated with extended Hückel method (EHCO). The original parameters [15] have been used, namely : K=1.75, I_{H1s}=1.2, I_{C2s}=21.4 eV, I_{C2p}=11.4 eV and $\zeta_{C2s} = \zeta_{C2p}$ = 1.625. Bond lengths are 1.54 Å and 1.08 Å respectively for C-C and C-H; angles around each carbon are tetrahedric.

Table 1 shows the evolution of P_{C-C} and P_{C-H}, the total overlap populations between the C-C and C-H nuclear centers going from ethane to the infinite polyethylene. Corresponding values for molecules of increasing size converge well to the infinite polymer values.

Table 1. P_{C-C} and P_{C-H} values for ethane, butane and infinite polyethylene. Central C-C bond has been chosen for molecules

Bond	C_2H_6	C_4H_{10}		$(C_2H_4)_\infty$
C-C	.7184	.7447	...	.7790
C-H	.8079	.8178	...	.8595

Atomic populations calculations give the following values :

$$P_{C_{2s}} = 1.20 \; ; \; P_{C_{2p}} = 2.90 \text{ and } P_C = 4.10$$

The carbon electronic configuration is then $2s^{1.20}\ 2p^{2.90}$ to be compared with $2s^2 2p^2$ for the isolated carbon in its ^3P_o electronic state. There is an important electronic transfer from C_{2s} orbital to C_{2p}; hybridization index is $sp^{2.42}$.

Additional and even finer knowledge on the polymer orbitals, $\psi_n(k,\vec{r})$, can be obtained by inspection of $P^n_{p\alpha q\beta}$ quantities which distribute into bonding, non bonding and antibonding groups. This analysis can be conducted with profit on series of similar compounds to appreciate modifications in the energy spectrum in terms of bonding character. Correlation of atomic charges with ESCA chemical shifts has been successfully applied to a series of linear fluoropolymers [16] and polymer orbitals interpretation has also been conducted[17] using overlap populations. Other examples and uses can be found in the litterature.

III.3. *Electronic Density of States Distribution*

In polymeric materials, like in three-dimensional crystals, the number of atoms is very large and the eigenvalues are very dense and bounded; it is therefore more convenient to deal with energy levels distribution functions than with individual values. There are definite advantages to studying density of states functions :

- it provides a synthetic description of one electron levels
- it is closely related to experiment.

III.3.1. Electronic density of states distribution concept

Density of states is defined as the number of allowed levels per unit energy

$$\frac{dm}{dE} = \mathcal{D}(E) = \frac{a}{\pi} \sum_{n=1}^{n_o} \left| \frac{dk}{dE_n(k)} \right|_{E_n(k)=E} \qquad (25)$$

A simple way to get to this result is to consider a one energy band system as shown in Figure II. For each non degenerate band there are two electrons with opposite spin for each primitive cell of the polymer or for each allowed wave number k. Indeed, there are as many allowed wave numbers k in the Brillouin zone as we find unit cells constituting the polymer. Wave number k is defined as $\frac{2\pi}{a}\frac{j}{N}$, j = 0,1, ... N; when N goes up to infinite, k becomes a continuous variable defined in the Brillouin zone $[-\pi/a,+\pi/a]$, of length, L_{RC}, $2\pi/a$. The length in k-space per allowed wave number is $2\pi/aN$ so we have $aN/2\pi$ allowed k points per unit length of reciprocal space. We are now in a position to relate the wavenumber, k, to the number of electronic states, m

$$dm = 2 \times \frac{a}{2\pi} dk \qquad (26)$$

and for the one band system we verify the number of states within the Brillouin zone :

$$m = \frac{a}{\pi} \int_{L_{RC}} dk = 2 \qquad (27)$$

The density of states itself is defined as dm/dE, to understand the meaning we now turn to Figure II and use an increment form for the variation of m and k :

$$\Delta m = \frac{a}{\pi} \Delta k \tag{28}$$

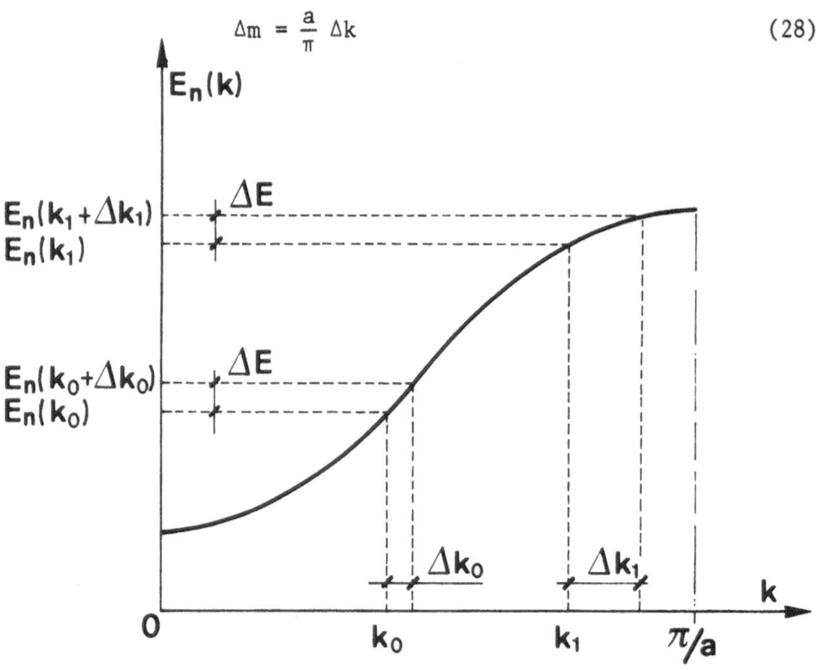

Figure II: Variation of the density of states contribution, Δk_i, within an energy band

We write down the definition of the density of states function using quantities shown in Figure II.

$$\frac{\Delta m}{\Delta E} = \frac{a}{\pi} \left| \frac{\Delta k}{E_n(k_o+\Delta k) - E_n(k_o)} \right|_{E_n(k) = E} \tag{29}$$

and let ΔE tend to zero to finally end up with the expected expression

$$\frac{dm}{dE} = \frac{\pi}{a} \left| \frac{dk}{dE_n(k)} \right|_{E_n(k) = E} \tag{30}$$

For a many bands system we just have to sum over all doubly occupied bands and we obtain the first expression of this paragraph, equation (25).

III.3.2. Evaluation of electronic density of states distributions

Though conceptually simple, electronic density of states distributions are not straighforward to evaluate numerically. Whenever the first derivative of $E_n(k)$, $E_n'(k)$, goes to zero there will be a large contribution to the density of states giving rise to singularities in the spectra. We know in advance there are at least three values of the wave number k where $E_n'(k)$ vanishes identically, i.e. $k = 0$ and $k = \pm \frac{\pi}{a}$. However other extrema can occur but their number and location cannot be predicted a priori. When implementation on computer is to be considered, one has to turn to new function, $D(E_i)$, the density of electronic states histogram. It exhibits the same behaviour as $\mathcal{D}(E)$ but fortunately gives rise to finite peak's height only. It is an average of $\mathcal{D}(E)$ over some energy interval ΔE

$$D(E_i) = \frac{1}{\Delta E} \int_{E_i - \frac{\Delta E}{2}}^{E_i + \frac{\Delta E}{2}} \mathcal{D}(E)dE \qquad (31)$$

$$E_i = E_o + i\Delta E;\ i=0,1,2,\ldots;\ \Delta E > 0$$

The value of ΔE is chosen according to a particular experiment to interpret. $\mathcal{D}(E_i)$ is obviously the limiting function of $D(E_i)$ when ΔE goes to zero. $D(E_i)$ is the height of the ith rectangular box centered at E_i; it represents an element of the histogram of the density of states.

Various approaches to compute those $D(E_i)$ have been proposed. The histogram method involves generating the eigenvalues, $E_n(k)$, of the system at a large number of points in the Brillouin zone and then counting the results into boxes that subdivide the energy range [18]. Another approach, more effective in the sense of saving computing time and numerical stability is currently prefered. Basically it consists of an analytical or continuous integration over small intervals $[k_i - \Delta k/2,\ k_i + \Delta k/2]$ located at wave number values for which a diagonalization has been requested. Figure III depicts schematically the way the histogram $D(E_i)$ is constructed. Energy bands in each of the above mentionned intervals, $k_i \pm \Delta k/2$, are approximated by quadratics. Detailed procedure can be found in reference [19].

Related and prior to any density of states calculations is the band indexing problem. Due to prohibitive computing time associated with diagonalization, the energy bands are known at discrete wave number values only, $E_n(k_i)$. Bands are usually given an index,n, increasing in order of increasing energy, but difficulties arise because the rule breaks down at degeneracy points. Figure IV with two energy bands illustrates the situation. For this set of known eigenvalues two possibilities exist : crossing or non crossing.

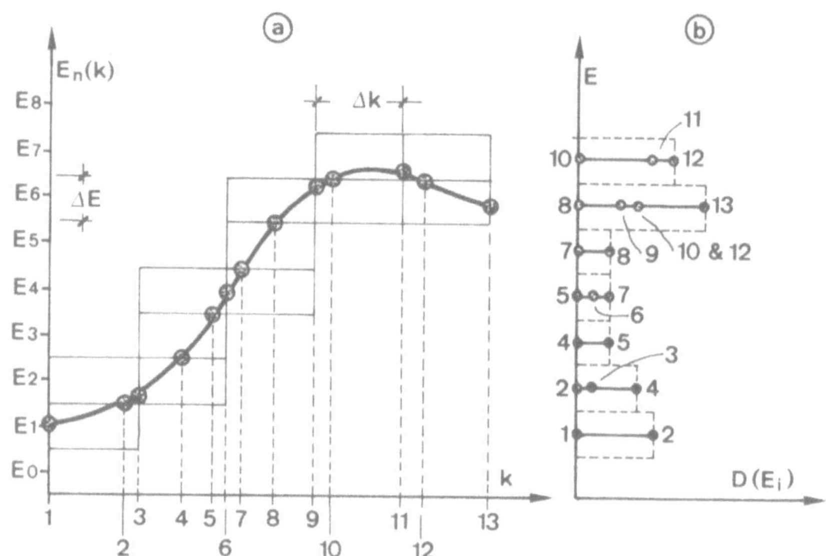

Figure III : a) Schematic band structure and its contributions to the density of states, 1-2, 2-3, ... 12-13.
b) Resulting density of states.

To these two alternatives correspond to entirely different physical realities. For cases where point symmetry exists within the unit cell the problem can be solved using group theory but when there is a low symmetry and a large number of bands this becomes practically untractable. For present time we solve the band indexing problem using graphic interactive program [20] with the knowledge of local first derivatives, $E_n'(k)$ of the energy eigenvalues. They are computed according to [21]

$$E_n'(k) = \underline{C}_n^+ (k) \, [\underline{H}'(k) - E_n(k) \, \underline{S}'(k)] \, \underline{C}_n (k) \qquad (32)$$

$\underline{H}'(k)$ and $\underline{S}'(k)$ are respectively the first derivatives of the Hamiltonian and overlap matrices; they are easily computed since they are expressed as trigonometric series. Work is actually in progress to avoid any human intervention in the routines string from diagonalization to final property evaluation; in this particular indexing problem perturbation theory seems to offer an elegant wayout[22].

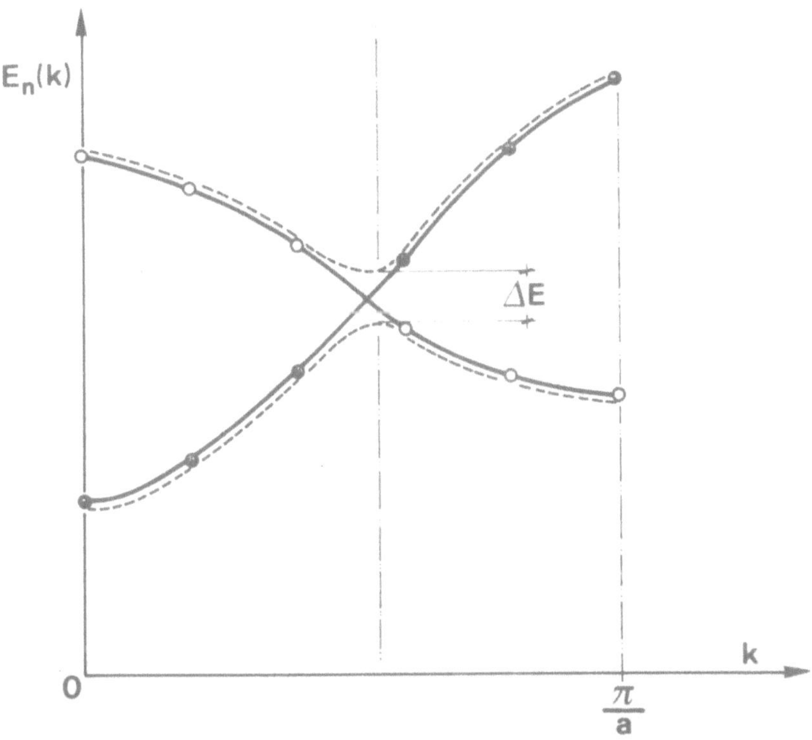

Figure IV : A two bands structure where curves can equally cross or avoid each other

III.3.3. Applications to density of states

Density of electronic states distributions can be used to test results on energy band structure obtained from various methods. Figures V and VI reproduce band structures and their electronic density of states distributions on polyethylene calculated with ab-initio[23], FSGO[5], LCLO[24], CNDO/2[25], CNDO/W[25]and EHCO[25]methods. Band structures graphs and energy scale occupation appear similar except for CNDO/2 which deviates strongly CNDO/W, a CNDO/2 method with special parameters[26] well suited for hydrocarbon studies, requires density of states comparison to reveal discrepancy in the peak structure. It is known that derivatives enhance variations of a function therefore no wonder density of states distribution is a more sensitive test on energy

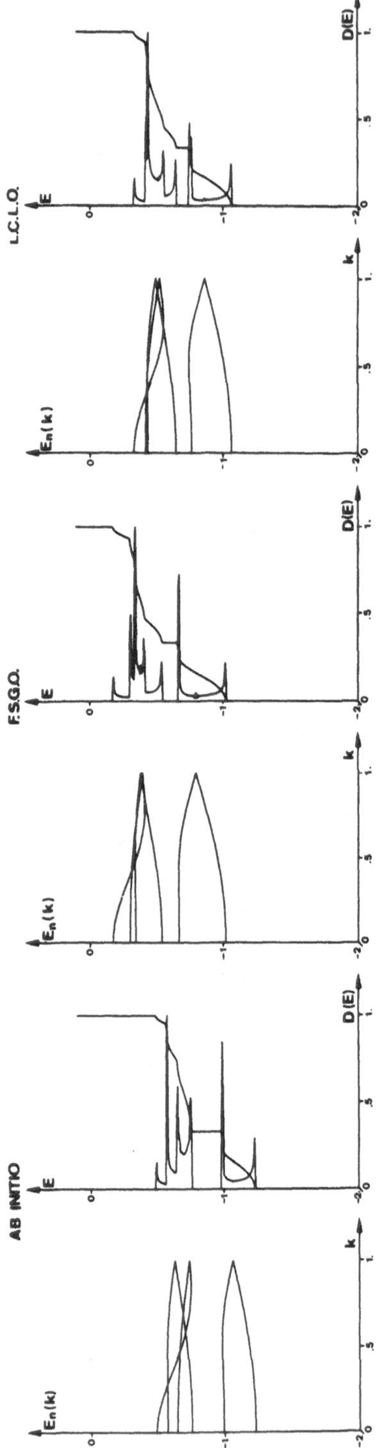

Figure V : Band structure, $E_n(k)$, and density of states, D(E), of the infinite polyethylene calculated with ab initio, FSGO, LCLO methods

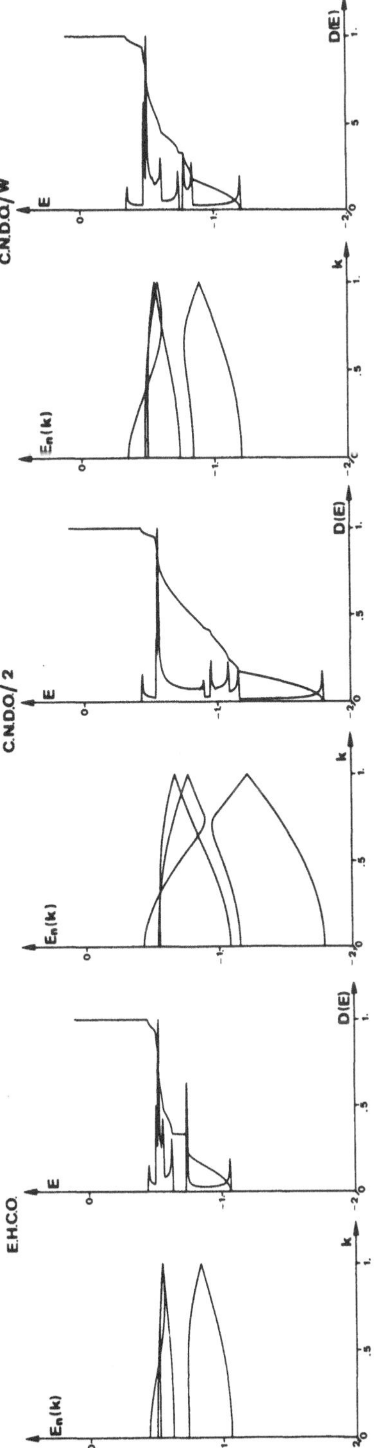

Figure VI : Band structure, $E_n(k)$, and density of states, D(E), of the infinite polyethylene calculated with EHCO, CNDO/2, CNDO/W methods

bands since it involves inverse of first derivatives, $E'_n(k)$.

Experimental XPS spectra are not directly comparable to density of states, but two additional effect can be introduced in the calculations to produce curves with almost the same type of content as in the experiment. First one has to take photoionization cross-section into account. It induces a sensible modification on the general shape of the density of states. Until now the calculations of the XPS intensities have been made [27] mainly according to the empirical model of Gelius :

$$I_n(k) \sim \sum_p C^*_{np}(k) \left[\sum_q S_{pq}(k) \, C_{nq}(k) \right] \sigma_p \qquad (33)$$

The σ_p's are the relative photoionization cross-sections refering to a particular atomic subshell; approximate values have been obtained by Gelius [28] by applying the intensity model to small molecules with known orbital assignments. A typical $I_n(k)$ graph is shown in Figure II of reference [27] To obtain the density of electronic states distribution modulated by the XPS cross-sections, $\overline{D}(E_i)$, each energy eigenvalue contribution is weightend by $I_n(k)$:

$$\overline{D}(E_i) \sim \frac{1}{\Delta E} \sum_{n=1}^{n_o} \int_{E_i-\Delta E/2}^{E_i+\Delta E/2} \{ |E_n^1(k)|^{-1} \times I_n(k) \} \, dE \qquad (34)$$

Second, XPS spectrometers have an inherent limitation in their resolution. For example, the HP.5950 A ESCA spectrometer shows a resolution ranging from 0.5 to 0.7 eV. It is found the experimental resolution function is reasonably well fitted by a gaussian function of full width, Γ, at half maximum of 0.7 eV. The final theoretical simulation, $D_s(E_i)$ of the experimental spectrum is obtained after the convolution of $\overline{D}(E_i)$ by the gaussian fitting curve has been made :

$$D_s(E_i) \sim \int_{-\infty}^{+\infty} \overline{D}(E) \exp\left[-(E-E_i)^2/2\sigma^2 \right] dE \qquad (35)$$

where $\Gamma = 2.345\sigma$.

Comparison between experimental and simulated XPS spectra will be discussed at length in a forthcoming lecture [29].

REFERENCES

[1] "Proceedings of the NATO ASI on the Electronic Structure of Polymers and Molecular Crystals", J.M. André and J. Ladik editors, B9, Plenum Press (1975)
[2] G. Del Re, J. Ladik, G. Biczo, Phys. Rev., 155, 997 (1967)
[3] J.M. André, J. Chem. Phys., 50, 1536 (1969)
[4] J.M. Ziman, in "Principles of the theory of solids", 2d ed., pp. 15-19 (1972)
[5] J.M. André, J. Delhalle, Ch. Demanet, M.E. Lambert-Gérard, Internat. J. Quantum Chem., S10, 99 (1976)
[6] F.E. Harris, J. Chem. Phys., 56, 4422 (1972)
[7] J. Delhalle, F.E. Harris, submitted to Theoretica Chimica Acta
[8] E. Zakrajsek, J. Zupan, Annales Soc. Scient. Bruxelles, 89, 337(1975)
[9] J.M. André, in "Proceedings of the NATO ASI on the Electronic Structure of Polymers and Molecular Crystals", J.M. André and J. Ladik editors, pp. 1-21, B9, Plenum Press (1975)
[10] J. Ladik, "Electronic Structure and Transport Properties of Biopolymers", NATO ASI, Namur (Belgium) 1977
[11] J. Ladik, "Energy Band Structures of Highly Conducting Polymers", NATO ASI, Namur (Belgium) 1977
[12] P. Schüster, "Hydrogen Bonds and Intermolecular Interactions on Polymers", NATO ASI, Namur (Belgium) 1977
[13] J.M. André, Computer Physics Comm., 1, 391 (1970)
[14] R.S. Mulliken, J. Chem. Phys., 23, 1833 (1955)
[15] R.Hoffmann,J. Chem. Phys., 39, 1397 (1963)
[16] J.M. André, J. Delhalle, Chem. Phys. Letters, 17, 145-149 (1972)
[17] J. Delhalle, S. Delhalle, J.M. André, J.J. Pireaux, J. Riga, R. Caudano, J.J. Verbist, to appear in J. Electron Spectroscopy
[18] J. Delhalle, Bull. Soc. Chim. Belges, 84, 135 (1975)
[19] J. Delhalle, S. Delhalle, Internat.J. Quantum Chem., 11, 349 (1977)
[20] J. Delhalle, D. Thelen, J.M. André, Computers and Chemistry, submitted
[21] J.M. André, J. Delhalle, G. Kapsomenos, G. Leroy, Chem.Phys. Lett., 14, 485 (1972)
[22] J. Delhalle, F.E. Harris, work in progress
[23] J.M. André, G. Leroy, Chem. Phys. Letters, 5, 71 (1974)
[24] J. Delhalle, J.M. André, S. Delhalle, C. Pivont-Malherbe, F. Clarisse, G. Leroy, D. Peeters, Theoretica Chim. Acta (Berl.), 43, 215 (1977)
[25] J. Delhalle, J.M. André, S. Delhalle, J.J. Pireaux, R. Caudano, J.J. Verbist, J. Chem. Phys., 60, 595 (1974)
[26] J.M. Sichel, M.A.Whitehead, Theoretica Chimica Acta (Berl.), 11, 220 (1968)

[27] J. Delhalle, S. Delhalle, J.M. André, Chem. Phys. Letters, 34, 430 (1975)
[28] U. Gelius, in "Electron Spectroscopy", D.A. Shirley editor, pp. 311-334, North-Holland, Amsterdam (1972)
[29] J. Verbist, "Photoelectron Spectroscopy as Applied for Determining Band structures of Polymers", NATO ASI, Namur (Belgium) 1977.

THE EVALUATION OF ELECTRON REPULSION INTEGRALS

A. Veillard

E.R. n° 139 du C.N.R.S.
Université L. Pasteur, BP 296 R8, 67000 Strasbourg

I. INTRODUCTION

One major problem in the ab initio electronic structure calculations is the evaluation of the interelectronic repulsion integrals

$$[pq|rs] = \int\int \chi_p^{*(1)} \chi_q^{(1)} (1/r_{12}) \chi_r^{*(2)} \chi_s^{(2)} dv_1 dv_2 \quad (1)$$

where the χ denote the basis functions which are used for the expansion of the molecular orbitals φ

$$\varphi_i = \sum_{p=1}^{m} C_{ip} \chi_p \quad (2)$$

The same problem arises also in the crystal orbital calculations. In fact, it is the relative ease of computation of these electron repulsion integrals which determines the choice of the basis functions. Presently, ab initio calculations for molecules other than linear are almost exclusively based on the use of the Gaussian type functions (GTF) which were introduced by Boys in 1950 [1]. The unnormalized GTF, usually centered on the atoms, are of the type

$$r^{n-1} e^{-\zeta r^2} Y_{lm}(\theta,\varphi) \quad (3)$$

The ease of integral evaluation for Gaussian-type functions comes from the fact that, for n-l-1 restricted to zero or even-integer values, the product of two Gaussians on different centers is equivalent to a single Gaussian on a new-center [2]. An integral [pq|rs] in which the four basis functions are Gaussian centered on

J.-M. André et al. (eds.), Quantum Theory of Polymers, 23-30.

different atoms then reduces immediately to a two-center integral which can be evaluated explicitly in terms of known functions.

The main disadvantage in using a Gaussian basis set is that a relatively large number of Gaussian functions is required to produce the same accuracy achieved for instance with a relatively small number of Slater functions. For first-row atoms, a 7s, 3p gaussian basis set (namely 7 basis functions of s symmetry, 3 functions of symmetry p_x, 3 of symmetry p_y, ...) is about as good as a minimal basis set 2s, 1p of Slater functions and a (10s, 6p) gaussian basis set is required to produce what is called a "double-zeta" accuracy (namely the accuracy achieved with a Slater basis set 4s, 2p). Thus a calculation for a "large" molecule usually requires a basis set made of several hundred of Gaussians. For instance a minimal basis set calculation for a model of oxyhemoglobin, $Fe(N_4C_{20}H_{12})(NH_3)O_2$, requires a basis set of 529 gaussians and a somewhat better calculation used a basis set of 662 functions. The number of electron repulsion integrals, for a basis set of m basis functions, is roughly $m^4/8$. The calculation for the model of oxyhemoglobin requires the evaluation of approximately 10^{10} integrals. Six or seven years ago, the average time needed to compute an electron repulsion integral on the fastest computers was of the order of 100 μs and, with this standard, the above calculation would have required about 300 hours. The actual calculation with the small basis set required in fact 6 hours of CPU time on a Univac 1110. A new generation of computer programs has been developped in several laboratories over the past few years, with an increased efficiency in the evaluation of the electron repulsion integrals. This efficient evaluation is a achieved essentially through :

i) the "batch" technique which evaluates simultaneously a batch of integrals $[G^AG^B | G^CG^D]$ where G^A, G^B, G^C and G^D denote a group of basis functions with a common exponent and located on the same center (for instance a set of three functions p_x, p_y and p_z or a set of six d functions d_{xx}, d_{xy}, d_{xz}, d_{yy}, d_{yz} and d_{zz}). This is advantageous since the distinct integrals of the batch $[G^A_pG^B_q|G^C_rG^D_s]$ possess many common factors which in this way need to be evaluated only once (with the previous generation of computer programs, integrals were computed independantly one after the other and in this way, these common factors had to be computed several times). This grouping was first introduced in the GAUSSIAN 70 program of Pople <u>et al</u> for the s and p-type functions [3] (with the restriction that the s and p functions share the same exponent).

ii) the use of the symmetry properties of the molecule in order to avoid either the calculation of integrals which are zero by symmetry or the redundant computation of integrals which are equal by symmetry (to within a sign). This is not a new feature since the first program developped in the early sixties for molecular calculations, POLYATOM [4], already took advantage of the symmetry properties by generating a list of integral labels that has no integrals in it which are zero by symmetry, and with the integrals equal by symmetry (to within a sign) grouped together, so that only the first member of the group needs evaluation. This list of integral labels was later processed for integral evaluation.

II. BATCH EVALUATION OF ELECTRON REPULSION INTEGRALS

II.1. Integrals over basis functions of s- and p-type

The electron repulsion integral over Gaussian functions is defined as

$$\mathrm{ERI} = \iint G(\alpha_1,A,l_1,m_1,n_1)G(\alpha_2,B,l_2,m_2,n_2)r_{12}^{-1}\, G(\alpha_3,C,l_3,m_3,n_3) \, G(\alpha_4,D,l_4,m_4,n_4)\, dv_1 dv_2 \tag{4}$$

where

$$G(\alpha,A,l,m,n) = x_A^l\, y_A^m\, z_A^n \exp(-\alpha r_A^2) \tag{5}$$

denotes the unnormalized Gaussian function, l, m and n are integers positive or zero, $\vec{r}_A$ is the vector of component x_A, y_A, z_A joining the point A to the point defining the location of electron 1 for the left-hand side of r_{12}^{-1} and of electron 2 for the right-hand side of r_{12}^{-1} . The expression of the electron repulsion integral may be reduced to a triple summation (the mathematical analysis is originally due to Wright [5], a detailed treatment may be found in References [6] and [7])

$$\sum_{I=0}^{L}\sum_{J=0}^{M}\sum_{K=0}^{N} C_I\, C_J\, C_K\, F_{I+J+K} = \sum_{\gamma}\left\{\sum_{I,J,K}^{I+J+K=\gamma} C_I\, C_J\, C_K\right\} F_\gamma \tag{6}$$

with

$$L = l_1+l_2+l_3+l_4 \quad M = m_1+m_2+m_3+m_4 \quad N = n_1+n_2+n_3+n_4 \tag{7}$$

F_γ denotes the incomplete gamma function and is a function of the coordinates of the four centers and of the four exponents. The coefficients C_I, C_J, C_K are functions of the coordinates of the four centers, of the four exponents and of the integers l_1 l_2 l_3 l_4 (for C_I), m_1 m_2 m_3 m_4 (for C_J) and n_1 n_2 n_3 n_4 (for C_K).

Batch processing of the integrals is based on the following assumption : at each basis function of the p_x type on a given center (with an exponent α) correspond one p_y and one p_z functions on the same center with the same exponent (similarly the basis functions of the d-type correspond to sets of six d functions d_{xx}, d_{xy}, d_{xz}, d_{yy}, d_{yz} and d_{zz} on the same center and sharing the same exponent).

Integrals over s and p functions may be classified into 7 families, each family including several types of integrals (Table I).

	nb. of types
ssss	1
psss	3
ppss	9 if $p_1 \neq p_2$
	6 if $p_1 \equiv p_2$
psps	9 if $p_1 \neq p_3$
	6 if $p_1 \equiv p_3$
ppps	27 if $p_1 \neq p_2$
	18 if $p_1 \equiv p_2$
pspp	27 if $p_3 \neq p_4$
	18 if $p_3 \equiv p_4$
pppp	81 if $p_1 \neq p_2 \neq p_3 \neq p_4$
	54 if $p_1 \equiv p_2$ or $p_3 \equiv p_4$
	45 if $p_1 \equiv p_3$ and $p_2 \equiv p_4$
	36 if $p_1 \equiv p_2$ and $p_3 \equiv p_4$
	21 if $p_1 \equiv p_2 \equiv p_3 \equiv p_4$

Table I. The 7 families of integrals over s- and p-functions and the corresponding number of types.

For instance integrals psss correspond to the three possible types $p_x$sss, $p_y$sss and $p_z$sss. A set of four p-type functions p_1, p_2, p_3 and p_4 (namely p_{1x}, p_{1y}, p_{1z}, p_{2x}, ...) on different centers generates 81 integrals. Each of these integrals requires the calculation of seven C terms, for instance, for the integral $p_{x1}\, p_{x2}\, p_{x3}\, p_{x4}$ (L = 4, M = 0, N = 0)

$$\text{ERI} = \sum_{I=0}^{4} \sum_{J=0}^{0} \sum_{K=0}^{0} C_I\, C_{J=0}\, C_{K=0}\, F_{I+J+K}$$

$$= {}^{y}C_0\, {}^{z}C_0\, ({}^{x}C_0F_0 + {}^{x}C_1F_1 + {}^{x}C_2F_2 + {}^{x}C_3F_3 + {}^{x}C_4F_4) \quad (8)$$

(we use the notation ^{x}C, ^{y}C and ^{z}C for C_I, C_J, C_K when we specify the value of I, J and K). One can see easily that the number of C terms will be the same, namely 7, for every [pp|pp] integral. If the 81 integrals are computed independantly, this will require the computation of 567 C terms (many C terms being recomputed several times). On the contrary, batch processing of the integrals insures that all the integrals of a given family, for instance p_1 p_2 p_3 p_4 are computed simultaneously. This is done by computing and storing first all the C terms needed for the given batch of integrals, next by carrying the summation over the different C terms in formula (6) for the integrals of the batch. For a given batch of integrals the coordinates of the centers and the exponents are fixed and the only parameters left are (l_1, l_2, l_3, l_4) for C_I, (m_1, m_2, m_3, m_4) for C_J, (n_1, n_2, n_3, n_4) for C_K. Since l, m, n = 0, 1, there are 2^4 = 16 C_I (denoted C_{0000}, C_{1000}, C_{01000} ... according to the values of l_1, l_2, l_3 and l_4) and the total number of C terms to be computed is now 48 instead of 567 when the integrals are processed sequentially (each C term being now computed only one whereas, in the sequential processing it was computed several times for different integrals).

The batch processing may be implemented in the following way. The program will loop over the "formal" basis functions of the "reduced" basis set. The formal functions are the basis functions without angular component and define a reduced basis set of M functions (M = a+b+c) with a functions of s-type, b of the p-type, and c of the d-type. Each "formal" basis function generates m_i "real" basis functions with m_i = 1 for s functions, 3 for p functions and 6 for d functions (the total number of basis functions being a + 3b + 6c). The organization of the program may be represented by the following statements

```
DO  100  I = 1,M
DO  100  J = 1,I
DO  100  K = 1,I
LIM = K
IF (I.EQ.K) LIM = J
DO  100  L = 1,LIM
DO   90  I1 = 1,ICP1
DO   90  J1 = 1,ICP2
DO   90  K1 = 1,ICP3
DO   90  L1 = 1,ICP4
```

The four inner loops correspond to the circular permutations which generate the m_i basis functions (for instance, if the four indices I, J, K, L define an integral psss, then ICP1 = 3 ICP2 = ICP3 = ICP4 = 1 and this generates the batch of there integrals $p_x$sss, $p_y$sss and $p_z$sss). In practice, rather than looping over the indices of these circular permutations, one prefers a sequential structure for the program, with different branchings for the different families of integrals.

II.2. Integrals over basis functions of s-, p- and d-types.

A sequential structure of the batch program allows an important saving in computation time and requires only a moderate programming effort in the case of s- and p-type functions since the number of types of integrals for a given family (and for a given batch) remains relatively small. However such a sequential organization is ruled out for the integrals involving d-functions, since the number of types becomes exceedingly high (1296 for the dddd integrals). However the programming effort can be greatly reduced by the fact that the 1296 formulas for the dddd integrals can be reduced to a much smaller number [7]. Once the whole set of C terms needed for a given batch has been computed (the computation of the C terms for the integrals involving d-functions has been described by Benard [7]), one is left with the summation over the different C terms within a given batch. The batch dddd may be divided into several sets of integrals such as

$$
\begin{array}{l}
xy\ xy\ xy\ y^2 \\
xy\ xy\ xy\ yz \\
xy\ xy\ xy\ z^2 \\
xy\ xy\ xz\ y^2 \\
\cdots\cdots\cdots
\end{array}
$$

namely the set of integrals of the type $x^1\ x^1\ x^1\ x^0$ ($l_1 = 1$ $l_2 = 1$ $l_3 = 1$ $l_4 = 0$). The expansion of the terms in ${}^xC$ will be identical for all these integrals and will involve a number of terms such as

$$ {}^xC^0_{1110}\ F_0 + {}^xC^1_{1110}\ F_1 + {}^xC^2_{1110}\ F_2 + {}^xC^3_{1110}\ F_3 \qquad (9) $$

$$ {}^xC^0_{1110}\ F_1 + {}^xC^1_{1110}\ F_2 + {}^xC^2_{1110}\ F_3 + {}^xC^3_{1110}\ F_4 \qquad (10) $$

(we use the notation ${}^xC^I_{l_1l_2l_3l_4}$ with $I = 1$ to $l_1 + l_2 + l_3 + l_4$).

In this way the 1296 formulas for a dddd integral can be reduced to a small number of types, for instance :

- integrals of the type $a^2a^2a^2a^2$, namely $x^2x^2x^2x^2$ or $y^2y^2y^2y^2$ or $z^2z^2z^2z^2$ which satisfy the following conditions over the powers of x or y or z

$$ \Sigma n_a = 8 \quad \Sigma n_b = 0 \quad \Sigma n_c = 0 \quad (a,b,c = x \text{ or } y \text{ or } z) \qquad (11) $$

Then

$$ {}^bC^0_{0000} = 1 \qquad {}^cC^0_{0000} = 1 \qquad (12) $$

and these integrals reduce to

$$ERI = {}^{a}C^{0}F_0 + {}^{a}C^{1}F_1 + \ldots + {}^{a}C^{8}F_8 \quad (13)$$

- integrals of the type $a^2a^2a^2ab$ or $a^2a^2aba^2$ or $a^2aba^2a^2$ or $aba^2a^2a^2$, namely

$$\Sigma n_a = 7 \quad \Sigma n_b = 1 \quad \Sigma n_c = 0 \quad (14)$$

The development of the corresponding integrals is of the form

$$ERI = {}^{b}C^{0}({}^{a}C^{0}\,F_0 + {}^{a}C^{1}\,F_1 + \ldots + {}^{a}C^{7}\,F_7) + {}^{b}C^{1}({}^{a}C^{0}\,F_1 + {}^{a}C^{1}\,F_2 + \ldots + {}^{a}C^{7}\,F_8) \quad (15)$$

This formula will apply for 24 types of integrals (the four types written above with six possibilities over the choice of a and b as x or y or z).

These algorithms have been implemented in the system of programs Asterix [8] and lead to a very efficient computing of the integrals over d-type functions. Representative timings have been reported by Benard [7]. The method has been extended recently to f-orbitals [9].

III. MISCELLANEOUS

The use of molecular symmetry has been discussed by Saunders [6b]. A test on the magnitude of the integrals, in order to avoid the unnecessary computation of small integrals (usually less than 10^{-7}), is a common feature of most of the integral programs presently used (see for instance Reference [10]). The calculation of the incomplete gamma functions has been discussed by several authors [2,11,12].

REFERENCES

[1] S.F. Boys, Proc. Roy. Soc. A200, 542 (1950)

[2] See for instance : I. Shavitt, in "Methods in Computational Physics", edited by B. Alder, S. Fernbach and M. Rotenburg (Academic, New-York, 1963), Vol. 2, p. 1.

[3] M.D. Newton, W.A. Lathan, W.J. Hehre and J.A. Pople, J. Chem. Phys., 51, 3927 (1969).

[4] I.G. Csizmadia, M.G. Harrison, J.W. Moskowitz, S. Seung, B.T. Sutcliffe and M.P. Barnett, "The Polyatom System", Technical Note 36, Cooperative Computing Laboratory, MIT.

[5] J.P. Wright, "Quarterly Progress Report", Solid State and Molecular Theory Group, MIT, October 1963, p. 35.

[6a] H. Taketa, S. Huzinaga and K.O-Ohata, J. Phys. Soc. Jap., 21, 2313 (1966).

[6b] V.R. Saunders, in "Computational Techniques in Quantum Chemistry and Molecular Physics", G. Diercksen, B.T. Sutcliffe and A. Veillard ed., D. Reidel, Dordrecht (Holland), 1975, p. 347.
[7] M. Bénard, J. Chim. Phys., 1976, p. 413.
[8] M. Bénard, A. Dedieu, J. Demuynck, M.M. Rohmer, A. Strich and A. Veillard, "Asterix": a system of programs for the Univac 1110", unpublished work.
[9] M. Barry and M. Bénard, unpublished results.
[10] R. Ahlrichs, Theoret. Chim. Acta, 33, 157 (1974).
[11] L.J. Schaad and G.O. Morrell, J. Chem. Phys., 54, 1965 (1971).
[12] J. Arents, Chem. Phys. Let., 12, 489 (1972).

X-RAY PHOTOELECTRON SPECTROSCOPY AND ITS APPLICATION TO POLYMER ELECTRONIC STRUCTURE STUDIES

Jacques J. VERBIST
Facultés Universitaires Notre-Dame de la Paix,
Laboratoire de Spectroscopie Electronique,
61, rue de Bruxelles, B-5000 - Namur (Belgium)

I. INTRODUCTION

The purpose of these lectures is to introduce an experimental method which has become increasingly important over the last decade in electronic structure studies of molecules, solids and surfaces. Perhaps because of the length of its explicit name, X-ray photoelectron spectroscopy has become popular under the acronym E.S.C.A. (Electron Spectroscopy for Chemical Analysis).

Our discussion will progress along the following headlines : first, we will attempt to define why E.S.C.A. appears so appropriate for our purpose, as an experimental method. We will describe how to obtain, and also how to interpret spectra. Finally, we will survey in some detail the most recent results in the field of electronic structure and properties of polymers.

II. E.S.C.A. AN EXPERIMENTAL METHOD

II.1. Generalities

Experimental methods for the study of matter have one thing in common : you have to do something to the sample, something you can describe as well as possible; then depending on the way the sample reacts to the perturbation, you try to learn as much as you can from what you observe.

J.-M. André et al. (eds.), Quantum Theory of Polymers, 31-48.

This statement is not as trivial as it may look. Any use of experimental data should be done, keeping this well in mind. It is the guideline to the following sections.

We first need to choose an appropriate way of perturbing the sample. In spectroscopy, an incident beam of electromagnetic radiation - or eventually particles - will be used. Then, we need to understand the physics of the interaction. Finally, the resulting spectrum needs to be decoded, and this will be the object of Section III.

Two important facts of general interest are to be underlined here :

1° as the experiment consists of disturbing the sample, the spectrum always reflects both the initial (fundamental, unperturbed) state, but also the final state properties of the sample.

2° any sample must be real, and will therefore behave more or less differently from what is expected for the ideal system.

II.2. *The Photoelectric Effect*

A wide range of photon energies, and a variety of possible projectile-particles are available to the spectroscopist. The choice will be based on a number of practical considerations :

- cost and difficulty of obtaining the incident beam
- amount and presentation of sample
- need and quality of special conditions, such as vacuum
- nature and properties of the detected particles, defining the analyzing system and detector.

The physical chemist or chemical physicist will also have to name what he wants to know about the sample; when it comes to study electronic structure and properties, it soon becomes clear that methods based on the photoelectric effect, such as photoelectron spectroscopy, are particularly well suited for this purpose.

The incident beam will be at least in the 10 eV energy-range, so as to override the electron energy eigenvalues in a solid or molecule. As a result of the interaction, electrons will have a given probability of being kicked out from their bound state, leaving a photo-ion behind them.

In this process, the electron carries away a positive kinetic energy, E_k, which is the balance between the initial (left-hand side in equation (1)) and final state (right-hand side) total energies, E_i and E_f, of the system :

$$E_i + h\nu = E_f + E_k \qquad (1)$$

This energy-conservation equation can be rearranged :

$$E_k = h\nu - (E_f - E_i) \qquad (2)$$

in order to display the role of the total energy difference, which is usually defined as the "binding energy".

$$E_b = E_f - E_i \qquad (3)$$

This value is commonly used as horizontal scale in E.S.C.A. spectra.

It should be stressed upon that it is not the measured quantity : the instrument provides an estimation of the electron kinetic energy, E'_k, which is different from the theoretical E_k by a value, C, which under good conditions is a constant for a given experiment :

$$E'_k = E_k - C \qquad (4)$$

This sets the so-called "calibration" problem.

We will now look in some detail how photoelectron spectroscopy works, and what the general aspects of a spectrum is like.

II.3. *Photoelectron spectroscopy*

II.3.1. Initial state description

Before photoemission, the sample is supposed to be in its equilibrium fundamental state; its total energy results from electronic, vibrational, and other contributions, the first two being of most interest to us.

The photon is usually in the UV or soft X-ray range. It is obvious from equation (2), that we need a monochromatic source (Table 1) : He lamps, and Al or Mg Kα X-ray sources are the most common ones at constant wavelength. It is noteworthy that synchrotron radiation, now available in a number of research centers (see Ref. [6] page 29) is a powerful tunable X-ray source, of which the advantage will be discussed about photoemission cross-sections.

Table 1. Characteristic data for common photon sources in photoelectron spectroscopy

Source	hν (eV)	Natural linewidth (eV)	λ(Å)
He (I)	21.22	∿ 0.002	584.3
He (II)	40.8	∿ 0.002	303.8
Mg $K\alpha_{1,2}$	1253.6	0.8	9.9
Al $K\alpha_{1,2}$	1486.6	0.9	8.3
Synchrotron	tunable		tunable

II.3.2. Photoemission

The interaction of electromagnetic radiation and charged particles in matter leads to a variety of effects. An X-ray beam can be transmitted, elastically or inelastically scattered, or ionise the sample. The relative probabilities of these phenomena depend on the incident beams frequency, direction of incidence, and polarization; but also on the electron transition rates between initial and final states, which can be evaluated theoretically.

Physically, the whole process takes place in a very short time, of the order of 10^{-16} sec. However, depending on the electronic properties of the system, some rearrangements can occur so fast, that they can be considered simultaneous with photoemission. This is the case for easily polarizable systems; the final state will in that case need to take the electronic "relaxation" in account.

II.3.3. Final state

The outgoing photoelectron, with positive kinetic energy, can be described by a plane wave perpendicular to its direction vector.

The final state of the sample, however, is more complicated to describe. In the simplest case, it is identical to the initial state, except for the missing electron. We can then apply Koopmans Theorem [1] and the frozen-core approximation. As pointed out before, however, there are cases where electronic rearrangements of different types occur so rapidly, that the final state description needs to take them into account : all other electrons in the molecule (or solid) will readjust to the creation of a positive hole; the system will thereby be stabilized, and the excess energy carried away by the photoelectron, according to equation (1).

In some cases, a valence electron can be excited (shaken up) or even ejected (shaken off) upon photoemission of a first electron. These processes, again, occur with a relative probability within that of photoelectron emission.

Finally, it must be noted that the existence of non-zero spin for a molecule or species under study creates an internal reference for unpaired core electrons left by photoemission. In that case, depending on the exchange interaction, several final states, more or less different in energy, can result (multiplet splitting).

II.3.4. The spectrum

The experimental arrangement for photoelectron spectroscopy consists typically of :

- an X-ray source (or U.V. lamp), possibly followed by a monochromator to reduce the natural linewidth (see Table I) with Kα X-rays;
- a sample inlet and handling system, fitted on a high vacuum (pressure $\leqslant 10^{-6}$ Torr) chamber, because of the extreme reactivity of electrons with gas molecules. The samples can be in the gas phase, but here we will only discuss the case of solids, such as polymers;
- an electron energy analyzer, magnetic or electrostatic, based on the bending of the charged particle motion;
- an electron multiplier, connected to an appropriate amplification circuit and to a multichannel analyzer or other data acquisition and handling system.

The use of an X-ray monochromator, eventually as part a dispersion compensation system, not only improves the instrumental resolution, but also [2] eliminates some spurious satellites resulting from other X-ray wavelengths emitted by the source, and drastically reduces the background cause by *Bremsstrahlung* photoelectrons.

When a spectrum is finally recorded, it shows the number of photoelectrons *versus* measured kinetic energy (E'_k in equation (4)). A typical wide-range spectrum is reproduced in Fig. 1. Its anatomy will be discussed in Section III. Usually, E'_k is converted into E_b scale, but care has then to be taken of determining the calibration constant C in equation (4) above.

Further details about the E.S.C.A. method will be found in the literature, *e.g.* in the benchmark books by Siegbahn et al [3,4], and in more recent references ([2,5,6] in English; [7] in German; [8] in French).

III. E.S.C.A. IN PRACTICE

Deleting strictly instrumental aspects of the experiment, which can be found elsewhere [6] , we will focus here on the analysis and interpretation of a spectrum, once it has been obtained as in Fig. 1, and the meaningful details have been magnified. From now on, we will also concentrate on X-ray excited photoelectron spectra, as only these contain informations on core and valence levels.

III.1. *Spectrum analysis*

We will distinguish three categories of details in a wide-range E.S.C.A. spectrum :

- sharp peaks, which usually include the strongest, but sometimes also very weak features,
- the valence region in the 0-30 eV binding energy range, normally weak compared to the core level peaks,
- a typical background, of which the intensity increases stepwise with every strong peak, going to higher E_b (lower E_k). This is the result of inelastic collisions, in arbitrary number, leading to energy loss by photoelectrons travelling through the sample.

The sharp peaks can be classified as follows :

1° genuine core level photoelectron peaks
2° shake-up or shake-off satellites
3° fine structure peaks (multiplet splitting)
4° quantified energy loss peaks, such as plasmons.

The following characteristics may be measured on the spectrum for every peak :

a. the binding energy
b. the intensity (peak surface or peak height)
c. the linewidth or lineshape (in case of asymmetry).

These measured values depend on three categories of factors :

- fundamental : related to the laws of physics.
- instrumental : inherent to the design and operating conditions of the real apparatus.
- "experimental" : we will restrict this adjective to a given experiment, with a real sample.

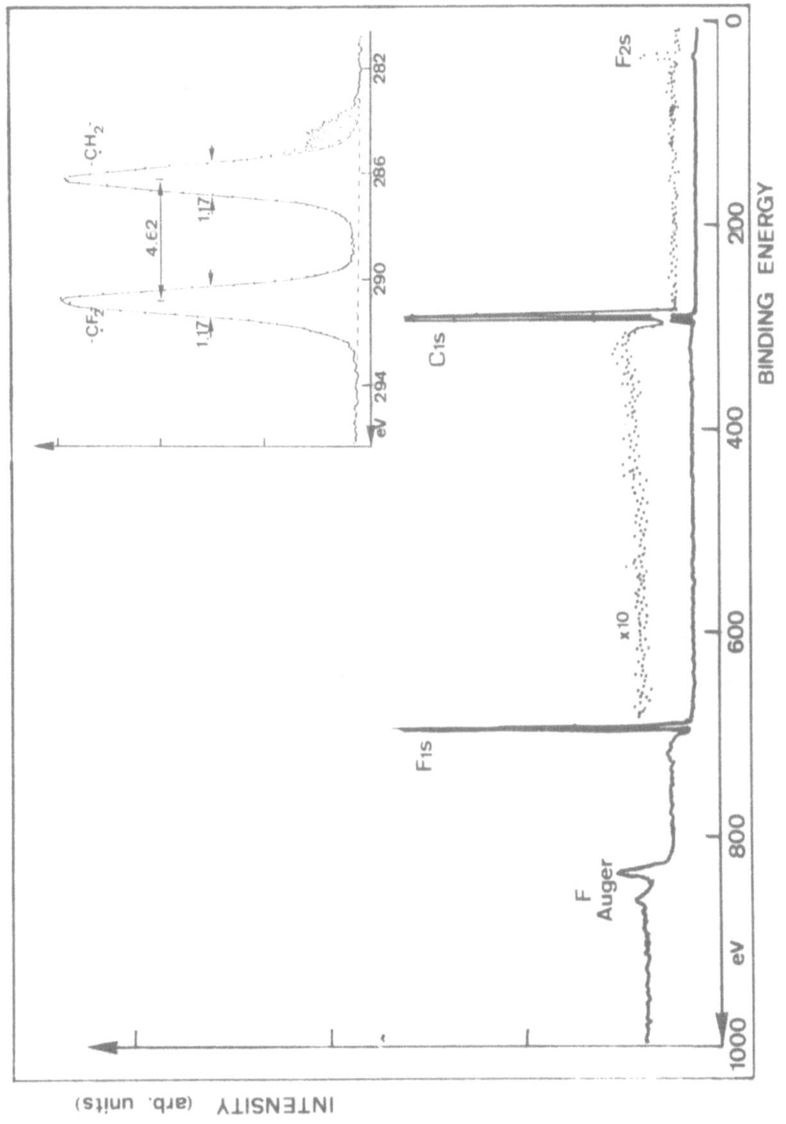

Figure 1 : Wide-range E.S.C.A. spectrum of an ethylene - tetrafluoroethylene copolymer.

The three types of contributions to any measured value on a spectrum should be well understood, in order to correctly distinguish the fundamental data which one is to deduce about the electronic structure of the sample. We will briefly survey them here :

III.1.1. The binding energy of a peak depends on the $(E_f - E_i)$ value, as defined in equation (3), but this can only be obtained from the observed E'_k if we know the calibration constant C in equation (4). Its value will depend in part of the instrument design, sample and spectrometer work functions, but also, in the case of insulating solid samples, on the electron retardation, an artifact resulting from positive charge accumulation on the surface. This will be an experimental factor, as the charging effect is a function of sample thickness, X-ray beam power, vacuum conditions, and eventually of the use of a low-energy electron "flood -gun" to compensate the effect.

Calibration of a spectrum is done by setting $E_b = 0$ at the Fermi level, when this can be observed, *i.e.* with conducting samples. Otherwise, external references are to be used, with procedures which have to be adjusted to every particular problem. The procedure used should be indicated in every paper, in order to allow valid comparisons between experiments on different instruments and/or under different conditions. Valence band structures are treated in the same way as peaks.

III.1.2. Peak intensity is fundamentally the result of photoelectric cross section, which is a function of the electron orbital and of the incident radiation parameters, but also of the photoelectron escape probability from the solid. Because of this factor, the effective investigation depth is reduced to ∿ 20 Å below the sample surface; E.S.C.A. is therefore very appropriate for studying surface states selectively. When this is not the purpose, care has to be taken to obtain a clean surface.

Instrumental and experimental factors of intensity include the spectrometer luminosity, the amount of material and atomic concentrations in the sample, the X-ray beam flux, and the detector efficiency.

III.1.3. The minimum line width or line shape is set by the uncertainty principle, in view of the very short characteristic time of the photoemission process. However, other contributions result from the instrumental resolution, including the possible lack of monochromaticity of the exciting radiation. A typical line in X-ray excited E.S.C.A. is 1 eV wide at half-height; the typical precision on the line position is then 0.1 eV.

The spurious charging effect, mentionned above for insulating samples, also leads to line broadening : this is the main reason for attempting to cancel it by the use of an electron gun located near the sample surface. The quality of sample preparation should also not be neglected, when good resolution is requested.

Asymmetric peaks, with a tail towards lower kinetic energies, have been observed, due to vibrational structure or to very low energy losses, when the resolution is too low to resolve individual components.

III.2. *Spectrum interpretation*

We will now define and illustrate the different types of photoelectron peaks which were listed in the preceding section.

III.2.1. Core and valence level photoelectron peak positions

On a correctly calibrated spectrum, the binding energy value represents the difference in total energy between final and initial states of the sample; in our case, this consists dominantly of electronic energy, with some possible vibrational fine structure, which is occasionally observed in E.S.C.A. as a line broadening.

Both the initial and final states are affected by molecular structure; electron eigenvalues will be tuned to their immediate surroundings in the molecule or solid. This is the basis of substitutional effect on the core electrons of a given atom, better known as the E.S.C.A. chemical shifts. We can relate this to permanent polarisation effects or inductive effects. A good example is provided by the polymers derived from polyethylene (Table II). Other popular correlations are between E_b and atomic charges, molecular potential, group electronegativities, *etc.* [9].

When the sample is highly polarizable, a fast relaxation process occurs, and the final state will be of lower absolute energy; the electron kinetic energy then becomes higher, shifting the E.S.C.A. peaks to lower binding energy. A remarkable correlation has been reported between polarizability, as observed by E.S.C.A. and in protor affinities. Relaxation effects can be different for the core and valence levels within the same molecule, and also between gas and solid of the same compound : as illustration, Table III summarizes a recent application for benzene, where a comparative study has led to the determination of the relaxation energies.

Table II. C_{1s} chemical shift increment for various substituents attached to the carbon atom (eV)

Polymer	Substituent	Shift
Polyethylene	- H	0.0
Poly(vinyl fluoride)	- F	3.1
Poly(vinyl chloride)	- Cl	1.8
Poly(vinyl hydroxide)	- OH	1.9
Poly(propylene)	- CH_3	- 0.1
Poly(1-butene)	- CH_2CH_3	- 0.2
Poly(styrene)	- C_6H_5	- 0.6
Poly(vinyl cyclohexane)	- C_6H_{11}	- 0.25

Table III. Summary of molecular and solid state electronic properties determined by E.S.C.A. for benzene (eV units).

Work function	4.1
First ionization energy	7.95
Energy gap	7.8 ± 0.2
Molecular relaxation energies:	
Core hole (C_{1s})	15.7
Extramolecular relaxation energies :	
- Core hole (C_{1s})	2.00
- Valence hole	1.35(*)

(*) extrapolated from A.I. BELKIND and V.V. GRECHOV, Phys. Stat. Sol. A26, 377 (1974).

III.2.2. Shake-up satellites of core level peaks

In the photoemission process, several final states can occur, in which a valence electron is simultaneously excited. Due to the higher final state energy, the kinetic energy of the photoelectron will then be a few eV lower. The relative probability of these phenomena is usually very low, but there are examples, such as in transition-metal compounds 2p spectra, where they are up to 60 % [11]

The shake-up satellites are found to obey the monopole selection rule, and their separation from the "main" core level peak is a valuable indication of the energy separation of valence levels, the upper one being uncompletely filled or completely unoccupied, and in that case non-observed in the E.S.C.A. valence band spectra.

An example of shake-up satellite interpretation is available in the case of acenes [12] ; a typical C_{1s} spectrum with satellites is reproduced in Fig. 2.

III.2.3. Multiplet splitting; plasmon energy loss peaks

These features, briefly defined above, will not be discussed because of their limited application in polymers.

III.2.4. Intensities

a. Core levels.

Photoemission probability from an inner shell is little affected by chemical bonding. The observed core level intensities are therefore useful either for determining polymer composition, or surface modifications. Relative photoelectric cross-sections have been calculated for Al Kα and Mg Kα - the most common X-ray photon sources [13].

When the surface composition of the polymer is different from the bulk, the situation gets more complicated. One has in that case to take into account the thickness of the surface layer, as well as mean free path informations, in order to completely interpret the intensity of core levels.

b. Valence levels.

The situation is rather different here. Molecular orbital cross-sections, σ_j, reflect their atomic origin and composition, through a formalism proposed by GELIUS [14] , and which has been widely verified :

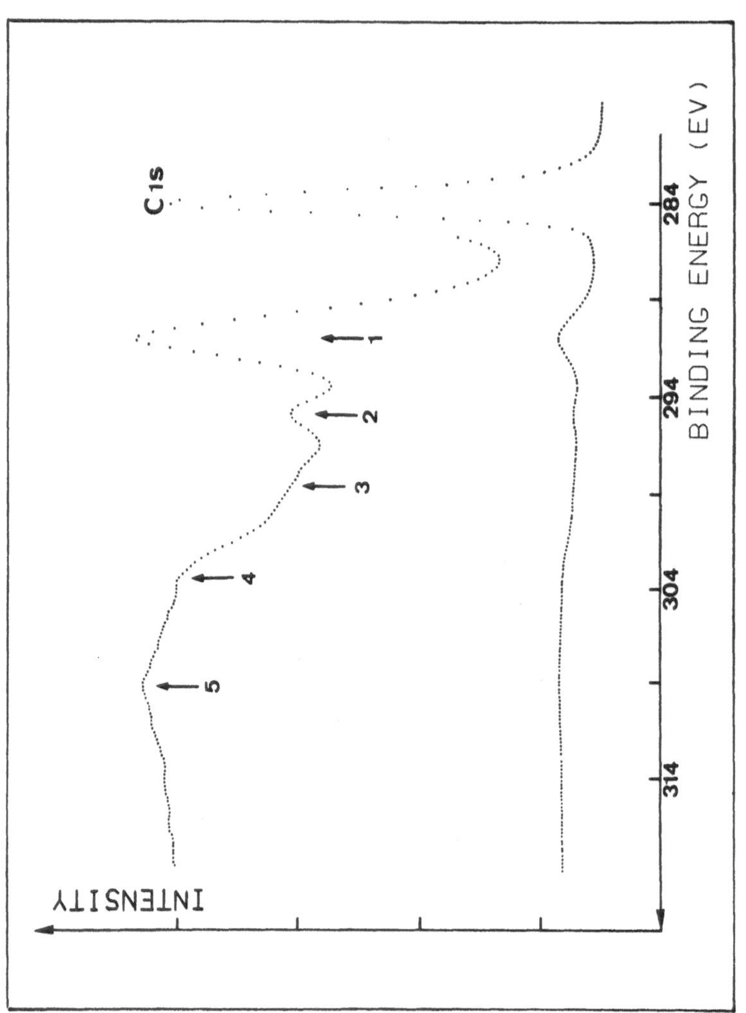

Figure 2 : Core level with shake-up satellites in solid benzene.

$$\sigma_j = \sum_{A,\lambda} P_{A\lambda_j} \sigma_{A\lambda} \quad (5)$$

Here j denotes a particular molecular orbital
A and λ define a given orbital of a particular atom
P is a linear combination factor related to electron population

so that the sum extends over all atomic orbitals of all atoms involved in the molecule. The validity of this empirical formula is limited to X-ray excitation, as all valence photoelectrons are then assumed to have the same final state.

IV. E.S.C.A. APPLIED TO ELECTRONIC STRUCTURE AND PROPERTIES OF POLYMERS

IV.1. *Atomic charges*

This theoretical concept, reflecting the electronic spatial distribution in a chemical system, can be defined in a number of ways, depending on the definition of the "atom" in a molecule. Core level photoelectron spectra, especially of light atoms (N, C, O ...) provide,an excellent way of testing any charge distribution model, as they are a sensitive probe of the real situation.

In a point charge model, core level chemical shifts for a given orbital are linearly related to the electrostatic potential[15]:

$$\Delta E_b = k\ \Delta Q_A + \Delta V_A \quad (6)$$

where ΔE_b is the chemical shift of a given core level, relative to some reference
k a proportionality factor
ΔQ_A the charge difference on the atom with respect to the reference
ΔV_A the potential energy created by the surrounding atoms.

The standard example for polymers is the effect of fluoro-substitution in polyethylene [16] : every additional fluorine atom, replacing hydrogen, induces an increment of 3.1 eV on the C_{1s} level to which it is bonded, and 0.8 eV to the neighbouring one.

IV.2. *Electronic relaxation*

Besides and in addition to the charge distribution effect, the chemical shifts can be affected by the electronic relaxation; this effect is particularly high when the molecule is highly polarizable. A remarkable illustration of this effect is found in the series of alkanes [17] as a function of increasing chain length. Polarizability increases with the size of the molecule, but rapidly levels off to an asymptotic value.

The origin of electronic relaxation can be both molecular and extra-molecular. Comparison of the gas and solid phase measurements (Fig. 3) has allowed to determine such contributions individually, particularly in the case of alkanes [17] and benzene [18] (Table III).

Relaxation can also affect differently core and valence levels. In the case of benzene, it has been shown that extramolecular relaxation energies are 2.0 eV for the C_{1s} level, and 1.35 eV for valence levels [18] .

IV.3. *Solid state parameters*

Two solid state characteristic parameters can be efficiently determined from a detailed analysis of E.S.C.A. spectra : the work function, and the energy gap. The deduction of these values is based on an accurate calibration of the spectra, and better, on comparison of gas and solid phase data, with detailed understanding of the contributions. This has been recently demonstrated in the case of benzene and other acenes (naphtalene, anthracene, tetracene, [12,18] .

IV.4. *Valence levels of homopolymers*

The most informative part of an E.S.C.A. spectrum is, without any doubt, the valence region, roughly located from 0-30 eV in binding energy. Two difficulties arise however, which can be overcome by careful experimental work <u>and</u> by comparison with theoretical models : first, intensities of valence bands in X-ray photoelectron spectra are very weak; second, the valence bands are very difficult to interpret without theoretical basis.

For this reason, theoretical work in the field is valuable to further experimental studies of polymers, including industrial applications.

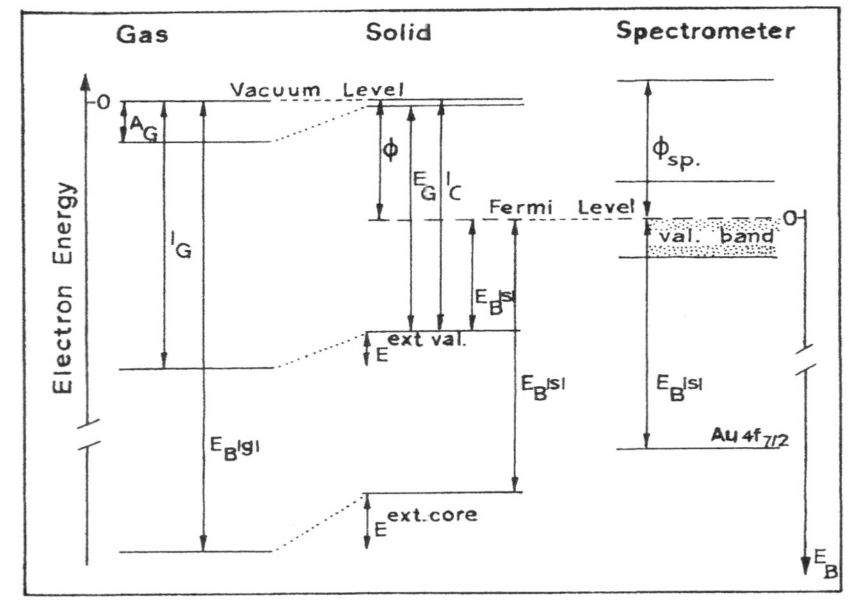

Φ	work function
E_G	energy gap
I_C	first ionization potential of the solid
I_G	first ionization potential of the gas
A_G	electron affinity (gas)
$E_B(g)$	binding energy (gas)
$E_B(s)$	binding energy (solid)
$E^{ext\ val}$	extramolecular relaxation energy (valence hole)
$E^{ext\ core}$	extramolecular relaxation energy (core hole)

Figure 3 : Relation between molecular and solid-state parameters expressed in energy units.

The effect of fluoro-substitution in polyethylene [19] and that of other simple substituents [20] have been recently described. Particularly in the case of polymers containing only carbon (polypropylene, polystyrene ...), the valence band is almost the only source of information, and it is established that extremely detailed conclusions can be drawn about the nature and the structure of the molecule.

It has been recently emphasized that cross-sectional effects, described above, have to be taken into account, when theoretical models of the electron density of states are compared to experimental data. Omission of this factor can lead to erroneous interpretations of one or more valence peaks[19].

Finally, the interpretation of core level shake-up satellites provides an experimental method of locating excited electron energy levels, which are important in the study of electronic properties. Again, this is illustrated as a function of molecular size by the series of acenes [12] . Shake-up satellites have been observed in polyethylene [21] , and also in other polymers [22] .

IV.5. *Copolymer valence bands*

Many real polymers do not result from a simple monomer, but from copolymerization of two monomers; the properties of such materials can be blended to almost any desirable result. This is however an active research field, and there is a strong need for characterizing the polymers which are prepared under different conditions.

E.S.C.A. valence band studies are very promising in that respect. A recent example is provided by an ethylene-tetrafluoroethylene copolymer [23] , for which the experimental valence band proves to be very sensitive to structural effects.

CONCLUSIONS

E.S.C.A. appears as a powerful method for investigating electronic structure and properties of polymers, provided that the necessary patience is taken for obtaining high quality experimental data, and that all the available informations from the spectra are correctly and completely understood. The attention is drawn to valence band structures, core level peaks, but also to their satellites. Peak positions should be interpreted in terms of total energy differences between initial and final states of the sample; intensities have to be related in particular to photoemission cross-section, and not only to densities of states or level populations Finally, it is reminded that the experiment itself requests a perturbation of the system, and that E.S.C.A., as an experimental method, has to do with real (not always close to ideal) samples.

ACKNOWLEDGMENT

The author wishes to thank his coworkers, whose work during five years has produced most of the experimental results illustrating this text, for the spirit in which research is continuing at the lab.

REFERENCES

[1] T. Koopmans, Physica, 1, 104 (1933).

[2] K. Siegbahn, D. Hammond, H. Fellner-Feldegg, E.F. Barnett, Science, 176, 245 (1972).

[3] K. Siegbahn, C. Nordling, A. Fahlman, R. Nordberg, K. Hamrin, J. Hedman, G. Johansson, T. Bergmark, S.E. Karlsson, I. Lindgren, B. Lindberg. E.S.C.A. Atomic, Molecular and Solid State Structure studied by means of Electron Spectroscopy. Almqvist and Wiksells AB, Uppsala, 1967.

[4] K. Siegbahn, C. Nordling, G. Johansson, J. Hedman, P.F. Heden, K. Hamrin, U. Gelius, T. Bergmark, L.O. Werme, R. Manne, Y. Baer. E.S.C.A. Applied to Free Molecules. North-Holland Publishing Co., Amsterdam, 1969.

[5] R. Caudano, J. Verbist (editors). Electron Spectroscopy - Progress in Research and Applications. (Proceedings of an International Conference held in Namur, 1974). Elsevier Scientific Publishing Co., Amsterdam, 1974.

[6] T.A. Carlson. Photoelectron and Auger Spectroscopy. Plenum Press, New York and London, 1975.

[7] H. Fellner-Feldegg, Wiss. Z. (Leipzig, Math-Naturwiss. Reike), 25, 355 (1976).

[8] J. Tousset, Analusis, 3, 221 (1975).

[9] D.A. Shirley, Advances in Chemical Physics, 23, 85 (1973). (I. Prigogine, S.A. Rice, editors). Wiley-Interscience, New York.

[10] R.L. Martin, D.A. Shirley, J. Amer. Chem. Soc.,96, 5299 (1974).

[11] a. S.K. Sen, Ann. Soc. Scient. Brux., 90, 125 (1976).
b. S. Asada, S. Sugano, J. Phys. Soc. Japan, 41, 1291 (1976).

[12] J. Riga, J.J. Pireaux, R. Caudano, J.J. Verbist, Physica Scripta, in press (1977). Report LSE 77-7-39.

[13] J.H. Scofield, J. Electr. Spectr. Rel. Phenom., 8, 129 (1976).

[14] U. Gelius, in Electron Spectroscopy, edited by D.A. Shirley, North-Holland, Amsterdam, 311 (1972).

[15] U. Gelius, Phys. Scripta, 9, 133 (1974).

[16] J.J. Pireaux, J. Riga, R. Caudano, J.J. Verbist, J.M. André, J. Delhalle, S. Delhalle, J. Electr. Spectr. Rel. Phenom., 5, 531 (1974).

[17] a. J.J. Pireaux, S. Svensson, E. Basilier, P-Å Malmqvist, U. Gelius, R. Caudano, K. Siegbahn, Phys. Rev. A14, 2133 (1976).
b. J.J. Pireaux, R. Caudano, Phys. Rev. B15, 2242 (1977).

[18] J. Riga, J.J. Pireaux, J. Verbist, Mol. Phys. 34, 131 (1977).

[19] J. Delhalle, S. Delhalle, J.M. André, J.J. Pireaux, J. Riga, R. Caudano, J. Verbist, J. Electr. Spectr. Rel. Phenom., in press (1977). Report LSE 76-6-18.

[20] J.J. Pireaux, J. Riga, R. Caudano, J.J. Verbist, J. Delhalle, S. Delhalle, J.M. André, Y. Gobillon, Physica Scripta, in press (1977). Report LSE 77-6-38.

[21] J.J. Pireaux, R. Caudano, J.J. Verbist. J. Electr. Spectr. Rel. Phenom., 5, 267 (1974).

[22] D.T. Clark, A. Dilks, J. Polym. Science (Chemistry Ed.), 14, 533 (1976).

[23] J.J. Pireaux, J. Riga, R. Caudano, J. Verbist, Y. Gobillon, J. Delhalle, S. Delhalle, J.M. André, submitted to J. Polym. Science. Report LSE 76-11-23.

CORRELATION ENERGY IN SOLIDS

N.H. March

Theoretical Chemistry Department,
University of Oxford, Oxford OXI 3TG, England

ABSTRACT. The problem of the correlation energy in solids is discussed via three principal approaches:

(i) Density functional theory. Here the main basis is the many-body study of interacting electrons in jellium.

(ii) Strong correlation in narrow energy bands. The transition metals afford the main area of interest here. The work of Gutzwiller is given some prominence.

(iii) Bond localization of electrons. While this approach is difficult to make quantitative, it is, it appears, a powerful way of simulating electron correlation effects.

1. INTRODUCTION

The customary definition of correlation energy is given by the difference between the exact energy E and the Hartree-Fock energy E_{HF},

$$E_{corr} = E - E_{HF} \tag{1.1}$$

It is worth dwelling, for a moment, before turning to the solid state, on the gross features of neutral atoms, with atomic number Z. Thus, we know from the Thomas-Fermi statistical theory that the non-relativistic total binding energy is proportional to $Z^{7/3}$. The correction due to exchange (analogue of Hartree-Fock method) is proportional to $Z^{5/3}$, and approximately the correlation energy is proportional to Z. Evidently, as we go to larger and larger atomic numbers, the exchange, and even more so, the correlation energy, become smaller and smaller fractions of the total binding energy,

J.-M. André et al. (eds.), Quantum Theory of Polymers, 49–73.

which becomes dominated by the Hartree terms. In atoms, therefore, we can summarize by saying that the correlation energy is quantitatively important for light atoms, but becomes less interesting with increasing atomic number.

The above argument is gross, in the sense that we have (in essence) been neglecting un-closed shell effects. One manifestation of correlation effects in atoms is, in fact, Hund's rule, which tells us that the state of maximum multiplicity is favoured. We shall return to this point briefly below, but it is more interesting to turn from the single-centre atomic problem to the multicentre problem of molecules and solids.

2. ELECTRON CORRELATION IN H_2 MOLECULE

Let us begin with the simplest multicentre problem involving correlation, the ground state of the H_2 molecule. We refer to the two classic approaches to building the symmetrical spatial ground state wave function* $\Psi(1,2)$, namely :

(i) The molecular orbital approach
(ii) The Heitler-London method

2.1. Linear combination of atomic orbitals

We shall use (i), by way of illustration, in the special, though widely used, approximation in molecules, in which the molecular orbital Ψ_{mo} holding both electrons (with, of course, opposed spins) is constructed as a linear combination of atomic orbitals (LCAO). Thus, we build from the ls orbital centred on nuclei A and B respectively, the approximate spatial wave function

$$\Psi(12) = \psi_{mo}(1)\psi_{mo}(2) = N[\psi^A_{1s}(1)+\psi^B_{1s}(1)]\,[\psi^A_{1s}(2)+\psi^B_{1s}(2)] \quad (2.1)$$

where N is a normalization factor. If we expand this out we can write

$$\begin{aligned}\Psi(12) &= N[\psi^A_{1s}(1)\psi^B_{1s}(2) + \psi^B_{1s}(1)\psi^A_{1s}(2)] \\ &+ N[\psi^A_{1s}(1)\psi^A_{1s}(2) + \psi^B_{1s}(1)\psi^B_{1s}(2)]\end{aligned} \quad (2.2)$$

It is now quite clear from (2.2) that the LCAO MO wave function (2.1) has two parts :

---

* The spin wave function is $\alpha(1)\beta(2) - \alpha(2)\beta(1)$ and is of course antisymmetric.

(a) The usual Heitler-London wave function, (given by the first term on the RHS of (2.2)).
(b) The ionic configurations in which both electrons are on either nucleus A or nucleus B (second term on the RHS of (2.2)).

It is now clear that as one increases the internuclear separation $R_{AB} \equiv R$, the ionic configurations, which from (2.2) are given equal weight with the Heitler-London form in the approximate LCAO MO wave function (2.1), must, in reality, get less and less important. The conclusion is that for large R, the LCAO MO method becomes inappropriate, the Heitler-London function becoming exact in the limit R tends to infinity.

2.2. Coulson-Fischer treatment of H_2

The above point can be highlighted by recalling the arguments of Coulson and Fischer (1949), who enquired, in the light of the difficulty with the MO function (2.1) as to the separation $R_{critical} = R_c$ say, at which (2.1) becomes seriously in error. These workers argued that, as the H-H distance increases, one of the two molecular orbitals (assumed identical in writing (2.1)) tends to concentrate more round nucleus A and the other round nucleus B. This situation could be represented by replacing $\psi_{mo}(1)$ and $\psi_{mo}(2)$ in (2.1) by

$$\psi_{1s}^{A} + \lambda\,\psi_{1s}^{B} \qquad \text{and} \qquad \psi_{1s}^{B} + \lambda\,\psi_{1s}^{A}$$

where λ is now a parameter not necessarily equal to unity. To be more precise, one can proceed to work with a variational wave function

$$\Psi_{\lambda}(12) = [\,\psi_{1s}^{A}(1)+\lambda\psi_{1s}^{B}(1)][\,\psi_{1s}^{B}(2)+\lambda\psi_{1s}^{A}(2)] \tag{2.3}$$

and determine λ by minimizing the total energy of the H_2 molecule (wave function (2.1) is evidently the special case of (2.3) when $\lambda = 1$).

The finding of Coulson and Fischer, on carrying through this calculation, was that for a H-H separation less than 1.6 times the equilibrium separation in H_2, λ is exactly unity. But when this separation is exceeded, λ rapidly falls to zero, clearly indicating a severe breakdown of the simple MO wave function (2.1).

If we symmetrize the above wave function (2.3), then we obtain

$$\Psi(1,2) = \Psi_{\text{Heitler-London}} + \mu\,\Psi_{\text{ionic}} \tag{2.4}$$

which was first employed by Weinbaum (1933). Of course, if this is minimized with respect to μ, and the result converted to λ, it again follows that the naive MO approximation, with $\lambda = 1$, exaggerates the importance of the ionic terms.

We have here been speaking of the role of electron correlation. In elementary terms, at sufficiently large R, the electrons go back onto their own atoms, i.e. the H_2 molecule dissociates into two neutral H atoms. It is very unfavourable, energetically, to include configurations representing one proton and one hydrogen negative ion.

3. CORRELATION IN MULTI-CENTRE PROBLEMS

Here, then we see that the role of electron correlation is much more profound in a multi-centre problem than in the single-centre atomic case. What is at stake, in simple pictorial terms, is whether electrons belong to the 'molecule' (which includes solids as the limit of very large molecules) as a whole, i.e. the electrons are delocalized, or whether electrons go back on to their 'own' atoms, i.e. are localized. This is what the problem of correlation energy is all about in multicentre systems (molecules and solids).

3.1. Examples of localized versus delocalized electrons

There seem to be cases where the answer is quite clearcut. Let us consider a few examples :

(a) The 3s valence electrons in body-centred-cubic sodium metal. Here, the evidence is overwhelming that these electrons are delocalized, i.e. move in molecular (Bloch) orbitals.

(b) The f electrons in rare earth metals. Here, in contrast to (a), the evidence is strongly in favour of the localized picture.

(c) The d electrons in transition metals. Here, it would seem that we have an intermediate case. There appear to be phenomena usefully talked about in terms of d bands, and yet electron correlation clearly cannot be ignored. We shall return to this case therefore later in the lectures.

As a final example, more closely related to the philosophy underlying this School, we take covalently bonded solids, e.g.

(d) The bonding electrons in the diamond structure. Evidence here is for highly directional charge distributions, in which electrons are 'localized' in bonds. Again, this very language is really about aspects of electron correlation.

In principle, in each of the four examples (a) - (d) above, we would like to be able to write down the total wave function of the ground state of the solid: the form of which must naturally reflect the 'localization' or otherwise of electrons. In practice, even when we can do so (approximately, of course), the wave function will be so complex as to be almost prohibitively difficult to use in calculations. Though we shall follow this route in case (c) below, this is the point at which we must discuss how recent trends in treating electron correlation have moved the emphasis away from the many-electron wave function, towards conceptually simpler quantities. An obvious quantity, which we shall use a great deal in these lectures, is the ground-state electron density $\rho(\underset{\sim}{r})$ in the molecule or solid under discussion.

3.2. Ground-state electron density

This quantity $\rho(\underset{\sim}{r})$, the number of electrons per unit volume at $\underset{\sim}{r}$ in the ground state, can naturally be calculated if we can get the N electron wave function Ψ from Schrödinger's equation. It has the great merit that it is an observable, and in practice X-ray scattering from a crystalline solid can determine it to useful accuracy.

But here, something of a dilemma arises. Even if we could get $\rho(\underset{\sim}{r})$ exactly from the N electron wave function (which we cannot!), it would be delocalized and there is no unambiguous way in which, say in example (d) above, we could extract the electron density in a Si-Si 'bond'.*

Thus, though the electron density is an observable, it cannot, of itself, tell us about localization of electrons. Nevertheless, the electron density $\rho(\underset{\sim}{r})$, in spite of its belonging to the molecule or crystal as a whole, is connected with electron correlation energy in a manner which it will be fruitful, at this point, to discuss.

4. GROUND-STATE ENERGY AS A FUNCTIONAL OF ELECTRON DENSITY

While Hartree, Fock and Slater were pioneering the self-consistent field calculation of one-electron wave functions, a parallel, though cruder, approach was being emphasized by Thomas, Fermi and later Dirac. It is this second approach, in which, originally, atoms and later molecules (Hund) and solids (Slater and Krutter) were described by means of the electron density that we focus on at this point.

The idea of Thomas and, independently, Fermi was to treat the electron cloud in an atom as an inhomogeneous Fermi gas to which free electron relations could be applied locally. In its simplest form, the argument went as follows. Whereas in the free Fermi gas, with wave functions which are simply plane waves, the constant electron density, ρ_o say, is related to the Fermi momentum p_f by the usual free electron equation

$$\rho_o = \frac{8\pi}{3h^3} p_f^3 \tag{4.1}$$

---

* Though this is, in general, true, in one or two exceptional cases, such as an ionic crystal Na Cl say, the X-ray scattering shows the electron density drops to (practically) zero between atoms and we can allocate charge without ambiguity to Na and to Cl.

in an inhomogeneous electron gas we have

$$\rho(\underset{\sim}{r}) = \frac{8\pi}{3h^3} p_f^3(\underset{\sim}{r}) \tag{4.2}$$

it being assumed thereby that (4.1) is applicable locally at $\underset{\sim}{r}$. If the electrons move in a one-body potential energy $V(\underset{\sim}{r})$ then the (classical) energy equation for the fastest electron takes the form

$$E_f \equiv \mu = \frac{p_f^2(\underset{\sim}{r})}{2m} + V(\underset{\sim}{r}) \tag{4.3}$$

In eqn (4.3), m is evidently the electronic mass, while the Fermi energy E_f must be independent of $\underset{\sim}{r}$ for otherwise electrons could redistribute to lower the total energy. E_f, or its equivalent in the above case, the chemical potential μ, minus the potential energy $V(\underset{\sim}{r})$ can be used to eliminate $p_f(\underset{\sim}{r})$ in eqn. (4.2), to yield

$$\rho(\underset{\sim}{r}) = \frac{8\pi}{3h^3} (2m)^{\frac{3}{2}} [\mu - V(\underset{\sim}{r})]^{\frac{3}{2}} \tag{4.4}$$

If the potential energy $V(\underset{\sim}{r})$ varies sufficiently slowly in space, then (4.4) must be, to adequate accuracy, the same density that one would obtain for an N electron assembly by finding the lowest N eigenfunctions ψ_i of the Hamiltonian $\frac{-\hbar^2}{2m}\nabla^2 + V(\underset{\sim}{r})$ and forming

$$\rho = \sum_{i=1}^{N} \psi_i^2(\underset{\sim}{r}) \tag{4.5}$$

Notice now that while the above argument seems to be a 'one-body' theory, we can measure the (many-body) electron density $\rho(\underset{\sim}{r})$ by X-ray scattering. Hence, in a system with slowly varying density $\rho(\underset{\sim}{r})$, we could extract, operationally, a one body potential $V(\underset{\sim}{r})$ from (4.4) which we could then use to determine the one-electron wave functions ψ_i in (4.5), and hence to reconstruct the density.

Here then we have a pointer to the fact that, at least in slowly varying electron densities, we can construct a many-body $\rho(\underset{\sim}{r})$ from a one-body potential. This result, which was recognized by many earlier workers, was finally proved by Kohn and his co-workers (Hohenberg and Kohn, 1964; Kohn and Sham, 1965), this work essentially completing the Thomas-Fermi theory. This approximate theory, leading to eqn. (4.4) can be derived from a variation principle for the total energy E, namely

$$E = C_k \int \rho^{5/3} d\underset{\sim}{r} + \int \rho V_N d\underset{\sim}{r} + \frac{1}{2} \int \rho V_e d\underset{\sim}{r} + U_{nn} \tag{4.6}$$

where the terms on the right hand side are the kinetic energy, the electron-nuclear potential energy, the electrostatic interaction of the charge cloud $\rho(\underset{\sim}{r})$ with itself and the nuclear-nuclear interaction energy. If we make $\delta E/\delta\rho = 0$, subject to

$$\int \rho \, d\underset{\sim}{r} = N \tag{4.7}$$

we regain eqn. (4.4) with μ playing now the role of the Lagrange multiplier introduced to take care of the subsidiary condition (4.7).

Dirac later introduced the exchange energy by adding a term (essentially) $-c_e \int \rho^{4/3} \, d\underset{\sim}{r}$ to the RHS of eqn (4.6), the corresponding Euler equation being the Thomas-Fermi-Dirac equation. Gombas and others added a term for the correlation energy on to eqn. (4.6) in addition to Dirac's exchange energy.

The work of Kohn and co-workers enables us to write an exact counterpart of equation (4.6), namely

$$E = \int t_{\underset{\sim}{r}}[\rho] \, d\underset{\sim}{r} + \int \rho V_N d\underset{\sim}{r} + \frac{1}{2}\int \rho V_e d\underset{\sim}{r} + \int \varepsilon_{xc}[\rho] \, d\underset{\sim}{r} + U_{nn} \tag{4.8}$$

Here $t_{\underset{\sim}{r}}[\rho]$ is the single-particle kinetic energy density, while $\epsilon_{xc}[\rho]$ is the exchange and correlation energy density. The corresponding Euler equation is readily found to be

$$\mu = \frac{\delta t_{\underset{\sim}{r}}}{\delta \rho} + V_N + V_e + \frac{\delta \epsilon_{xc}}{\delta \rho} \tag{4.9}$$

Since $t_{\underset{\sim}{r}}$ is a single-particle kinetic energy density, it is readily seen (cf Kohn and Sham 1965) that a one-body potential $V(\underset{\sim}{r})$ can be defined through

$$V(\underset{\sim}{r}) = V_N + V_e + \frac{\delta \epsilon_{xc}}{\delta \rho} \tag{4.10}$$

Finding the one-body eigenfunctions of $\frac{-\hbar^2}{2m}\nabla^2 + V(\underset{\sim}{r})$ and inserting them in (4.5) leads to the exact density $\rho(\underset{\sim}{r})$. But in practice, exact knowledge of ϵ_{xc} in eqn (4.10) would be tantamount to having an exact solution of the N-body problem, and this is out of the question. Therefore, the widely adopted procedure is to approximate $\delta\epsilon_{xc}/\delta\rho$ in eqn (4.10), the best known procedure being to write

$$\epsilon_{xc} = -c_e \, \rho^{4/3} \tag{4.11}$$

as in the Dirac modification of Thomas-Fermi theory. Slater has argued that correlation can be simulated by multiplying (4.11) by a factor α , and this is then the Dirac-Slater form of one-body potential theory.

We can therefore regard the basic result of eqn (4.10) as telling us that the one-body potential to be used to construct the (in principle) exact many-body $\rho(\underset{\sim}{r})$ has two parts, $V_N + V_e \equiv V_{Hartree}$ (though to be calculated with the exact $\rho(\underset{\sim}{r})$) plus an exchange and correlation potential. In general the one-body eigenfunctions of $\frac{-\hbar^2}{2m}\nabla^2 + V(\underset{\sim}{r})$ will be delocalized, i.e. will belong to the whole nuclear framework, in accord with our assertion above that $\rho(\underset{\sim}{r})$ will be delocalized. Nevertheless, the form of ϵ_{xc} in eqn (4.10) will depend on whether electrons are localized by Coulomb correlations, or whether the delocalizing effect of the kinetic energy outweighs Coulomb repulsions.

4.4. Local density approximation

In eqn (4.11) we indicated one possible, and widely useful, approximation to the exchange and correlation energy density ϵ_{xc}, namely to use the free electron exchange energy density locally. As remarked, Slater proposed to take at least some account of correlation by multiplying this by a scale factor, to be chosen in some way to simulate correlation repulsions.

4.4.1. Sommerfeld model

Pressing the argument, it is natural to study the correlation energy of the uniform background (Sommerfeld) model as a function of the (constant) density $\rho_0=3/4\pi r_s^3$, r_s being the mean interelectronic spacing. For this case, the result referred to above that the ground state energy is a unique functional of $\rho(\underline{r})$ reduces to the (almost trivial) statement that $E = E(r_s)$.

(a) High density

For this problem, the one-electron wave functions are plane-waves, and the Hartree-Fock determinant of these leads very directly to the result (energies in Rydbergs when r_s is in units of the Bohr radius a_0).

$$\frac{E_{HF}}{N} = \frac{2.21}{r_s^2} - \frac{0.916}{r_s} \tag{4.12}$$

with N electrons. The first term is kinetic energy (mean Fermi energy) while the second is the expectation value of $\sum_{i<j} e^2/r_{ij}$, the Coulomb repulsion term, with respect to the unperturbed Slater determinant of plane waves (i.e. $-\frac{0.916}{r_s}$ is the exchange energy, calculated by first-order perturbation theory). This term is easily used to regain eqn (4.11). However, as Macke showed, the second order perturbation term diverges and this heralds the presence of a term proportional to $\ln r_s$ in the total energy, i.e.

$$\frac{E}{N} = \frac{2.21}{r_s^2} - \frac{0.916}{r_s} + A\ln r_s + \ldots \tag{4.13}$$

The constant A was first obtained by Gell-Mann and Brueckner. Thus, in the small r_s (high density) limit, the leading term in the correlation energy is $A\ln r_s$. In case it looks very surprising to see such a term, one might rewrite the virial theorem for the Sommerfeld (jellium) model as (see March 1958)

$$2T + U = -r_s\frac{dE}{dr_s} = -\frac{dE}{d(\ln r_s)} \tag{4.14}$$

and it is not unnatural to see either r_s, $\ln r_s$ or both appear in high density expansions. The next term (independent of r_s) in the high density expansion of the total energy is also known.

Another way of expressing all this is to focus on the pair correlation function $g(r)$ between electrons : in the uniform gas this is the simplest non-trivial quantity, the density being constant. Sitting on an electron at the origin $\rho_o\,[g(r) - 1]$ gives the density of other electrons at distance r from it, and we can get the mean potential energy/electron by considering the potential due to the 'hole' $\rho_o\,[g(r) - 1]$ round it, at the origin. Thus

$$\frac{U}{N} = -\frac{1}{2}\int \rho_o \frac{g(r)-1}{r}\, d\underset{\sim}{r} \tag{4.15}$$

the $\frac{1}{2}$ entering to avoid counting electron-electron interactions twice over.

From the Hartree-Fock determinant of plane waves, $g(r)$ is readily calculated as

$$g(r) = 1 - \frac{9}{2}\left(\frac{j_1(k_f r)}{r}\right)^2 \tag{4.16}$$

where $j_1(x)$ is the first-order spherical Bessel function$(\sin x - x\cos x)/x^2$. This is the so-called Fermi hole, representing the statistical correlation between parallel spin electrons in a Fermi gas. The value $g(0) = \frac{1}{2}$ reflects the fact that in this approximation antiparallel spin electrons are not correlated : so that when Coulomb repulsions between such electrons are switched on we must have $g(0) < \frac{1}{2}$.

b) Low density limit

Having discussed the energy, and pair correlation function in the high density limit $r_s \to 0$ we now turn to the strong interaction regime of very large r_s. Then, we see from the Hartree-Fock energy (4.12) that in the limit $r_s \to \infty$, the potential energy U/particle, proportional to $1/r_s$, will dominate the kinetic energy. As can be anticipated from our arguments above, and as was first pointed out by Wigner (1934, 1938) electrons will become localized and form an electron 'crystal', thereby avoiding each other in the most efficient manner. Of the various crystal lattices examined to date, the body-centred cubic has the lowest Madelung energy and one finds as $r_s \to \infty$

$$\frac{E}{N} \simeq \frac{-1.80}{r_s} \tag{4.17}$$

leading to a correlation energy of $0.88/r_s$ from (4.12) and (4.17). The Hartree-Fock approach is seen to be almost a factor of 2 in error, the delocalized determinant of plane waves being therefore a quite inappropriate wave function.

Various authors have developed interpolation schemes to bridge the gap between the high and low density results for the correlation energy in jellium discussed above.

The reason we have gone into all this is to emphasize that we now know, not by any means exactly, but to useful accuracy, the correlation energy of jellium through the entire density range. Converting this to an energy density, we can combine it with exchange and then we can regard $\epsilon^{o}_{xc}(\rho_o)$ as known.

4.4.2. Form of $\epsilon_{xc}[\rho]$ from jellium results

The most widely used local density approximation is then to write for the as yet unknown functional $\epsilon_{xc}[\rho]$ the approximate local density result, based on our knowledge of jellium

$$\epsilon_{xc}[\rho] \doteq \epsilon^{o}_{xc}(\rho(\underset{\sim}{r})) \qquad (4.18)$$

Then we know the one-body potential $V(\underset{\sim}{r})$ and hence can proceed with calculations of one-electron functions.

The above expression for the exchange and correlation energy appears to be valuable in dealing with valence electrons, such as those in Na metal, example a above, which are certainly highly uniform.

It is hardly reasonable to expect the above approximation to be valid in very inhomogeneous electron gases, though gradient corrections to eqn (4.18) have been examined by Herman and co-workers and look promising (see also Stoddart et al who have summed subseries of gradient terms to all orders and obtained numerical results for Be metal).

For the f electrons in rare earth metals, in contrast, an atomic-like correlation problem presents itself and is presently not well understood.

Thus, having indicated the potential offered for the future by the density functional method, it must be said that reliable results for $\epsilon_{xc}[\rho]$ are only as yet available for slowly varying densities. Therefore, in the remainder of these notes we shall focus on :

(i) Strong correlations in narrow energy bands (relevant, among other examples, to case (c) above)

(ii) Exchange and correlation in the presence of chemical bonding (our example (d) above).

These will also have the merit of widening the above discussion, though (i) above will, as we shall see, lead us back towards the Coulson-Fischer work. In (ii), while in principle we should be able to work via $\epsilon_{xc}[\rho]$, in practice we shall follow Inkson and work with the self-energy (see also Collings here)

Before turning to these cases, we conclude the discussion by remarking once again that as we go from the high density limit, with a determinant of plane waves, and a sharp Fermi surface, eventually electron-electron interactions create particle-hole pairs which so change the character of the assembly that, at some critical density, all semblance of a Fermi surface is lost and we pass to an insulating state. It needs emphasising that this metal-insulator transition, the so-called Wigner transition, taking place in jellium, is quite different from that taking place as we bring a lattice of H atoms together from large lattice spacing. Eventually, in this latter one-electron/atom case, we pass to a metallic state, the assembly now having undergone a so-called Mott transition. Thus electron correlation can, in admittedly extreme circumstances, cause dramatic effects as we go through a delocalized electron-localized electron transition. We shall be even more explicit about this below in case (i) to which we now turn.

5. STRONG CORRELATIONS IN NARROW ENERGY BANDS

A quite different regime from that of the homogeneous electron gas (jellium) discussed above occurs when we have tight-binding bands. One example, no doubt among many, is our example (c) for the d electrons in transition metals. Then our previous arguments can hardly be expected to remain appropriate.

We shall give some prominence, in this regime, to the approximate variational calculation of Gutzwiller (1965). Though we shall write down below a model Hamiltonian in the language of field theory, let us introduce his method by listing two basic assumptions of his treatment :

(1) Electrons of one spin direction are assumed fixed when one studies the motion of an electron with opposite spin; a reasonable and usual assumption for narrow bands.

(2) The correlations between electrons of opposite spin are assumed to reduce the amplitude of the one-electron wave function on the sites already occupied by an electron of opposite spin, but are otherwise assumed to produce no scattering.

Let η be the average reduction in amplitude of a one-electron wave function on a doubly occupied site. Clearly we must have η in the range $0 \leqslant \eta \leqslant 1$. For the one-electron wave function with wave vector $\underset{\sim}{k}$ and spin σ one can write approximately

$$\phi_{\underset{\sim}{k}\sigma}(r) \rightarrow \frac{1}{[1-(1-\eta^2)N_{-\sigma}]^{1/2}} \sum_{\underset{\sim}{l}} \exp(i\underset{\sim}{k}.\underset{\sim}{l}) \, [1-(1-\eta)n_{\underset{\sim}{l}}-\sigma] \times w_\sigma(\underset{\sim}{r}-\underset{\sim}{l}) \tag{5.1.}$$

where $N_{-\sigma}$ is the number of electrons with spin $-\sigma$, $\underset{\sim}{l}$'s denote the direct lattice vectors, $n_{\underset{\sim}{l}-\sigma}$ is the number of electrons at site $\underset{\sim}{l}$ with spin $-\sigma$, while $w_\sigma(r)$ is the Wannier function representing the single-band under discussion.

Two results now follow for $\eta \neq 1$:

(a) The concentration ν of doubly occupied sites is reduced by a factor η^2 .

The law of mass action can be written for 2 singly occupied sites A and B with opposite spins → 1 empty plus 1 doubly occupied site

$$c_o \nu = \eta^2 c_\uparrow c_\downarrow \tag{5.2}$$

where $c_o = 1 - n + \nu$ is the concentration of empty sites and

$$c_\uparrow = c_\downarrow = \tfrac{1}{2} n - \nu \tag{5.3}$$

is the concentration of single occupied ones.

(b) Transfer integrals for an electron involving a hop from or to a doubly occupied site will be lowered by a factor η. Optimization with respect to the total number of doubly occupied atoms ν then gives the ground state energy.

It will be clear to the reader that the wave function (5.1) is closely related to that of Coulson and Fischer discussed earlier in section 2.2.

To formalize this, let us write down, using Gutzwiller's notation, the model Hamiltonian which represents a single tight-binding band with only intra-atomic Coulomb interactions between the electrons. It takes the form

$$\begin{aligned} H = & \sum_{\underset{\sim}{k}} \epsilon_{\underset{\sim}{k}} [a^+_{\underset{\sim}{k}\uparrow} a_{\underset{\sim}{k}\uparrow} + a^+_{\underset{\sim}{k}\downarrow} a_{\underset{\sim}{k}\downarrow}] \\ & + C \sum_{\underset{\sim}{g}} a^+_{\underset{\sim}{g}\uparrow} a^+_{\underset{\sim}{g}\downarrow} a_{\underset{\sim}{g}\downarrow} a_{\underset{\sim}{g}\uparrow} \end{aligned} \tag{5.4}$$

Here $a^+_{\underset{\sim}{k}}$ and $a^+_{\underset{\sim}{g}}$ are the creation operators for electrons in the Bloch state $\underset{\sim}{k}$ and the Wannier state g respectively, C is the intra-atomic Coulomb repulsion and $\epsilon_{\underset{\sim}{k}}$ is the kinetic energy, with the zero of energy chosen so that

$$\sum_{\underset{\sim}{k}} \epsilon_{\underset{\sim}{k}} = 0 \tag{5.5}$$

5.1. Results for one electron per atom

Gutzwiller used his results to obtain a criterion for itinerant ferromagnetism: we shall here choose to discuss instead the more recent results of Brinkman and Rice (1970), who applied Gutzwiller's method to the problem of the metal-insulator transition.

For other than one electron per atom, Gutzwiller's variational state is inevitably metallic, so we restrict ourselves to this electron/atom ratio.

Gutzwiller's work then yields expressions for η, and for q, the discontinuity in the single particle occupation number $\langle n_{\underset{\sim}{k}} \rangle$ at the Fermi surface, in terms of $\bar{\nu}$ defined by

$$\bar{\nu} \equiv \langle n_{i\uparrow} n_{i\downarrow} \rangle_{\eta} \tag{5.6}$$

as

$$\eta = \bar{\nu}/(\frac{1}{2} - \bar{\nu}) \tag{5.7}$$

$$q = 16\,\bar{\nu}\,(\frac{1}{2} - \bar{\nu}) \tag{5.8}$$

Minimizing the energy with respect to $\bar{\nu}$ one then obtains

$$\bar{\nu} = \frac{1}{4}\,(1 - \frac{C}{C_o}) \tag{5.9}$$

$$q = 1 - (\frac{C}{C_o})^2 \tag{5.10}$$

while the expectation value of the energy in the (paramagnetic) ground state is given by

$$\langle H \rangle = \bar{\epsilon}\,(1 - C/C_o)^2 \tag{5.11}$$

Here

$$\bar{\epsilon} = 2 \sum_{k<k_f} \epsilon_{\bar{k}} < 0 \tag{5.12}$$

is the average energy without correlation and $C_o = -8\bar{\epsilon}$

Thus at a critical value of the interaction strength $C = C_o$, the number of doubly occupied sites and the discontinuity in the single-particle occupation number at the Fermi surface go to zero. The value of the energy also approaches zero, the expectation value of the energy of a paramagnetic localized insulating state.

However, it must be that some magnetically ordered insulating ground state will have a lower energy than the paramagnetic insulating state and a transition to an insulating magnetically ordered ground state will occur for a value of $C < C_o$.

In spite of this circumstance, it is of interest to calculate the properties of the Gutzwiller trial wave function in the metallic state.

If, following Brinkman and Rice (1970) it is assumed that the effective-mass renormalization m*/m is due to the frequency dependence of the self-energy only, as for example, in the electron-phonon and paramagnon problems (see Schrieffer 1964; Berk and Schrieffer, 1969; Doniach and Engelberg, 1969) then m*/m is immediately calculable from the discontinuity q as

$$(m^*/m) = q^{-1} = [\,1 - (C/C_o)^2\,]^{-1} \tag{5.13}$$

showing that the effective mass diverges as $C \to C_o$.

In Gutzwiller's work, the minimum energy occurs for states with differing numbers of upward and downward spin electrons. The spin susceptibility χ can thus be calculated as

$$\chi^{-1} = \frac{1-(C/C_o)^2}{\rho(E_f)} \left[1 - \rho(E_f) C \left\{ 1 + \frac{(C/2C_o)}{1+(C/C_o)^2} \right\} \right] \tag{5.14}$$

which again diverges as $C \rightarrow C_o$, in the same manner as the effective mass.

It should be noted that, in this case, the Stoner enhancement factor $[1- \rho(E_f)C]^{-1}$ does not appear : it is the effective mass enhancement that is basic here.

5.2. Correlation effects in transition metals

The Gutzwiller approximation discussed above has been employed by Friedel and Sayers (1977) to study the influence of correlation effects on the bulk modulus and equilibrium lattice spacing of the transition metals.

The empirical facts on the bulk modulus are as follows :

(i) The bulk modulus in the 4d and 5d transition metal series varies regularly with the filling of the band, with a maximum in the middle of the series (Gschneider, 1964).

(ii) In the 3d series, however, there is a marked deviation from this behaviour, the bulk modulus of Gr, Mn, Fe and Co being smaller than the values expected from the trend in the 4d and 5d series.

Actually, a similar anomaly occurs in the equilibrium atomic volume of the 3d series (Gschneider 1964), the atomic volume of Fe and Co being larger than that of Ni, in contrast to the behaviour in the 4d and 5d series where the atomic volume varies regularly with the filling of the d band. Mn is notable for its large atomic volume and small bulk modulus.

In terms of Gutzwiller's theory, Friedel and Sayers (1977) have shown that, to second-order in the electron-electron interaction U, the part of the d electron energy dependent on the bandwidth W is

$$E_B = - AW - B/W \tag{5.15}$$

where

$$A = \frac{z}{20}(10 - z) \tag{5.16}$$

$$B = 45\left[\frac{z}{10}\left(1 - \frac{z}{10}\right) U\right]^2 \tag{5.17}$$

* Essentially the interaction energy C used above.

z being the average number of electrons/atom in the d band. Here, five overlapping rectangular d bands have been assumed, in the interests of simplicity.

The term - AW represents the band energy of the d electrons in the absence of correlations and leads to good agreement with the measured cohesive energy in the 4d and 5d series (Friedel, 1969).

The second term in E_B, namely - B/W represents the first-order contribution of correlations to the energy and reduces the electron interaction energy $9z^2U/20$ in the solid by preventing the electrons hopping on to the same atom.

Friedel and Sayers now combine the above expression for E_B with a Born-Mayer repulsive term

$$E_R = C \exp(-pR) \tag{5.18}$$

where R is the Wigner-Seitz radius (cf Ducastelle 1970), and they use the approximation

$$W = W_o e^{-qR} \tag{5.19}$$

for the band width.

Neglecting correlation (i.e. B = 0) they regain the result of Ducastelle (1970) for the equilibrium Wigner-Seitz radius

$$R_o = \frac{1}{p-q} \ln \frac{pC}{qAW_o} \tag{5.20}$$

Now, the correlation effects we are interested in here lead, to first order in the correlation, to the relative change in the Wigner-Seitz radius due to correlations

$$\frac{R-R_o}{R_o} \simeq \frac{q}{(p-q)R_o} \frac{z}{10}\left(1 - \frac{z}{10}\right)\frac{U^2}{W^2} \tag{5.21}$$

The effect of correlations is thus seen to be a lattice expansion, the effect being greatest for a half filled d band (assuming one can neglect the dependence of U, W, pR_o and qR_o on z).

Using the values $pR_o \simeq 6$, $qR_o \simeq 3$ used by Ducastelle and the average values U ≃ 3 ev, W ≃ 6 ev deduced for the 3d series by Friedel and Sayers (1977) a 20% increase in the Wigner-Seitz radius in the middle of the 3d series due to correlation effects results. This is somewhat larger than the observed anomaly. Friedel and Sayers suggest that inclusion of higher-order terms in the expansion in U/W would serve to reduce this correction.

Bulk modulus

The bulk modulus K is given by

$$K = \frac{Vd^2E}{dV^2} = \frac{R^2}{qV}\frac{d^2E}{dR^2} \tag{5.22}$$

Following Ducastelle, for the case B = 0 one obtains

$$K_o = \frac{q(p-q)}{qV} R_o^2 AW \simeq \frac{AW}{V} \tag{5.23}$$

The change in bulk modulus is given, to first order in the correlation, by

$$\begin{aligned} K - K_o &= \frac{-q^2R^2}{V} 5\left[\frac{z}{10}\left(1 - \frac{z}{10}\right)\right]^2 \frac{u^2}{W} \\ &\simeq \frac{-45}{V}\left[\frac{z}{10}\left(1 - \frac{z}{10}\right)\right]^2 \frac{u^2}{W} \end{aligned} \tag{5.24}$$

with q R ≃ 3.

The detailed form of band width is not very important, altering only the coefficient 45 by some 10% probably.

Correlation effects are thus seen to reduce the bulk modulus, the reduction being greatest in the middle of the series. For z = 5, U = 3 ev and W = 6 ev, this reduction is around 50% and is in good agreement with the size of the measured anomaly in the 3d series.

Friedel and Sayers also discuss the effect of magnetism in Fe, Co and Ni on the above results. They conclude that the effect of magnetization is rather smaller than that of correlation; both effects acting to decrease the bulk modulus and increase the Wigner-Seitz radius due to transfer of electrons into orbitals of high kinetic energy.

Having discussed various aspects of the physical effects of strong correlations in narrow energy bands, we turn finally to example (d), namely correlations in chemically bonded solids.

6. EXCHANGE AND CORRELATION POTENTIAL IN CHEMICALLY BONDED SOLIDS (ESPECIALLY SI)

In solids in which the electron density varies strongly in space, one should not expect any local approximation to the exchange and correlation potential to be valid, as we remarked in section 4.4.

Kane (1971) has shown that in Si any local potential will leave a discrepancy of about 15 to 20% between band gaps and effective masses. It is not possible to fit both to experiment: and the reason is that in the diamond-like structure we are dealing with highly directional and inhomogeneous electron density clouds.

Using the formalism of Hedin (1965; see also Hedin & Lundqvist S., 1969) with a static screened interaction Kane (1972) calculated the variation in the electron self-energy through the valence and conduction bands and showed that this variation was incompatible with an $\alpha\rho^{\frac{1}{3}}$ potential.

Bennett and Inkson (1976) have extended Kane's calculation for the valence electrons using a model dielectric function due to Inkson (1972). In this way, it is possible to write down a dynamic screened interaction W. Following Kane, these workers have employed pseudowavefunctions in calculating the screened exchange part of the self-energy. They also calculate the Coulomb hole part using the above dynamic interaction and a free-electron Green function.

They find only a weak energy dependence in the Coulomb hole term; it leads to a flattening of the total self-energy across the band of about 1 ev.

Kane demonstrated that the principal band gaps are not unduly affected by using a static instead of a dynamic dielectric function and the results of Bennett and Inkson for the principal points at which Kane uses a dynamic function agree with his. However, the self-energy at these points is found to be sensitive to whether or not a static approximation is invoked.

Bennett and Inkson argue that the static approximation is inappropriate for screened exchange energies in the conduction bands, however.

6.1. Screened exchange part of self-energy

The method employed by Bennett and Inkson (BI) to obtain the screened exchange part of the self-energy was to find the Fourier components $\Sigma\ (\underset{\sim}{k} + \underset{\sim}{G},\ \underset{\sim}{k} + \underset{\sim}{Q}\ ;\ E(\underset{\sim}{k})\ )$ of the self-energy potential $\Sigma(\underset{\sim}{r}\underset{\sim}{r}'\ ;\ E\ (\underset{\sim}{k})\ )$. Their model for these Fourier components is diagonal and depends only on $|\underset{\sim}{k} + \underset{\sim}{G}|$ and on the band index.

The dynamic self-energy, following Hedin (1965) is approximated by the form

$$\Sigma(\underset{\sim}{r}\underset{\sim}{r}'\omega) = \frac{i}{2\pi}\int_{-\infty}^{\infty} d\omega'\ G(\underset{\sim}{r}\underset{\sim}{r}'\ \omega-\omega')\ W(\underset{\sim}{r}\underset{\sim}{r}'\ \omega')\ \exp(-i\delta\omega') \tag{6.1}$$

G being the interacting Green function, W the dynamic screened interaction and δ a positive infinitesimal. The presence of δ ensures that unoccupied states do not contribute to the screened exchange part of the self-energy.

BI use the standard form

$$G(\underset{\sim}{r}\underset{\sim}{r}'\omega) = \sum_{k} [\ \phi_{\underset{\sim}{k}}(r)\phi^{*}_{\underset{\sim}{k}}(r')] / [\ \omega-E(\underset{\sim}{k}) + i\delta(\mu-\omega)] \tag{6.2}$$

for the Green function, the Φ_k's being a complete orthonormal set of eigenfunctions for the system, having eigenvalues $E(\underset{\sim}{k})$. They take these in their work as an empirically fitted set of pseudowavefunctions. [Here, we follow them by using $\underset{\sim}{k}$ in the extended zone scheme; unless we label states $(\underset{\sim}{k},\ \nu)$

when k is in the first Brillouin zone and $\nu = 1,2,3,4$ is a band index] . In the equation for G, μ is the chemical potential.

Now, following BI the screened interaction is assumed diagonal: and thus one can write

$$W(\underset{\sim}{q}\ \underset{\sim}{q}'\ \omega') = (2\pi)^3\,\delta\,(\underset{\sim}{q}-\underset{\sim}{q}')\,U(q)/\,\varepsilon(q,\omega') \tag{6.3}$$

where $U(q)$ is simply the Coulomb interaction $4\,\pi\,e^2/q^2$.

Using the fact that the $\phi_{\underset{\sim}{k}}$'s are Bloch functions we may write their Fourier components as

$$<\underset{\sim}{q}|\phi_{\underset{\sim}{k}}(r)> = (2\pi)^3 \sum_{\underset{\sim}{G}} \delta\,(\underset{\sim}{k}+\underset{\sim}{G}-\underset{\sim}{q})\ a_{\underset{\sim}{G}}(\underset{\sim}{k})/\sqrt{V} \tag{6.4}$$

V being the volume of the crystal, and as usual the $\underset{\sim}{G}$'s are reciprocal lattice vectors.

Taking the double Fourier transform of Σ we obtain then

$$\Sigma(\underset{\sim}{k}+\underset{\sim}{G},\ \underset{\sim}{k}+\underset{\sim}{Q},\omega) = \frac{i}{2\pi}\int_{-\infty}^{\infty} d\omega' \sum_{\underset{\sim}{l},\underset{\sim}{S}} \frac{a_{\underset{\sim}{S}+\underset{\sim}{G}}(\underset{\sim}{l})\ a_{\underset{\sim}{S}+\underset{\sim}{Q}}(\underset{\sim}{l})}{\omega-\omega' - E(\underset{\sim}{l}) + i\delta\,(\mu-\omega-\omega')}$$

$$\times\ v(|\underset{\sim}{l}+\underset{\sim}{S}-\underset{\sim}{k}|)/\epsilon(|\underset{\sim}{l}+\underset{\sim}{S}-\underset{\sim}{k}|,\omega')\ \exp(-i\delta\omega') \tag{6.5}$$

the $\underset{\sim}{S}$'s being reciprocal lattice vectors and the $\underset{\sim}{l}$ labelling the eigenstates. As usual, translational symmetry of the lattice has ensured that only Fourier components whose arguments differ by reciprocal lattice vectors may be non-zero.

Following Hedin and Lundqvist (1969) it is useful to split up the contour integration in 6.5 into two parts

$$\begin{aligned} \Sigma &= [\ \text{Residues of } G \times W(\omega - E(\underset{\sim}{l}))\,] \\ &\quad + [\ G(\omega_p) \times \text{Residues of } W\ ] \\ &= \text{Screened exchange part} + \text{Coulomb hole part} \end{aligned} \tag{6.6}$$

Then one obtains for the screened exchange contribution Σ_{SX} of the self energy

$$\Sigma_{SX}(\underset{\sim}{k}+\underset{\sim}{G},\ \underset{\sim}{k}+\underset{\sim}{Q};\ E(\underset{\sim}{k})) = -\sum_{\underset{\sim}{l}}^{occ}\ \sum_{\underset{\sim}{S}} a_{\underset{\sim}{S}+\underset{\sim}{G}}(\underset{\sim}{l})\ a^{*}_{\underset{\sim}{S}+\underset{\sim}{Q}}(\underset{\sim}{l})$$

$$U(|\underset{\sim}{l}+\underset{\sim}{S}-\underset{\sim}{k}|)/\epsilon(|\underset{\sim}{l}+\underset{\sim}{S}-\underset{\sim}{k}|;\ E(\underset{\sim}{k})-E(\underset{\sim}{l})) \tag{6.7}$$

only states below the Fermi level being included in the l sum. The coefficients $a_G(\underset{\sim}{l})$ are taken to be those of pseudowavefunctions; this is fairly good since the overlap with the core states is not large.

For the dielectric function BI employ the Inkson (1972) model, namely

$$\epsilon_I(q,\omega) = 1 + (\epsilon_o-1)/(1 + \frac{q^2}{q_{TF}^2}(\epsilon_o-1) - \frac{\omega^2}{\omega_p^2}(\epsilon_o-1)) \quad (6.8)$$

where q_{TF} is the Thomas-Fermi wave number, ω_p is the plasma frequency and ϵ_o is the static dielectric constant.

The expectation value of the screened exchange is given by

$$\langle\Sigma_{SX}(\underset{\sim}{k})\rangle = \int d\underset{\sim}{r}d\underset{\sim}{r}' \, \phi^*_{\underset{\sim}{k}}(\underset{\sim}{r}) \, \Sigma_{SX}(\underset{\sim}{rr}';E(\underset{\sim}{k})) \, \phi_{\underset{\sim}{k}}(\underset{\sim}{r}')$$

$$= \sum_{\underset{\sim}{G}\,\underset{\sim}{Q}} a^*_G(\underset{\sim}{k}) \Sigma_{SX}(\underset{\sim}{k}+\underset{\sim}{G},\ \underset{\sim}{k}+\underset{\sim}{Q};\ E(\underset{\sim}{k}))\ a_{\underset{\sim}{G}}(\underset{\sim}{k}) \quad (6.9)$$

Inserting Σ_{SX} into this equation, one arrives at a first-order estimate of the screened exchange energy.

6.2. Coulomb hole term

To evaluate the Coulomb hole term, the dielectric function is usefully rewritten as

$$\epsilon_I^{-1}(q\omega) = (\omega_r(q)^2 - \omega_p^2 - \omega^2)/(\omega_r^2(q)-\omega^2) \quad (6.10)$$

where

$$\omega_r(q)^2 = \omega_p^2\frac{\epsilon_o}{\epsilon_o-1}\left(1 + \frac{q^2}{q_{TF}^2}\,\frac{\epsilon_o-1}{\epsilon_o}\right) \quad (6.11)$$

Thus, the dynamic screened interaction has poles at $\omega = \pm\ \omega_r(q)$ with residues $\pm\ (\omega_p^2/2\omega_r(q)) \times U(q)$. BI choose the contour of integration so as to include only the positive frequency pole. Since the energies of interest in the integral are now well above the Fermi level, one can approximately treat the electrons as free and replace $a_{\underset{\sim}{G}}$ by δ_{Go} and $E(\underset{\sim}{l})$ by l^2. Then one obtains for the Coulomb hole part Σ_{CH} of the self-energy the result

$$\Sigma_{CH}(\underset{\sim}{k}, E(\underset{\sim}{k})) =$$

$$\frac{-\omega_p^2}{4k(2\pi)^2} \int_0^\infty q\,dq \, \frac{U(q)}{\omega_r(q)} \ln \left| \frac{\omega_r(q) + 2kq + q^2}{\omega_r(q) - 2kq + q^2} \right| \tag{6.12}$$

where units have been used in which $\hbar^2 = 1$, $2m_e = 1$ and $2\pi/a = 1$, where a is the lattice constant : 5.43 A in Si.

In the static approximation, $\omega_r(|\underset{\sim}{\ell}-\underset{\sim}{k}|)$ is assumed to increase so rapidly with $|\underset{\sim}{\ell} - \underset{\sim}{k}|$ that in comparison $E(\underset{\sim}{k}) - E(\underset{\sim}{\ell})$ will be negligible. Then the simpler formula, corresponding to constant Coulomb hole over the zone results; namely

$$\Sigma_{CH} = -\frac{1}{2}\omega_p^2 \Sigma_{\underset{\sim}{q}} U(q)/\omega_r(q)^2 \tag{6.13}$$

We shall not go into detail on the numerical calculations of BI for Si. However, as an example, we show their results for the Fourier components of the screened exchange energy at X_4 ; $\underset{\sim}{k} = (110)$ in the Table below

$\underset{\sim}{G}$	$\underset{\sim}{Q}$	$\lvert\underset{\sim}{G} - \underset{\sim}{Q}\rvert$	Σ_{SX} in eV
000	000	0	- 4.62
000	111	3	0.57
000	-111	3	- 0.79
000	-1-1-1	3	0.80
000	- 200	4	- 0.0002
000	200	4	- 0.0001

BI found that, at least out to the free-electron Fermi surface, the Coulomb hole term was fairly constant (it changed from - 6.5 to -7.2 ev).

We summarize their findings by giving the exchange and correlation energy (in ev) in various approximations at principal points in the valence band

	$\Gamma_{25'}$	$L_{3'}$	X_4
Bennett-Inkson (1977)	-10.9	-11.2	-11.5
Dynamic (Kane 1972)	-11.0	-11.1	-10.9
Static (Kane 1972)	-12.1 -12.6	-12.1 -12.6	-12.1 -12.6

The agreement between BI and Kane's dynamic calculation is seen to be good.

Additional problems arise in extending the work to the conduction bands: we refer the reader to BI for a discussion of this.

7. ADDITIONAL NOTES

We shall conclude with some additional comments on the various parts of these lectures.

7.1. Jellium Model

There are a number of examples in which one can make contact between the theory based on this model and experiment: we mention two below.

a) In highly compensated n-InSb, donor orbits which overlap to form an impurity band, with metallic type conducting properties can be localized by applying a magnetic field. A rather dramatic metal-insulator transition occurs at a critical field. This transition has some remarkable similarities to the Wigner transition. But we need to observe Bragg diffraction to prove finally the existence of an electron crystal. This is a difficult experiment, especially as the melting temperature of the electron crystal would be very low.

b) Meier has very recently reported a possible test of exchange and correlation energy in rather low jellium. The system investigated was EuO which in pure form has an empty conduction band. Doping with >1 atomic % trivalent element leads to metallic behaviour with the conduction electrons usefully described by the jellium model.

The localized 4f electrons magnetically order below a temperature T_c and the conduction electrons polarize so that only states of one spin direction are occupied, in contrast to the paramagnetic situation above T_c.

It turns out that the threshold energy which is the minimum photon energy required to remove a 4f electron into the vacuum, depends strongly on magnetization. Meier shows how this can be related to the exchange and correlation energy in jellium. There is satisfactory agreement between theory and experiment, which is very encouraging at the low electron densities (strong correlations) involved in Meier's metallic state.

7.2. Bonding in Solid Si

Recent work by Stenhouse et al. (1977) has demonstrated that the electron density in one Si-Si bond, built up from sp^3 atomic hybrids by the LCAO method, gives an accurate electron density in the solid, by superposition. This is true in both crystalline and amorphous Si, as is demostrated by comparison of this model with electron and X-ray diffraction intensities.

There is little doubt that this is a manifestation of electron correlation leading to localization of electron density in bonds. The work of Bennett and Inkson (1977; see also the lectures of T. Collins in this Volume) should eventually throw more light on this, and provide a basic justification for this superposition of bonds' model.

Professor Koutecký has also drawn my attention during the Institute to the success of LCAO descriptions of Si band structure, based on sp^3 hybridization, with a limited parametrization. It would be interesting if such parameters could now be related to the non-locality and energy dependence of the self energy Σ discussed in sec.

7.3. Narrow Bands with Strong Interaction

a) Three dimensions

Continuing the discussion of section 5.2, we should record that Sayers (1977) has applied Gutzwiller's method to explain the difference between the 3d series of transition metals on the one hand, and the 4d and 5d metals on the other. As remarked in section 5.2, using a simple band model, Friedel (1969) was able to explain the parabolic variation of the cohesive energy in the 4d and 5d metals, peaking at the refractory metals in the middle of the series. Sayers demonstrates that the inclusion of the Gutzwiller-Hubbard interaction produces a dip near the half filled band in the 3d series. Comparison with experiment then makes it clear that correlation is important in the 3d series, but of small consequence in the 4d and 5d series.

In connection with the results (5.13) and (5.14) Brinkman and Rice (1970) point out that whereas in conventional Stoner enhancement theory, the spin susceptibility of a metal is much more strongly modified from the band result $\propto N(E_f)$, the density of states at the Fermi surface, than the electronic specific heat, the experimental properties of the metallic state of V_2O_3 show that the spin susceptibility and electronic specific heat are enhanced by roughly the same amount. But though the example is interesting, it is not decisive as V_2O_3 is surely a complicated many band material.

b) One-dimensional Gutzwiller-Hubbard model

We record here a result of some interest in polymer problems, namely that in one dimension the Hamiltonian (5.4) can be solved exactly (Lieb and Wu, 1968).

For the details, we must refer the reader to the original paper. However the allowed energies E take the form

$$E \propto \sum_j \cos k_j = \int \varrho(k,C) \cos k \, dk \tag{7.1}$$

where the quasi-momenta k_j are determined by a somewhat complicated set of equations. As the length of the chain tends to infinity, and for the case of a half filled band, the ground state energy takes the form

$$E = -4N \int_0^\infty \frac{J_0(\omega)J_1(\omega)\, d\omega}{\omega[1+\exp(1/2\omega C)]} \tag{7.2}$$

where J denotes a Bessel function of the first kind.

For small C, one regains the Hartree-Fock result

$$\frac{E}{N} = -\frac{4}{\pi} + \frac{1}{4} C + \dots \tag{7.3}$$

but at higher order in C there are terms which cannot be expanded in powers of C.

The full expression for E has then been evaluated numerically as a function of the Coulomb interaction strength C by Johansson and Berggren (1969), and hence for this problem the correlation energy in now known exactly.

In addition to the ground state, the electronic excitation spectrum is also known for this model and appears highly relevant to that of a one-dimensional chain with conjugated bonds.

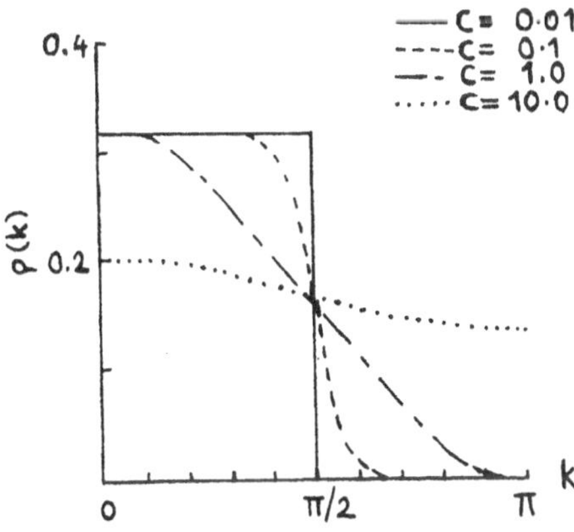

Fig. 1

We conclude by reporting calculations (Grant and March, 1977, to be published) of $\rho(k,C)$ for a half filled band. These are shown in the Figure where ρ is plotted as a function of k for various values of the interaction strength C from 0 to 10. Evidently the step function is blurred out as soon as the interaction $C \neq 0$, which, as pointed out by Lieb and Wu (1968), implies the absence of a Mott transition for a half filled band in one dimension. Calculations of the occupation number n(k,C) of the Bloch states are currently in progress as well as a study of the relation between the first and second order density matrices from the Lieb-Wu ground state wave function.

ACKNOWLEDGMENTS

It is a pleasure to thank Colin Sayers for helpful discussions of Gutzwiller's method and Peter Grout for his collaboration on the calculations for the Lieb-Wu model.

REFERENCES

1. M. Bennett and J.C. Inkson, J. Phys. C10, 987 (1977).
2. N.F. Berk and J.R. Schrieffer, Phys. Rev. Letts. 17, 433 (1969).
3. W.F. Brunnen and T.M. Rice, Phys. Rev. B2, 4302 (1970).
4. C.A. Coulson and I. Fischer, Phil. Mag. 40, 386 (1949).
5. P.A.M. Dirac, Proc. Camb. Phil. Soc. 26, 376 (1930).
6. S. Doniach and S. Engelberg, Phys. Rev. Letts. 17, 750 (1969).
7. F. Duscastelle, J. Physique 31, 1055 (1970).
8. J. Friedel in Physics of Metals I, Electrons, J.M. Ziman, ed. Cambridge Univ. Press, 1969.
9. J. Friedel and C.M. Sayers, J. Physique 38, 697 (1977). ibid 38, L263.
10. M. Gell-Mann and K.A. Brueckner, Phys. Rev. 106, 364 (1957).
11. K.A. Gschneider, Solid State Physics 16, 275, eds. F. Seitz and D. Turnbull, Acad. Press, New York (1964).
12. M.C. Gutzwiller, Phys. Rev. 137, A1726 (1965).
13. L. Hedin, Phys. Rev. 139, A796 (1965).
14. L. Hedin and S. Lundquist, Solid State Physics 23, 1, eds. F. Seitz, D. Turnbull and H. Ehrenreich, Academic Press, New York (1969).
15. F. Herman, J.P. van Dyke and I.B. Ortenburger, Phys. Rev. Letts. 22, 807 (1969).
16. P. Hohenberg and W. Kohn, Phys. Rev. 136, B864 (1964).
17. J.C. Inkson, J. Phys. C5, 2599 (1972).
18. B. Johansson and K.F. Berggren, Phys. Rev. 181, 855 (1969).
19. E.O. Kane, Phys. Rev. B4, 1910 (1971); Phys. Rev. B5, 1493 (1972).
20. W. Kohn and L.J. Sham, Phys. Rev. 140A, 1133 (1965).

21. E.H. Lieb and C. Wu, Phys. Rev. Letts. 20, 1445 (1968).
22. N.H. March, Phys. Rev. 110, 604 (1958).
23. C.M. Sayers, J. Phys. F7, 1157 (1977).
24. J.R. Schrieffer, Theory of Superconductivity, Benjamin Press, New York (1964).
25. J.C. Slater, Phys. Rev. 81, 385 (1951).
26. J.C. Slater and H.M. Krutter, Phys. Rev. 47, 559 (1935).
27. B. Stenhouse, P.J. Grout, N.H. March and J. Wenzel, Phil. Mag. (July number, 1977).
28. J.C. Stoddart, P. Stoney, N.H. March and I.B. Ortenburger, Nuovo Cimento, 23B, 15 (1974).
29. S. Weinbaum, J. Chem. Phys. 1, 593 (1933).
30. E.P. Wigner, Phys. Rev. 46, 1002 (1934), Trans. Far. Soc. 34, 678 (1938).

QUANTUM MECHANICAL TREATMENT OF EXCITED ELECTRONIC STATES OF POLYMERS

T.C. Collins *

Directorate of Physics, Office of Scientific Research, Bolling Air Force Base, Washington D.C. 200332, USA

I. INTRODUCTION

In this series of lectures, we will develop methods for obtaining the one electron like excitations of atoms, molecules, and solid systems to a high degree of accuracy. The starting point of the development is the eigenvalues and wave functions of the Hartree-Fock model. In this approximation the N lowest eigenvalues (of an N-electron system) of the Fock operator represent ionization energies according to Koopman's theorem. No relaxation of the change caused by removing one electron is included in the Hartree-Fock model. Likewise the virtual or unoccupied state solutions of the Fock operator gives eigenvalues which are related to electron affinities, namely the N+1 system. Thus one has solutions of the N+1 systems instead of excited states of the N system with no relaxation and correlation included.

In the first part of the lectures, the correction to the Fock operator will be made so that the virtual states will see the hole from which it was excited [1]. The second simple correction will be to include the long range correlation effects in a perturbative way [2], and the third is the inclusion of local field effects [3]. Then in the last part, the Green's function formalism will be used to construct a set of Bethe-Salpeter equations from which the excitation energies and corresponding Bethe-Salpeter amplitudes are deduced [4].

* Alexander von Humboldt Senior U.S. Scientist Awardee at the Chair of Theoretical Chemistry of the Friedrich Alexander University Erlangen-Nürnberg

J.-M. André et al. (eds.), Quantum Theory of Polymers, 75-101.

II. $\hat{O}\hat{A}\hat{O}$

In obtaining an excitation Hamiltonian the first step considered here is to make sure that the excited states or virtual states see the correct types of field. Since for atoms and molecules plus a large number of crystals (of which most of the organic polymers belong), the excitations are localized; one wants to have this effect included in the excitation Hamiltonian. A localizing operator for the occupied space orbitals which will give the same density as the Hartree-Fock charge density has been suggested by Adams [5] and Gilbert [6]. Namely, one has

$$A = \hat{\rho} \hat{A} \hat{\rho} \quad , \tag{2.1}$$

where ρ is the charge density operator of the Hartree-Fock operator and has the property

$$\rho^2 = \rho \quad . \tag{2.2}$$

$\hat{A}$ is an arbitrary operator which is chosen to help reduce the computation problem. If Ψ_i is a virtual orbital, one has

$$\hat{\rho} \hat{A} \hat{\rho} \Psi_i = 0 \quad . \tag{2.3}$$

On the other hand, if Ψ_i is an occupied orbital, one has

$$\hat{\rho} \hat{A} \hat{\rho} \Psi_i \neq 0 \quad . \tag{2.4}$$

The use of $\hat{\rho}$ as a projection onto the occupied space can be extended to form a projection operator onto the virtual space. One has

$$1 = \hat{\rho} + (1 - \hat{\rho}) = \hat{P} + \hat{O} \tag{2.5}$$

and one can form an operator B of the form

$$B = \hat{O}\hat{A}\hat{O} \quad . \tag{2.6}$$

Here again $\hat{A}$ is an operator which is at present an arbitrary operator. The objective is to formulate an $\hat{A}$ such that when $\hat{O}\hat{A}\hat{O}$ is added to the Fock operator, the resulting operator will have eigenvalues which are the excitations of the system being investigated The removal of an electron of a core state of an atom will leave behind a Coulomb potential of the form

$$V(\vec{r}_1) = - \int \Psi_c^*(\vec{r}_2) \frac{e^2}{|\vec{r}_1 - \vec{r}_2|} \Psi_c(\vec{r}_2) \, d\vec{r}_2 \quad . \tag{2.7}$$

There will be some change in the charge density. Since we are looking at core states, the attraction is the major force, and the removed orbital or hole will retain nearly the same shape. However, the outer shells may relax substantially. (One has also an exchange potential which will be discussed latter is this section.) The major change in the remaining electron charge density will come from the outer electron orbits. This gives the relaxation and causes the energy change between the ionized atom and the ground state to be less than the Hartree-Fock eigenvalue. We will return to this in a latter section.

We begin the mathematical formulation by defining a system Hamiltonian for a N-particle system in terms of general one-body operators f_i and two-body operators g_{ij} . We find

$$\mathcal{H} = \sum_{i=1}^{N} f_i + \tfrac{1}{2} \sum_{\substack{i,j=1 \\ i \neq j}}^{N} g_{ij} \quad . \tag{2.8}$$

Using a Slater determinantal-type wave function of the form

$$\Psi(x_1, \ldots x_N) = (N!)^{-1/2} \det\{\phi_i(x_i)\} = (N!)^{1/2} \hat{A}_N \prod_{i=1}^{N} \phi_i(x_i) \quad , \tag{2.9}$$

where $\hat{A}_N$ is the antisymmetrizing operator and the ϕ_i's are spin orbitals, the energy, E, is given as

$$E = \langle \Psi | \mathcal{H} | \Psi \rangle = \sum_{i=1}^{N} \langle i | f_1 | i \rangle + \tfrac{1}{2} \sum_{i,j=1}^{N} \left(\langle ij | g_{12} | ij \rangle - \langle ij | g_{12} | ji \rangle \right) . \tag{2.10}$$

If the orbitals are not permitted to relax, the energy needed to remove an electron in state ϕ_N is given as

$$\Delta E_N = \langle N | f_1 | N \rangle + \sum_{i=1}^{N} \left(\langle iN | g_{12} | iN \rangle - \langle iN | g_{12} | Ni \rangle \right. \quad . \tag{2.11}$$

Let us assume Ψ is the system ground state and the orbitals ϕ are chosen so that the system energy is stationary; one finds that the orbitals ϕ satisfy the Hartree-Fock equation

$$\left(f_1 + \int d\tau_2 \sum_{i=1}^{N} |\phi_i(x_2)|^2 g_{12} \right) \phi_j(x_1) - \sum_{i=1}^{N} \phi_i(x_1) \cdot \\ \cdot \int d\tau_2 \, \phi_i^*(x_2) \, g_{12} \, \phi_j(x_2) = \varepsilon_j \phi_j(x_1) \quad . \tag{2.12}$$

The solutions to (2.12) fall into two classes. One class is for $i \leq N$, where one finds that ΔE_i as given by (2.11) is exactly equal to ε_i given by (2.12). This is a statement of Koopmans' theorem. However, there are solutions to (2.12) for which $i > N$. For such solutions Koopmans' theorem is not satisfied, and the eigenvalues are not ionization energies of the N-particle system. These orbitals are termed virtual and are labeled by a, b, etc. from now in this lecture set. It is noted that the variable x includes space and spin degrees of freedom and that integration implies summation on spin variables.

Consider a ϕ_a solution to (2.12) and also the energy needed to remove the electron in ϕ_a from an N-body system, where ϕ_a replaces ϕ_N [as in (2.9)] and the ϕ_i , where we adopt the convention that i,j are always less or equal to N and are orbitals occupied in the ground state plus are solutions to (2.12). We find

$$\Delta E_a = \langle a|f_1|a\rangle + \sum_{i=1}^{N-1} (\langle ai|g_{12}|ai\rangle - \langle ai|g_{12}|ia\rangle) . \tag{2.13}$$

However, the ε_a for (2.12) is given by

$$\varepsilon_a = \langle a|f_1|a\rangle + \sum_{i=1}^{N} (\langle ai|g_{12}|ia\rangle - \langle ai|g_{12}|ai\rangle) , \tag{2.14}$$

demonstrating the previously discussed failure of Koopmans' theorem for the virtual orbitals.

Consider now an operator of the form $\hat{O}\hat{A}\hat{O}$, where $\hat{A}$ is chosen to be

$$\hat{O}\hat{A}\hat{O} = \hat{O}[(-1)\langle N|g_{12} - g_{12}\hat{P}_{12}|N\rangle]\hat{O} . \tag{2.15}$$

$\hat{P}_{12}$ is defined such that

$$\langle a|\hat{O}\langle N|g_{12}\hat{P}_{12}|N\rangle\hat{O}|a\rangle = \langle aN|g_{12}|Na\rangle . \tag{2.16}$$

One may add $\hat{O}\hat{A}\hat{O}$ to (2.12) without disturbing the ground-state solution, so that the stationary condition is satisfied. The new equation of the orbitals is of the form

$$(\hat{F} + \hat{O}\hat{A}\hat{O})\phi = \varepsilon\phi . \tag{2.17}$$

For this equation the eigenvalue, ε_a, using (2.15) is

$$\varepsilon_a = \langle a|f_1|a\rangle + \sum_{i=1}^{N-1} (\langle ai|g_{12}|ai\rangle - \langle ai|g_{12}|ia\rangle) . \tag{2.18}$$

Thus for (2.17) and (2.15) one has a Koopmans' theorem for the virtual orbitals.

It is simple to show that in the unrelaxed orbital limit the difference in the eigenvalues of (2.17) using (2.15) are excitations of the N-particle system for single electrons. Assume the electron in the state N is excited to a state a. Call this energy for excitation ΔE_N^a . Using (2.10), if E_g is the total ground-state energy and E_N^a is the excited-state energy, one has

$$\begin{aligned} E_N^a - E_g = \Delta E_N^a \equiv \langle a|f_1|a\rangle + \sum_{i=1}^{N-1} (\langle ai|g_{12}|ai\rangle - \langle ai|g_{12}|ia\rangle) - \\ - \langle N|f_1|N\rangle - \sum_{i=1}^{N-1} (\langle Ni|g_{12}|Ni\rangle - \langle Ni|g_{12}|iN\rangle) \equiv \varepsilon_a - \varepsilon_N . \end{aligned} \tag{2.19}$$

Therefore the eigenvalue differences correspond to the excitation energies of the N-particle system for excitation from state $|N\rangle$. Please note that the choice of excitation orbital $|N\rangle$ is not special since any of the occupied states can be chosen to be $|N\rangle$.

This choice of $\hat{A}$ will lead to the same results for the triplet (the lowest excited state by Hund's rule); however, it will not reproduce the singlet case. For the triplet the exchange like term is zero and $\hat{A}$ is just the Coulomb-term. To obtain the spin singlet, the choice of $\hat{A}$ would be minus the Coulomb term plus twice the exchange term. Using (2.15) as $\hat{A}$ and with the spin of the substitute orbit the same as that of the removed orbit, one has $\hat{A}$ giving the average of the two energies. Two points should be made. First, in complicated systems, for which this development of the excitation operator is intended, S^2 is not a good quantum number in general. Second, one needs more than one determinant for a trial wave function in general even in simple systems to obtain correct spin symmetry. Thus in this work the pure-state requirement will be dropped. For the simple systems, such as the 1s2s excited state of He, an error of $\sim$0.4 eV is found, and in Be, for the $1s^2 2s3s$ excited state, one obtains an error of $\sim$0.3 eV.

Finally we consider the proper variational determination of the virtual Φa. We show that these are properly the solutions of (2.17) using (2.15) under the restriction of having only one Slater determinant. Consider a given set of orbitals ξ_i, $i \leq N-1$, for the self-consistent ground-state equation. Choose the ξ_a so that the energy of the state

$$\Psi_N^a(x_1,\dots x_N) = (N!)^{1/2}\hat{A}_N\Big(\xi_a(x_N)\prod_{i=1}^{N-1}\xi_i(x_i)\Big) \tag{2.20}$$

is stationary with respect to ξ_a, ξ_a^*. Requiring ξ_a to be orthogonal to the occupied orbitals leads to

$$\delta_{\xi_a^*}\Big[E_N^a - \sum_{i=1}^{N-1}\lambda\Big(\int \xi_a^*\xi_i\,d\tau - \delta_{ia}\Big)\Big] = 0 \tag{2.21}$$

with

$$E_N^a = \sum_{i=1}^{N-1}\langle i|f_1|i\rangle + \langle a|f_1|a\rangle + \tfrac{1}{2}\sum_{i=1}^{N-1}\sum_{j=1}^{N-1}\big(\langle ij|g_{12}(1-\hat{P}_{12})|ij\rangle + \sum_{i=1}^{N-1}\big(\langle ai|g_{12}(1-\hat{P}_{12})|ai\rangle\big) . \tag{2.22}$$

The variation produces for the ξ_a^* the equation

$$\Big(f_1 + \sum_{i=1}^{N-1}\int|\xi_i(2)|^2 g_{12}\,d\tau_2\Big)\xi_a(1) - \sum_{i=1}^{N-1}\xi_i(1)\int\xi_i^*(2)\xi_a(2)g_{12}\,d\tau_2 = \lambda_a\xi_a(1) . \tag{2.23}$$

We see that the form for the expectation value of this operator with ξ_a is the same as for (2.17) using (2.15) for Φ_a. Thus if we show Φ_a to be the same as ξ_a, the Φ_a of (2.17) and (2.15) are those which minimize the system energy for unrelaxed $\phi_i, i \leq N-1$.

The proof is to consider the matrix of the operator $\hat{B}$ defined by (2.23) with respect to the solution ϕ_a of (2.17) and (2.15). We need to evaluate $\langle \phi_a | \hat{B} | \phi_a \rangle$. Now the operator in (2.17) is given as

$$\hat{F} = \hat{B} + \int |\phi_N(2)|^2 g_{12}\, d\tau_2 - \phi_N(1) \int d\tau_2\, \phi_N^*(2)\, g_{12} \hat{P}_{12} . \tag{2.24}$$

Therefore

$$\begin{aligned} \langle \phi_a | \hat{B} | \phi_b \rangle &= \varepsilon_a \delta_{ab} - \langle aN | g_{12} | bN \rangle + \langle aN | g_{12} | Nb \rangle - \langle a | \hat{O}\hat{A}\hat{O} | b \rangle \\ &= \varepsilon_a \delta_{ab} - \langle aN | g_{12} | bN \rangle + \langle aN | g_{12} | Nb \rangle \\ &\quad + \langle aN | g_{12} | bN \rangle - \langle aN | g_{12} | Nb \rangle \equiv \varepsilon_a \delta_{ab} . \end{aligned} \tag{2.25}$$

Thus one finds that $\hat{B}$ is diagonal in the solution of (2.17) with (2.15); hence the energy is minimized as desired.

We also note that for configuration-interaction calculations based upon single and double replacements of orbitals in the ground-state eigenfunction, the orbitals defined by (2.15) and (2.17) should be optimal to a good degree. For one thing, the virtual orbitals minimize the energy of the excited wave functions, or, hence, minimize the energy difference between ground and excited wave functions. Thus in second-order perturbation theory the energy denominator is minimized. In addition the interaction integral in the numerator is enhanced since the virtual orbitals see a V^{N-1} potential rather than a V^N potential and are much more spatially localized than the usual Fock virtual orbitals. The spatial region of the virtual orbital is chosen to the occupied orbitals thus the interaction matrix elements are enhanced.

III. ELECTRONIC POLARON MODEL

In this section we will follow closely the work of Kunz given in Ref. (2). We will consider explicitly the case in which the highest valence band is p-type and the lowest conduction band is s-type or vice versa, a situation which holds for a large number of organic polymers. We will study the case of an electron in an otherwise empty band and a hole in an otherwise filled band. The procedure includes only that part of the electron-electron correlation which is contained in terms of simple single-particle excitations and two-particle excitations. It is assumed that the valence electrons are mostly responsible for the observed polarization properties. The results shall be obtained by second-order perturbation theory and all correlation between electrons and the crystal lattice shall be neglected. Our Hamiltonian is the same as (2.8) and we desire to have approximate eigenvalues for

$$\mathcal{H}^N \Psi_n^N(x_1 \ldots x_N) = E_n^N \Psi_n^N(x_1 \ldots x_N). \tag{3.1}$$

The case n=o refers to the ground state. Again if we ionize the system, we will always order things so that the electron in the Nth orbital is removed.

Let's define an approximate process for constructing $\Psi^{(N)}$ out of an expansion set of Slater determinants. Assume we have solved for the ground state of $\mathcal{H}^{(N)}$ in the Hartree-Fock limit, and this ground state is non-degenerate. Thus the various bands are either completely filled or empty in the ground state. This produces a single Slater determinant $\Psi_0^{(N)0}(x_1 \cdots x_N)$ which has an energy E^0 and is

$$\Psi_0^{(N)0}(x_1 \cdots x_N) = (N!)^{-1/2} \det(\phi_i(x_j))$$

or

$$|\Psi_0^{(N)0}\rangle \equiv |vac\rangle. \tag{3.2}$$

We order the orbitals such that $1 \leq i \leq N < a$.

We form additional Slater determinants using the virtual orbitals, where $\Psi_i^{(N)0a}$ is the determinant in which the ith occupied orbital has been replaced with the ath virtual orbital. Similarly, if orbitals i and j are replaced by orbitals a and b, the notation of this state is $\Psi_{ij}^{(N)0ab}$. Since the number of particles will be varied, second quantization will also be used. Using α^+ and α as fermion creation and annihilation operators, we have

$$|\Psi_0^{(N)}\rangle = |vac\rangle + \sum_{i,a} A_i^{(N)0a} \alpha_a^+ \alpha_i |vac\rangle + \sum_{ij,ab} B_{ij}^{(N)0ab} \alpha_a^+ \alpha_b^+ \alpha_i \alpha_j |vac\rangle. \tag{3.3}$$

In principle we need to determine the values of the A's and B's by the variational method. In the case of the nondegenerate ground state, the A's will be zero in the limit of second-order perturbation theory. We can construct approximate excited states of the N-body system by replacing an occupied orbital with a virtual orbital in $\Psi^{(N)0}$.

We also generate the ground state of the (N-1)-body system by removing the electron in the Nth orbital in the N-body system. This ground state is

$$|\Psi_{0N}^{(N-1)}\rangle = \alpha_N |vac\rangle + \sum_{i,a} A_i^{(N-1)a} \alpha_a^+ \alpha_i \alpha_N |vac\rangle + \sum_{ijab} B_{ij}^{(N-1)ab} \alpha_a^+ \alpha_b^+ \cdot \alpha_i \alpha_j \alpha_N |vac\rangle. \tag{3.4}$$

This wave function describes a nonmetallic solid with a hole produced in it.

The wave function for a conduction electron of the N+1 electron system in which we have added an electron in the ath state is

$$|\Psi_a^{a(N+1)}\rangle = \alpha_a^\dagger|vac\rangle + \sum_{i,b}{}' A_i^{(N+1)b}\alpha_b^\dagger\alpha_i\alpha_a^\dagger|vac\rangle + \sum_{ijbc}{}' B_{ij}^{(N+1)bc}\cdot \alpha_b^\dagger\alpha_c^\dagger\alpha_i\alpha_j\alpha_a^\dagger|vac\rangle . \tag{3.5}$$

The prime on the sum means that in summing over b and c, we omit state a and that in summing over i and j, we include a.

The Φ's in general satisfy the Hartree-Fock equation

$$F(\varrho)\Phi_i(x) = \varepsilon_i^0\Phi_i(x) \tag{3.6}$$

and the eigenvalues ε_i have the usual meaning given by Koopmans' theorem. The energy of the ground state N-body system and the N+1 body system are seen to be

$$E_0^{(N)} = \int \Psi_0^{(N)*}\mathcal{H}\Psi_0^{(N)}d\tau \Big/ \int \Psi_0^{(N)*}\Psi_0^{(N)}d\tau \tag{3.7}$$

and

$$E^{a(N+1)} = \int \Psi_0^{a(N+1)*}\mathcal{H}\Psi_0^{a(N+1)}d\tau \Big/ \int \Psi_0^{a(N+1)*}\Psi_0^{a(N+1)}d\tau , \tag{3.8}$$

and the ionization energy for the Nth electron is

$$E_N^{(N-1)} - E_0^{(N)} = -\varepsilon_N . \tag{3.9}$$

Now we define the correlation energy terms. We do this with respect to the energy of the base Hartree-Fock ground state. In so doing we rely on the content of Koopmans' theorem. The energy of the N-body ground state to the accuracy of second-order perturbation theory is given as

$$E_0^{(N)} = E^0 + \sum_{i,a}\frac{(V_i^{(N)0a})^2}{\varepsilon_i^0 - \varepsilon_a^0} + \frac{1}{2}\sum_{ij}\frac{(V_{ij}^{(N)0ab})^2}{\varepsilon_i^0 + \varepsilon_j^0 - \varepsilon_a^0 - \varepsilon_b^0 + v_{ij}^{ab}} . \tag{3.10}$$

Here the quantity V is defined as

$$V_i^{(N)0a} = \langle vac|\mathcal{H}\alpha_a^\dagger\alpha_i|vac\rangle$$

and

$$V_{ij}^{(N)0ab} = \langle vac|\mathcal{H}\alpha_a^\dagger\alpha_b^\dagger\alpha_i\alpha_j|vac\rangle .$$

Note, by Brillouin's theorem for our N-body ground state the V_i are zero. The quantities v_{ij}^{ab} are given as

$$v_{ij}^{ab} = -\langle ab|R|ab\rangle - \langle ij|R|ij\rangle + \sum_{k=i}^{j}\sum_{c=a}^{b}\langle ka|R|kc\rangle ,$$

$$R = (1-\hat{P}_{12})\, e^2/r_{12}$$

In general, for Bloch-like states the v's tend to go as 1/N and are hence usually neglected in solid state calculations.

From this we identify part of the correlation energy and call it a "pair correlation", $e_{ij}^{(N)}$, which is seen to be

$$e_{ij}^{(N)} = \sum_{a>N}\frac{(V_i^{(N)0a})^2}{\varepsilon_i^0-\varepsilon_a^0} + \sum_{a>N}\frac{(V_j^{(N)0a})^2}{\varepsilon_j^0-\varepsilon_a^0} + \sum_{a>N}\sum_{b>N}\frac{(V_{ij}^{(N)0ab})^2}{\varepsilon_i^0+\varepsilon_j^0-\varepsilon_a^0-\varepsilon_b^0} . \tag{3.11}$$

In terms of the "pair correlations" defined in (3.11) for the N-body case, which take on the appropriate form for the N±1-body cases, [that is, one simply changes the upper limit of the sums in (3.11)] we have expressions for the total enery which are

$$E_0^{(N)} = E^0 + \tfrac{1}{2}\sum_{i,j=1}^{N}{}' e_{ij}^{(N)} , \tag{3.12}$$

$$E_N^{(N-1)} = E^0 - \varepsilon_N^0 + \tfrac{1}{2}\sum_{i,j=1}^{N-1}{}' e_{ij}^{(N-1)} , \tag{3.13}$$

$$E^{a(N+1)} = E^0 + \varepsilon_a^0 + \tfrac{1}{2}\sum_{i,j=1}^{N+1}{}' e_{ij}^{a(N+1)} . \tag{3.14}$$

Here we add the superscript a to indicate that the orbital a has been occupied in (3.14). In the case where the electron in the Nth orbital is removed from the system, we define its energy as being the negative of the ionization potential, and hence find ε_N is given as

$$\begin{aligned}\varepsilon_N &= E_0^{(N)} - E_N^{(N-1)} \\ &= \varepsilon_N^0 + \tfrac{1}{2}\sum_{i,j=1}^{N-1}{}'(e_{ij}^{(N)} - e_{ij}^{(N-1)}) + \sum_{j=1}^{N-1} e_{Nj}^{(N)} .\end{aligned} \tag{3.15}$$

We define the energy of the conduction electron as being the electron affinity of the solid, and hence ε_a is found to be

$$\begin{aligned}\varepsilon_a &= E^{a(N+1)} - E_0^{(N)} \\ &= \varepsilon_a^0 + \tfrac{1}{2}\sum_{i,j=1}^{N}(e_{ij}^{a(N+1)} - e_{ij}^{(N)}) + \sum_{j=1}^{N} e_{(N+1)j}^{a(N+1)} .\end{aligned} \tag{3.16}$$

However, if we neglect the changes in correlation energy for the valence and core electrons due to the presence of the added electron in an empty band, one has

$$\varepsilon_\alpha = \varepsilon_\alpha^\circ + \sum_{j=1}^{N} e_{(N+1)j}^{\alpha(N+1)} \quad . \tag{3.17}$$

It is useful to define a self-energy for the conduction electron and for the valence hole. These are called Σ_e and Σ_h, respectively, and are given as

$$\Sigma_e = \sum_{j=1}^{N} e_{(N+1)j}^{\alpha(N+1)}$$

$$\Sigma_h = \tfrac{1}{2} \sum_{i,j=1}^{N-1}{}' \left(e_{ij}^{(N)} - e_{ij}^{(N-1)}\right) + \sum_{j=1}^{N-1} e_{Nj}^{(N)} \quad . \tag{3.18}$$

At this point it is useful to discuss the use of second-order perturbation theory with some hope of understanding its reliability. It is also useful to discuss types of configurations which one includes in evaluating the correlation energies. Note also that for our N$\pm$1 body systems, single-particle excitations are included. These are necessary for two reasons. First, these configurations are not closed-shell configurations so that Brillouin's theorem does not exactly apply, and second, the one-electron orbitals used in the N$\pm$1-body configurations do not satisfy exactly the Hartree-Fock equation for these systems but the N-body Hartree-Fock equation.

One of the limitations of using second-order perturbation theory is that only the interactions of the excited configurations with the ground-state configuration are included. Thus the mutual interaction of the excited configurations, which may lie very close together in energy, is neglected. This question has been considered for the present types of systems by Inoue et. al. [7] for the N+1 body case. They find the following:

(a) Many-body corrections are exactly zero for odd orders of perturbation;

(b) they give an explicit expression for fourth-order perturbation theory and evaluate this for a particular case.

It was found that this fourth-order term is several orders of magnitude smaller than the second-order correction. Thus it is reasonable at this point to stop with second-order.

One of the reasons that the higher-order perturbation-theory corrections are small is that some attention was given to the choice of how one constructs the excited configurations. The spectrum of virtual orbitals, as shown in the last section, is quite arbitrary and the choice of the ÔÂÔ leads to much faster convergence.

We now briefly discuss which configuations are to be included and how they are chosen. In doing this we restrict the explicit discussion to the N+1 body wave function, noting that the extension of the discussion to other cases is simply formal. The possibilities for single-particle excitation are twofold: (i) An electron can scatter internal to its own band; (ii) an electron can scatter to another band. Case (i) does not contribute to the correlation energy because the excited determinant belongs to a different irreducible representation of the translation group than the ground-state wave function and hence the natrix element coupling them is zero. Case (ii) we use the particular choice of excited orbital is an excitonic one rather than a band function. One can evaluate the specific contribution of this type of excitation, and it is small compared with the two electron excitations.

For the two-electron excitations one has three possibilities: (a) Two particles are scattered internal to their band; (b) two particles are excited to other states; (c) one particle is scattered internal to its band and one particle is excited to another state. Process (a) does not occur here since we have at most one particle free to scatter internal to its own band. Process (b) can be included, but an evaluation of this process in Ref. 2 shows that its influence is small compared to process (c). Thus we must consider process (c) carefully. If the configurations chosen are based upon the Bloch type of solution to Fock equation, one may anticipate difficulties in the use of second-order perturbation theory since there are clearly an infinite number of possible excited states nearly degenerate in energy which correspond to the same irreducible representation of the translation group as the base determinant and one may expect the mutual interaction of these states to be non-negligible. This difficulty is avoided making use of the arbitrariness of the Fock virtual orbitals. Following Toyozawa [8] one choses a Bloch function for the orbital of the electron scattered internal to its band. However, rather than represent the excited electron by a Bloch function one forms an excitonic wave function, in which both electron and hole are represented by localized functions in a given determinantal wave function, and linear combinations of such determinantal wave functions are formed which satisfy the periodic properties of the lattice. This appears to avoid the difficulties which could occur due to the neglect of the mixing of the excited configurations.

We will sketch the derivation of Ref. 8 emphasizing the physical arguments. Look at the self-energy term of the electron given in (3.18), that is, for example

$$\sum_{j=1}^{N} e^{\alpha\,(N+1)}_{(N+1)j} \; .$$

We restrict our attention to j being those electrons in the valence band or those electrons in the same band as the electron in question.

We restrict our attention to those configurations which lie closest in energy to the state of interest.

We wish to consider the interaction between the ith electron and the M electrons in the branch with the jth electron. Assume here the jth electron is in the valence band for simplicity. This interaction is

$$U(\vec{r}; \vec{r}_1, \dots \vec{r}_M) = \sum_{j=1}^{M} \frac{e^2}{|\vec{r} - \vec{r}_j|} \quad , \tag{3.19}$$

and the total Hamiltonian is

$$\mathcal{H} = \mathcal{H}_o(\vec{r}) + \mathcal{H}_c(\vec{r}_1, \dots \vec{r}_M) + U(\vec{r}; \vec{r}_1, \dots r_M) \; . \tag{3.20}$$

H_o includes the kinetic energy and the interaction of the electron with the core electrons and with the nuclei. H_c is the equivalent Hamiltonian for the valence electrons. The c system in the independent particle limit has excited states in which some electrons are in the conduction bands and some valence holes. In the next limit we find some excitons (i.e., spatially correlated electrons and holes). It is assumed the energy needed to create such an exciton is $\varepsilon_{n\vec{K}}$ and is independent of $\vec{K}$, the exciton translational quantum number. We specify the eigenstates of the system by the number of excitons of each $\vec{K}$ and n present (the excitons are considered to be bosons). This gives

$$\mathcal{H}_c \Psi(\cdots, n_{\vec{K}_i'}, \cdots n_{\vec{K}_j}, \cdots; \vec{r}_1, \dots \vec{r}_M) = (E_o + \sum_m \varepsilon_m \sum_{\vec{K}} n_{\vec{K},m}) \Psi \; . \tag{3.21}$$

(3.21) holds only if $\sum_m \sum_{\vec{K}} n_{\vec{K}m} \ll M$. We need the matrix elements of U with respect to the solutions of (3.21). The basis set we shall use is the linear-combination-of-atomic-orbitals (LCAO) method for the one-electron orbitals. The integral I which must be evaluate is

$$I = \langle \Psi(\cdots, n_{\vec{K}_i'}, \cdots n_{\vec{K}_j}, \cdots; \vec{r}_1 \cdots \vec{r}_M) | U | \Psi(\cdots, n_{\vec{K}_i}, \cdots n_{\vec{K}_j}, \cdots; \vec{r}_1 \cdots \vec{r}_M) \rangle \; . \tag{3.22}$$

This integral is nonzero in three cases. Case (1): n=n' for all $\vec{K}$ and j and the electrons excited coincide; case (2): n=n' for all $\vec{K}$ and j but 1 and here n+1=n' and the exited electrons coincide for the cases where n=n'; case (3): the reverse of case (2).

In investigating these cases, consider only changes in the total charge and in the dipole moment. In case (1), where there is neither change in total charge nor dipole moment, we can essentially use the Fock potential due to the valence electrons in the ground state. In this case I becomes

$$I = U_0(\vec{r}). \tag{3.23}$$

In cases (2) or (3), one obtains

$$I = e\Phi_{m'n}(\vec{r}) \tag{3.24}$$

where $\Phi_{m'n}(\vec{r})$ is the dipole potential due to the dipole moment μ on the m'th atom when it is excited to the state n;

$$\mu_n = \int \phi^{g*}(\vec{r})\, e\vec{r}\, \phi_n^e(\vec{r})\, d\vec{r} \tag{3.25}$$

where we use ground and excited atomic Slater determinants in the above.

In principle we must evaluate (3.25) for each excited state of the atom in question. However, we see that (3.25) is related to the transition oscillator strength for the transition 0→n of the atom. At this point we simplifiy the problem. We replace the spectrum by a single excited level which is ε above the groundstate. Since the main contribution is assumed to come from the first conduction band (or in this case exciton band), we give unit oscillator strength per electron for transitions from the ground to first excited state. Hence, we use the assumption to obtain a value for μ ; we drop the subscript since there is now only one excited state. As it turns out, the exact value in ε is not very critical.

Introducing boson creation and annihilation operators, $b^{\dagger}_{\vec{K}}$ and $b_{\vec{K}}$, for the exciton, one has

$$\langle \Psi_c | U | \Psi_c \rangle = U_0(\vec{r}) - e\Phi(\vec{r}) \tag{3.26}$$

and

$$\Phi(\vec{r}) = (\alpha / V^{1/2}) \sum_{\vec{K}} \frac{i}{|\vec{K}|} \left(b_{\vec{K}} e^{i\vec{K}\vec{r}} - b^{\dagger}_{\vec{K}} e^{-i\vec{K}\vec{r}} \right), \tag{3.27}$$

with

$$\alpha^2 = 2\pi\varepsilon(1 - 1/\varepsilon_\infty). \tag{3.28}$$

In the above V is the volume of the solid, and ε_∞ is the optical dielectric constant. Therefore

$$\mathcal{H}_{eff} = [\mathcal{H}_c + U_0(\vec{r})] + (E_0 + \varepsilon \sum_{\vec{K}} b^{\dagger}_{\vec{K}} b_{\vec{K}}) - e\Phi(\vec{r}) \; . \tag{3.29}$$

Here E_0 is the energy of the system when there is no conduction electron. It is the term $e\,\Phi(\vec{r})$ which contains the effect of pair correlation.

Using a Bloch function formed from localized functions as the one electron orbital for the electron in question, one has

$$\phi_{\vec{k}}(\vec{r}) = (N)^{-1/2} \sum_{\mu=1}^{N} e^{i\vec{k}\vec{R}_\mu} \phi(\vec{r}-\vec{R}_\mu) \tag{3.30}$$

We must evaluate $-e\,\Phi(\vec{r})$. To lowest order in interatomic overlap, one has

$$-e\,\Phi(\vec{r}) = \sum_{\vec{k}\vec{K}} V_{\vec{K}}(0)(b^+_{-\vec{K}} - b_{\vec{K}})\,\alpha^+_{\vec{k}-\vec{K}}\,\alpha_{\vec{k}} \tag{3.31}$$

where

$$V_{\vec{K}}(0) = e\left[\frac{2\pi\varepsilon}{V}\left(1-\frac{1}{\varepsilon_\infty}\right)\right]^{1/2} \frac{i}{|\vec{K}|}\int \phi^*_n(\vec{r}-\vec{R}_\mu)\,e^{i(\vec{k}+\vec{K})\vec{r}}\,\phi_m(\vec{r}-\vec{R}_\mu)\,d\vec{r}. \tag{3.32}$$

If the one-electron orbitals about site $\vec{R}_\mu$ are orthogonal and if $|\vec{K}|^{-1}$ is large compared to the size of an orbital ϕ , one finds

$$V_{\vec{K}}(0) = e\left[\frac{2\pi\varepsilon}{V}\left(1-\frac{1}{\varepsilon_\infty}\right)\right]^{1/2} \frac{i}{|\vec{K}|}\int \phi^*_n(\vec{r}-\vec{R}_\mu)\,e^{-i\vec{k}\vec{r}}\,\phi_m(\vec{r}-\vec{R}_\mu)\,d\vec{r}\,\delta_{nm}\,. \tag{3.33}$$

That is, the electron is scattered internal to its own band and this in part justifies our prior assumption relating to which configurations are needed. Now, subtracting out the energy of the ground state Hartree-Fock determinant, except for the conduction electron, the effective Hamiltonian becomes

$$\mathcal{H}_{eff} = \sum_{\vec{k}} \varepsilon_{\vec{k}}\,\alpha^+_{\vec{k}}\alpha_{\vec{k}} + \varepsilon \sum_{\vec{K}} b^+_{\vec{K}} b_{\vec{K}} + \sum_{\vec{k}\vec{K}} V_{\vec{K}}(0)(b^+_{-\vec{K}} - b_{\vec{K}})\,\alpha^+_{\vec{k}-\vec{K}}\,\alpha_{\vec{k}}\;. \tag{3.34}$$

In the ground state there are no excitons and hence only the first term remains which is the Hartree-Fock one for the conduction electron. The last term is the effect of correlating the electron with all of the valence electrons. Proceeding from here (to second-order perturbation theory) to find the self-energy for the electrons and the holes, one gets

$$\Sigma_e(\vec{k}) = \sum_{\vec{K}} \frac{|V_{\vec{K}}(0)|^2}{\varepsilon(\vec{k}) - \varepsilon - \varepsilon(\vec{k}-\vec{K})} \tag{3.35}$$

and

$$\Sigma_h(\vec{k}) = \sum_{\vec{K}} \frac{|V_{\vec{K}}(c)|^2}{\varepsilon(\vec{k}) + \varepsilon - \varepsilon(\vec{k}-\vec{K})}\;.$$

Again it should be pointed out that we have included only the lowest exciton band, and we restrict the scattering of the electron to its own band.

Several qualitative statements about (3.35) and (3.36) can be made. First, if the bands are flat, there is no dispersion in Σ_e or Σ_h and the expressions are independent of $\vec{k}$. Second, if the band width is less than ε, Σ_e is negative and Σ_h is positive. Finally, the expressions are such that the effect of correlation is to narrow the bands. This narrowing is proportional to the width of the band.

IV. LOCAL FIELD EFFECTS OR "SHORT RANGE" CORRELATION

There are several ways to get accurate evaluation of local field effects. The first is to look at the off-diagonal matrix elements of the K-dependent dielectric function,

$$\varepsilon(\vec{k}+\vec{K},\vec{k}+\vec{K}') \ .$$

One need to obtain ε^{-1} and calculate the energy shifts due to this term. Another way is to look at the perturbation energy shifts caused by the present of a localized hole and a localized electron. Using the definitions of ρ and $\hat{O}$ from section II one has a perturbation Hamiltonian of the form

$$\mathcal{H}' = \hat{O}\hat{A}\hat{\rho} + \hat{\rho}\hat{A}\hat{O} \tag{4.1}$$

where $\hat{A}$ is chosen to create the hole and put the added electron in its excited state.

One can also do cluster calculations to find the relaxation effects due to the presence of the hole [3]. First calculate the total energy of the N-body cluster self-consisting, then remove the electron and again calculate the total energy of the N-1-body cluster. The difference between the Hartree-Fock eigenvalue and the difference of the two-self-consistent calculations gives the relaxation effect due to the hole. The valence-electron cloud polarization due to the presence of the conduction electron can be estimated using a Mott-Littleton model. This is a first order perturbation correction. One could also estimate this effect by differences of self-consistent cluster calculations and using the $\hat{O}\hat{A}\hat{O}$ eigenvalues.

V. GENERAL STRUCTURE OF EXCITATIONS IN MANY-BODY SYSTEMS

We will give in this section a general description of excitations in a many-body system as contained in Ref. 4. The method used is to construct the particle-hole (PH) polarization propagator, and the poles of the PH polarization propagator correspond to single-particle excitations of the N-particle system. The propagator describes the density responce of the system to an external potential and thus may be used to construct the dielectric function. Thus, when we look at the PH polarization propagator, we not only get information about a system's excitation structure, but also about such quantities as collective modes and optical absorption structure.

The Green's function formalism is used to construct a set of Bethe-Salpeter equations from which we may deduce the excitation energies and corresponding Bethe-Salpeter amplitudes. From these we may construct the PH polarization propagator. The various terms of the Green's function expansion can easily be classified in such categories as hole-renormalization effects, particle-renormalization effects, and particle-hole interactions. Systematic analyses of the contributions to these categories are then possible.

The polarization propagator, as stated above, describes the response of the system to a perturbation of the form

$$\mathcal{H}^{ex}(t) = \int \hat{\Psi}^{+}(\vec{x})\, V^{ex}(x)\, \hat{\Psi}(\vec{x})\, d\vec{x} \tag{5.1}$$

where $V^{ex}(x)$ is an external perturbing potential, and $\hat{\Psi}(\vec{x})$ is a field operator of the form

$$\hat{\Psi}(\vec{x}, t=0) = \sum_{n} u_n(\vec{x})\, \hat{\alpha}_n(t=0) \quad . \tag{5.2}$$

Here $\{u_n(\vec{x})\}$ are a complete set of orthonormal spin orbitals, and $\{\hat{\alpha}_n(t)\}$ are Heisenberg operators obeying Fermi statistics.

The PH polarization propagator $\Pi(x,x')$ is given by

$$i\Pi(x,x') = \langle \Psi_0 | \hat{T}[\hat{\Psi}^{+}(x)\, \hat{\Psi}(x)\, \hat{\Psi}^{+}(x')\, \hat{\Psi}(x')] | \Psi_0 \rangle \tag{5.3}$$

where $|\Psi_0\rangle$ is the exact Heisenberg ground state of the system, and $\hat{T}$ is the time-ordering operator. Expanding in the complete, orthonormal set $\{u_n\}$,

$$i\Pi(x,x') = \sum_{\alpha\beta\lambda\mu} u^{*}_{\mu}(\vec{x})\, u_{\lambda}(\vec{x})\, u^{*}_{\alpha}(\vec{x}')\, u_{\beta}(\vec{x}')\, [\, i\Pi_{\lambda\mu;\alpha\beta}(t-t')\,] \quad . \tag{5.4}$$

In many cases the u's generated by the ÔÂÔ method give a more rapidly convergent many-body expansion than the Hartree-Fock orbitals.

As an aside, the relation between the PH polarization propagator and the inverse dielectric function is simply

$$\begin{aligned} \varepsilon^{-1}(x,x') &= \delta^{4}(x-x') + \int V(x-x'')\, \Pi(x'',x')\, dx'' \quad , \\ \varepsilon^{-1}_{\lambda\mu;\alpha\beta}(t-t') &= \delta(t-t')\,\delta_{\lambda\alpha}\,\delta_{\mu\beta} + \sum_{\gamma\delta} V_{\mu\lambda;\gamma\delta}\, \Pi_{\delta\gamma;\alpha\beta}(t-t') \quad . \end{aligned} \tag{5.5}$$

We will use these relations later to construct contributions to the particle and hole self-energies.

Now, if we define the frequency transform of $\Pi_{\lambda\mu;\alpha\beta}(t-t')$ as

$$\Pi_{\lambda\mu;\alpha\beta}(t-t') = \frac{1}{2\pi}\int d\omega\, \Pi_{\lambda\mu;\alpha\beta}(\omega)\, e^{-i\omega(t-t')} \tag{5.6}$$

we can construct the Lehmann representation as

$$\Pi_{\lambda\mu;\alpha\beta}(\omega) = \sum_n \Bigg(\frac{\langle\Psi_0|\hat{c}^+_\mu \hat{c}_\lambda|\Psi_n\rangle\langle\Psi_n|\hat{c}^+_\alpha \hat{c}_\beta|\Psi_0\rangle}{\omega-(E_n-E_0)+i\eta} - \frac{\langle\Psi_0|\hat{c}^+_\alpha \hat{c}_\beta|\Psi_n\rangle\langle\Psi_n|\hat{c}^+_\mu \hat{c}_\lambda|\Psi_0\rangle}{\omega+(E_n-E_0)-i\eta}\Bigg) \quad , \quad \eta = 0^+ \tag{5.7}$$

where the intermediate states $|\Psi_n\rangle$ refer to excited states of the N-body system. From (5.7) we see that the poles of $\Pi_{\lambda\mu;\alpha\beta}(\omega)$ give the energies of the excited states of the N-body system that can be reached by a density perturbation similar to that in (5.1).

Using the interaction representation, we can investigate the structure of $\Pi_{\lambda\mu;\alpha\beta}(\omega)$. The structure is depicted schematically as:

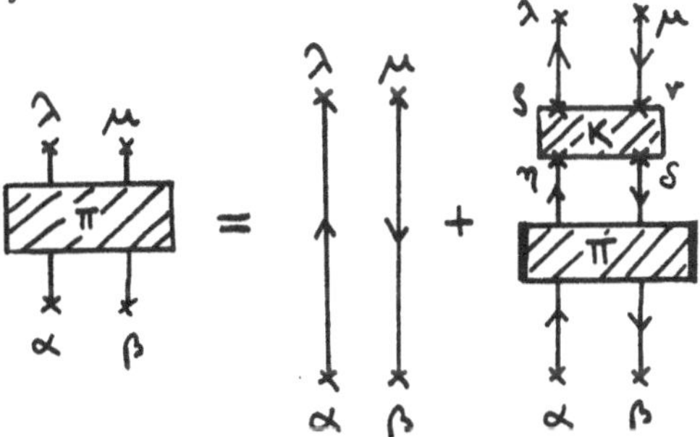

The directed lines are single-particle propagators which in general have self-energy renormalizations. $\Pi_{\lambda\mu;\alpha\beta}(\omega)$ also contains disjoint graphs, however, these terms are independent of (t-t') and contribute only at ω =o. Since we are only interested in $\omega \neq$ o, these terms will be omitted from further discussions. Then unsing the above structure, we may write the Bethe-Salpeter equations as

$$\Pi_{\lambda\mu;\alpha\beta}(\omega) = \Pi^0_{\lambda\mu;\alpha\beta}(\omega) + \sum_{\rho\nu,\eta\delta}\int \frac{d\omega_1 d\omega_2}{(2\pi)^2}\, \Pi^0_{\lambda\mu;\rho\nu}(\omega;\omega_1)\,\cdot$$

$$\cdot K_{\rho\nu\eta\delta}(\omega;\omega_1-\omega_2)\, \Pi_{\eta\delta;\alpha\beta}(\omega;\omega_2) \quad , \tag{5.8}$$

where we define

$$\Pi^{0}_{\lambda\mu;\alpha\beta}(\omega;\omega_1) \equiv -i\, G_{\lambda\alpha}(\omega_1)\, G_{\beta\mu}(\omega_1+\omega) \quad , \tag{5.9}$$

$$\Pi_{\lambda\mu;\alpha\beta}(\omega) \equiv \int \frac{d\omega_1}{2\pi}\, \Pi_{\lambda\mu;\alpha\beta}(\omega;\omega_1) \quad , \tag{5.10}$$

and

$$\Pi^{0}_{\lambda\mu;\alpha\beta}(\omega) \equiv \int \frac{d\omega_1}{2\pi}\, \Pi^{0}_{\lambda\mu;\alpha\beta}(\omega;\omega_1) \quad .$$

Note that $G_{\lambda\alpha}$ in (5.9) refers to the full one particle Green's function. Equ. (5.8) is an integral equation, but if the kernel K ($\omega;\omega_1-\omega_2$) dependend only on ω , the Bethe-Salpeter equation reduces to a Dyson-like equation:

$$\Pi_{\lambda\mu;\alpha\beta}(\omega) = \Pi^{0}_{\lambda\mu;\alpha\beta}(\omega) + \sum_{\rho\nu\eta\delta} \Pi^{0}_{\lambda\mu;\rho\nu}(\omega)\, K_{\rho\nu;\eta\delta}(\omega)\, \Pi_{\eta\delta;\alpha\beta}(\omega). \tag{5.11}$$

It is (5.11) that we will investigate later.

In matrix notation, (5.11) becomes

$$\underline{\Pi}(\omega) = \underline{\Pi}^{0}(\omega) + \underline{\Pi}^{0}(\omega)\,\underline{K}(\omega)\,\underline{\Pi}(\omega). \tag{5.12}$$

This equation is factorable, and the inverse is

$$\underline{\Pi}(\omega)^{-1} = \underline{\Pi}^{0}(\omega)^{-1} - \underline{K}(\omega). \tag{5.13}$$

The matrix elements of $\underline{\Pi}(\omega)$ have a pole at $\omega = E_n - E_0$ unless the numerators in (5.7) vanish; however, the structure of $\underline{\Pi}(\omega)$ in the vicinity of the pole is analytic, and for real ω ,

$$\underline{\Pi}^{+}(\omega) = \underline{\Pi}(\omega). \tag{5.14}$$

Thus, for real ω, $\underline{\Pi}(\omega)$ can be diagonalized with an unitary tranformation

$$\underline{U}(\omega)\,\underline{\Pi}(\omega)\,\underline{U}(\omega)^{-1} = \underline{\Pi}^{D}(\omega). \tag{5.15}$$

From the inverse

$$\underline{U}(\omega)\,\underline{\Pi}(\omega)^{-1}\,\underline{U}(\omega)^{-1} = \underline{\Pi}^{D}(\omega)^{-1} \tag{5.16}$$

we see that U(ω) also diagonalizes $\underline{\Pi}^{-1}(\omega)$. Since some of the diagonal matrix elements of $\underline{\Pi}^{D}(\omega)$ have poles at the exact excitation energies of the system, the corresponding diagonal matrix elements of $\underline{\Pi}^{\circ}(\omega)$ have zeros at the same points. Thus, the zero eigenvalues of $\underline{\Pi}^{-1}(\omega)$ correspond to the collective energy levels of the system, and we must solve the frequency-dependent eigenvalue problem

$$\underline{\Pi}^{-1}(\omega)\vec{C}(\omega) = \Lambda(\omega)\vec{C}(\omega) , \tag{5.17}$$

with

$$\Lambda(\omega) = 0 .$$

Combining (5.17) with (5.13) yields

$$[\underline{\Pi}^{\circ -1}(\omega) - \underline{K}(\omega)]\vec{C}(\omega) = \underline{0} \tag{5.18}$$

from which we can find the excitation energies ω and the corresponding eigenvectors $\vec{C}(\omega)$.

To proceed further, it is necessary to study the structure of $\underline{K}(\omega)$. The first-order expression for $\underline{K}(\omega)$ is

$$K^{(1)}_{\rho\nu;\eta\sigma} \equiv V_{\sigma\eta,\rho\nu} - V_{\rho\eta,\sigma\nu} , \tag{5.19}$$

where

$$V_{\sigma\eta,\rho\nu} = \int d\vec{r}_1 d\vec{r}_2\, u^*_\sigma(\vec{r}_1) u^*_\rho(\vec{r}_2) \frac{1}{r_{12}} u_\nu(\vec{r}_2) u_\eta(\vec{r}_1) . \tag{5.20}$$

Note that $K^{(1)}$ is independent of frequency so that (5.11) is correct to first-order in the electron-electron interaction. The Green's function lines may be written as

$$G_{\lambda\alpha}(\omega) = \sum_n \frac{A^n_\lambda(\omega) A^n_\alpha(\omega)^*}{\omega - \omega_n(\omega)} . \tag{5.21a}$$

The usual pole structure is implied with ω_n referring to the ionization energy of the (N±1) body system. The $A^n_\lambda(\omega)$ is a one-particle amplitude between the N- and N±1 body systems. We will return to the structure of the one-particle Green's function in more datail later. For free particles, (5.21a) becomes

$$G^{\circ}_{\lambda\alpha}(\omega) = \delta_{\lambda\alpha}\left(\frac{\Theta(\alpha - F)}{\omega - \omega^{\circ}_\alpha + i\eta} + \frac{\Theta(F-\alpha)}{\omega - \omega^{\circ}_\alpha - i\eta}\right) . \tag{5.21b}$$

Here F refers to the Fermi level and ω°_α to the eigenvalue of the free-particle Hamiltonian.

To obtain $\underline{\Pi}^0(\omega)$ use (5.9), (5.10) and (5.21a). After performing the frequency integration, we get

$$\Pi^0_{\lambda\mu;\alpha\beta}(\omega) = \sum_{m,n} \Big(\frac{g_m A^m_\beta(\omega_m) A^m_\mu(\omega_m)^* A^n_\lambda(\omega_m-\omega) A^n_\alpha(\omega_m-\omega)^*}{\omega - [\omega_m - \omega_n(\omega_m-\omega)]} - \\ - \frac{g_m A^n_\beta(\omega_m+\omega) A^n_\mu(\omega_m+\omega)^* A^m_\lambda(\omega_m) A^m_\alpha(\omega_m)^*}{\omega + [\omega_m - \omega_n(\omega_m+\omega)]} \Big). \tag{5.22}$$

Here and through this section, the index m refers to ionizations of the N+1 particle system, while l refers to ionizations of the N-1 one; all other indices are general. The factor g_m is

$$g_m = \Big(1 - \frac{\partial \omega_m(\omega)}{\partial \omega}\Big)_{\omega=\omega_m} . \tag{5.23}$$

If we use the free-particle Green's function in the expression for $\underline{\Pi}^0(\omega)$, we get

$$\Pi^{0,0}_{\lambda\mu;\alpha\beta}(\omega) = \delta_{\lambda\alpha}\delta_{\beta\mu}\Big[\frac{\Theta(\alpha-F)\,\Theta(F-\beta)}{\omega - (\omega^0_\alpha - \omega^0_\beta) + i\eta} - \\ - \frac{\Theta(F-\alpha)\,\Theta(\beta-F)}{\omega + (\omega^0_\beta - \omega^0_\alpha) - i\eta} \Big] . \tag{5.24}$$

Now inserting $\underline{K}^{(1)}(\omega)$ into (5.12) for $\underline{K}(\omega)$ yields the structure depicted below.

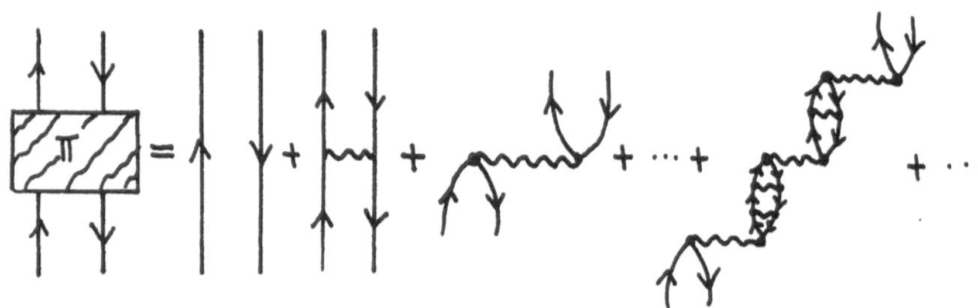

If we used $\underline{\Pi}^{0,0}(\omega)$ instead of $\underline{\Pi}^0(\omega)$ in (5.12), and $\underline{K}^{(1)}(\omega)$ in place of $\underline{K}(\omega)$, we would get the random-phase-approximation (RPA) equations for $\underline{\Pi}(\omega)$. In the RPA, (5.18) becomes

$$(\omega - \omega^0_{m\ell}) C^n_{m\ell} - \sum_{m'\ell'} \Big\{ [V_{m\ell,\ell'm'} - V_{mm',\ell\ell'}] C^n_{m'\ell'} + \\ + [V_{m\ell,m'\ell'} - V_{m\ell',m'\ell}] C^n_{\ell'm'} \Big\} = 0, \tag{5.25a}$$

$$-(\omega+\omega^{\circ}_{m\ell})C^n_{\ell m} - \sum_{m'\ell'} \{ [V_{\ell m,\ell' m'} - V_{\ell m',\ell' m}] C^n_{m'\ell'} + \\ + [V_{\ell m, m'\ell'} - V_{\ell\ell', m'm}] C^n_{\ell' m'} \} = 0 \tag{5.25b}$$

with

$$\omega^{\circ}_{m\ell} = \omega^{\circ}_m - \omega^{\circ}_\ell .$$

In the RPA we may make the identification

$$C^n_{\beta\alpha} = \langle \Psi_0 | C^+_\alpha C_\beta | \Psi_n \rangle \quad . \tag{5.26}$$

Using this equation, we may infer the correct spin structure for the excited state $|\Psi_n\rangle$ for closed-shell systems:

$$C^n_{\beta\alpha;S} = \frac{1}{\sqrt{2}}(C^n_{\beta\uparrow\alpha\uparrow} + C^n_{\beta\downarrow\alpha\downarrow})$$

$$C^n_{\beta\alpha;T} = \begin{cases} C^n_{\beta\uparrow\alpha\downarrow} \\ C^n_{\beta\downarrow\alpha\uparrow} \\ \frac{1}{\sqrt{2}}(C^n_{\beta\uparrow\alpha\uparrow} - C^n_{\beta\downarrow\alpha\downarrow}) \end{cases} \tag{5.27}$$

The indices S and T refer to singlet and triplet, and the spin is indicated explicitly. If we rearrange (5.25) with the help of (5.27) and assume our basis set is real, we get

$$\sum_{m'\ell'} (\Delta E_{m\ell,m'\ell'} C^n_{m'\ell'} + M_{m\ell,m'\ell'} C^n_{\ell' m'}) = \omega C^n_{m\ell}$$

and (5.28)

$$\sum_{m'\ell'} (\Delta E_{m\ell,m'\ell'} C^n_{\ell' m'} + M_{m\ell,m'\ell'} C^n_{m'\ell'}) = -\omega C^n_{\ell m}$$

where

$$\Delta E_{m\ell,m'\ell'} = \delta_{mm'}\delta_{\ell\ell'}(\omega^{\circ}_m - \omega^{\circ}_\ell) + 2\gamma V_{m\ell,m'\ell'} - V_{mm',\ell\ell'} \; ,$$

$$M_{m\ell,m'\ell'} = 2\gamma V_{m\ell,m'\ell'} - V_{m\ell',m'\ell} \quad ,$$

$$\gamma = \begin{cases} 1 & \text{for singlet,} \\ 0 & \text{for triplet.} \end{cases}$$

In addition, using restricted Hartree-Fock orbitals, $V_{m\ell,m'\ell'}$ is the matrix element over spatial orbitals independent of spin. The matrix $\underline{M}$ describes multipair excitations. If $\underline{M}=\underline{0}$, (5.28) are separable equations. Then their diagonal elements yield

$$\omega_T \approx \omega^{\circ}_{m\ell} - V_{mm,\ell\ell} \quad ,$$
$$\omega_S \approx \omega^{\circ}_{m\ell} - V_{mm,\ell\ell} + 2V_{m\ell,m\ell} \quad . \tag{5.29}$$

These matrix elements are equivalent to the first order $\hat{O}\hat{A}\hat{O}$ construction. In fact, the $\hat{O}\hat{A}\hat{O}$ equations derived from the $\hat{A}$ operator are equivalent to (5.28) if we set $\underline{M}=\underline{O}$ and $l=l'$ in ΔE. The open-shell result is more complicated than the above owing to the fact that the unperturbed ground state is multideterminantal. However, a construction can be made which gives equations similar to (5.28) with an approximate excitation spin structure, except that in this case the different spin excitations are coupled.

The structure of (5.18) would be similar to that of (5.28) if we retained $\underline{\Pi}^{\circ}(\omega)$ in (5.12) instead of using $\underline{\Pi}^{\circ,\circ}(\omega)$ but assumed $A^n_\alpha(\omega)$ was independent of ω. The latter is approximately true in most cases. The major difference would be that ω°_n would be replaced by ω_n in (5.28). Since ω_n differs from ω°_n by the renormalization effects coming from the particle or hole self-energies, an investigation of the self-energies would yield these effects. Also, it is clear from the structure of (5.28) and (5.29) that $\underline{K}(\omega)$ contains the particle-hole interactions. Thus we have a clear separation between structures which contribute to the relaxation and correlation of the hole or particle and those which contribute to particle-hole interactions.

To put this discussion on a firmer footing, make the replacements

$$A^n_\lambda(\omega) = A^n_\lambda \quad , \quad g_m = 1 \quad , \quad \omega_m(\omega) = \omega_m \quad .$$

We may then rewrite (5.22) as

$$\Pi^{\circ}_{\lambda\mu;\alpha\beta}(\omega) = \sum_{m,\ell} \frac{A^m_\beta A^{m*}_\mu A^\ell_\lambda A^{\ell*}_\alpha}{\omega - (\omega_m - \omega_\ell) + i\eta} - \sum_{m,\ell} \frac{A^\ell_\beta A^{\ell*}_\mu A^m_\lambda A^{m*}_\alpha}{\omega + (\omega_m - \omega_\ell) - i\eta} \quad , \tag{5.30}$$

where we make the replacement $\omega_\ell(\omega_n \pm \omega) \rightarrow \omega_\ell$. The poles of (5.22) occur at

$$\omega = \pm(\omega_m - \omega_\ell)$$

If the poles of $\underline{\Pi}(\omega)$ are not far from this value, then the above replacement is valid. Then if we use (5.30) and make the definition

$$\Pi^{0'}_{m\ell;m'\ell'}(\omega) = \delta_{mm'}\,\delta_{\ell\ell'} \sum_{\lambda\mu\alpha\beta} (A^{m}_{\beta})^{-1} (A^{m*}_{\mu})^{-1} \cdot \Pi^{0}_{\lambda\mu;\alpha\beta} (A^{\ell *}_{\alpha})^{-1} (A^{\ell}_{\lambda})^{-1} , \tag{5.31}$$

we can transform (5.12) to be

$$\underline{\Pi}'(\omega) = \underline{\Pi}^{0'}(\omega) + \underline{\Pi}^{0'}(\omega)\underline{K}'(\omega)\underline{\Pi}'(\omega) . \tag{5.32}$$

Here, $\underline{K}'(\omega)$ has the same structure as $\underline{K}(\omega)$ except it contains

$$V'_{no,pq} = \sum_{\sigma\eta\varrho\nu} A^{n*}_{\sigma} A^{o}_{\eta} V_{\sigma\eta,\varrho\nu} A^{p*}_{\varrho} A^{q}_{\nu} . \tag{5.33}$$

Thus we can solve equations for the renormalized excitations of the same form as the RPA equations

$$[\,\underline{\Pi}^{0'}(\omega) - \underline{K}^{(1)'}\,]\,\vec{C}(\omega) = 0 , \tag{5.34}$$

except that in (5.28), ω_n would replace ω_n^0 and $V'_{no,pq}$ would replace $V_{no,pq}$.

Finally, the construction of $\underline{\Pi}(\omega)$ using the Bethe-Salpeter amplitudes is

$$\Pi_{\lambda\alpha;\alpha\beta}(\omega) = \sum_n \left[\frac{C^n_{\lambda\mu}(\omega)\,C^{n*}_{\alpha\beta}(\omega)}{\omega - \omega^{ex}_n(\omega) + i\eta} - \frac{C^n_{\beta\alpha}(\omega)\,C^{n*}_{\mu\lambda}(\omega)}{\omega + \omega^{ex}_n(\omega) - i\eta} \right] . \tag{5.35}$$

We now turn to look at the self-energy structure and the general structure of the one-particle Green's function presented in (5.21). The structure is given by Dyson's equation.

$$G_{\lambda\alpha}(\omega) = G^0_{\lambda\alpha}(\omega) + \sum_{\eta\sigma} [\, G^0_{\lambda\eta}\, \Sigma_{\eta\sigma}(\omega)\, G_{\sigma\alpha}(\omega) \,] . \tag{5.36}$$

The self-energy $\Sigma_{\eta\sigma}(\omega)$ contains all many-body corrections to the one-particle Green's function. We seek solutions to the equation

$$\sum_{\alpha} [\, G^0_{\lambda\alpha}(\omega)^{-1} - \Sigma_{\lambda\alpha}(\omega) \,]\, A^n_{\alpha}(\omega) = 0 \tag{5.37}$$

which we used to construct (5.21a). The general structure of $G_{\lambda\alpha}(\omega)$ in the Lehmann representation is

$$G_{\lambda\alpha}(\omega) = \sum_n \Big[\frac{\langle\Psi_0|\hat{c}_\lambda|\Psi_n^{N+1}\rangle\langle\Psi_n^{N+1}|\hat{c}_\alpha^+|\Psi_0\rangle}{\omega - (E_n^{N+1} - E_0) + i\eta} + \frac{\langle\Psi_0|\hat{c}_\alpha^+|\Psi_n^{N-1}\rangle\langle\Psi_n^{N-1}|\hat{c}_\lambda|\Psi_0\rangle}{\omega - (E_0 - E_n^{N-1}) - i\eta} \Big] . \quad (5.38)$$

Thus the $\{\omega_n\}$ refer to ionizations of the (N$\pm$1) body system. Our interpretation of ω_ℓ (or ω_m) is to think of it as the energy necessary to create a hole among the occupied orbitals (or to fill a hole in the virtuals), and $\Sigma_{\lambda\alpha}(\omega)$ describes the relaxation and rearrangement of electrons around the hole.

The structure of $\Sigma_{\lambda\alpha}(\omega)$ may be determined using many-body perturbation theory. If one chooses SCF-RHF orbitals, all diagrams contributing to canonical Hartree-Fock are omitted. The second-order expression for $\Sigma_{\lambda\alpha}(\omega)$ is

$$\Sigma^{(2)}_{\lambda\alpha}(\omega) = \sum_{m\ell m'} \frac{V_{\lambda m',m\ell}(2V_{m\ell,\alpha m'} - V_{m'\ell,\alpha m})}{\omega + \omega^0_\ell - \omega^0_{m'} + i\eta} + \sum_{m\ell\ell'} \frac{V_{\lambda\ell',m\ell}(2V_{m\ell,\alpha\ell'} - V_{m\ell',\alpha\ell})}{\omega + \omega^0_m - \omega^0_\ell - \omega^0_{\ell'} - i\eta} . \quad (5.39)$$

A partial summation of third- and higher-order diagrams for $\Sigma_{\lambda\alpha}(\omega)$ is possible. The structure of these terms is given by the product of the appropriate term in (5.39) times an interaction matrix element devided by the same denominator and with appropriate sign. These elements form a geometric progression which may be summed to all orders by shifting the denominator in (5.39) to get:

$$\Sigma^{D}_{\lambda\alpha}(\omega) = \sum_{m\ell m'} \frac{V_{\lambda m',m\ell}(2V_{m\ell,\alpha m'} - V_{m'\ell,\alpha m})}{\omega + \omega^0_\ell - \omega^0_m - \omega^0_{m'} + V_{mm,\ell\ell} + V_{m'm',\ell\ell} - V_{mm,m'm'} + i\eta} + \sum_{m\ell\ell'} \frac{V_{\lambda\ell',m\ell}(2V_{m\ell,\alpha\ell'} - V_{m\ell',\alpha\ell})}{\omega + \omega^0_m - \omega^0_\ell - \omega^0_{\ell'} - V_{mm,\ell\ell} - V_{mm,\ell'\ell'} + V_{\ell\ell,\ell'\ell'} - i\eta} . \quad (5.40)$$

Note that higher-order diagonal exchange terms are not included in $\Sigma^{D}_{\lambda\alpha}(\omega)$. The non-diagonal terms omitted from $\Sigma^{D}_{\lambda\alpha}(\omega)$ tend to converge rapidly, and the major contributions from diagrams are included in (5.40).

To investigate further the constructions of $\Sigma_{\lambda\alpha}(\omega)$, note that $\Sigma_{\lambda\alpha}$ involve ionizations of the N$\pm$2 body systems as well as excitations of the N-body system. Thus, the correct first-order structure of these excitations is

$$\text{N+2:}\quad \begin{cases} \omega_m^0 + \omega_{m'}^0 + V_{mm,m'm'} \pm V_{mm',mm'} \,, & m \neq m' \\ 2\omega_m^0 + V_{mm,mm} \,, & m = m' \end{cases}$$

$$\text{N-2:}\quad \begin{cases} \omega_\ell^0 + \omega_{\ell'}^0 - V_{\ell\ell,\ell'\ell'} \mp V_{\ell\ell',\ell\ell'} \,, & \ell \neq \ell' \\ 2\omega_\ell^0 - V_{\ell\ell,\ell\ell} & \ell = \ell' \end{cases}$$

$$\text{N:}\quad \omega_m^0 - \omega_\ell^0 - V_{mm,\ell\ell} + \begin{cases} 2V_{m\ell,m\ell} \\ 0 \end{cases}$$

for spin-averaged orbitals, where the upper term refers to the singlet and the lower term the triplet. The partial summation in (5.40) may be modified to include averaged exchange interactions from the above excitations by shifting the denominators of (5.40) to read

$$\Sigma_{\lambda\alpha}^{DE}(\omega) : [\omega + \omega_\ell^0 - \omega_m^0 - \omega_{m'}^0 + V_{mm,\ell\ell} - V_{m\ell,m\ell} + V_{m'm',\ell\ell} - V_{m'\ell,m'\ell} - V_{mm,m'm'} + i\eta]$$

and

$$[\omega + \omega_m^0 - \omega_\ell^0 - \omega_{\ell'}^0 + V_{mm,\ell\ell} + V_{m\ell,m\ell} - V_{mm,\ell'\ell'} + V_{m\ell',m\ell'} + V_{\ell\ell,\ell'\ell'} - i\eta] \,. \tag{5.41}$$

Actually, calculations show little difference between the averaging used in (5.41) and the separation into singlet and triplet excitations.

Another set of higher-order diagrams which can be thought of as contributions to the correlated charge density in an HF-like model are included in Σ expressed as

$$\Sigma_{\lambda\alpha}^{(3)} = \sum_{\ell''\neq\mu} \{ [2V_{\ell\alpha,\mu\ell''} - V_{\lambda\mu,\alpha\ell''}] \Sigma_{\mu\ell''}^{(2)}(\omega_{\ell''}^0) + (2V_{\lambda\alpha,\ell''\mu} - V_{\lambda\ell'',\alpha\mu}) \Sigma_{\mu\ell''}^{(2)}(\omega_{\ell''}^0) \} - \sum_\mu (2V_{\lambda\alpha,\mu\mu} - V_{\lambda\mu,\alpha\mu}) \left. \frac{\partial \Sigma_{\mu\mu}^{(2)}(\omega)}{\partial\omega} \right|_{\omega=\omega_\mu^0} + \sum_{\mu\nu,m\ell\ell'} \frac{(2V_{\lambda\alpha,\mu\nu} - V_{\lambda\mu,\alpha\nu}) V_{\mu\ell',m\ell} (2V_{m\ell,\nu\ell'} - V_{m\ell',\nu\ell})}{(\omega_\ell^0 + \omega_{\ell'}^0 - \omega_m^0 - \omega_\mu^0)(\omega_\ell^0 + \omega_{\ell'}^0 - \omega_m^0 - \omega_\nu)} \,. \tag{5.42}$$

In order to study the effect of screening via the RPA dielectric function in the self-energy, we use (5.35) to construct

$$\Sigma^{SC}_{\lambda\alpha}(\omega) = \Sigma^{DE}_{\lambda\alpha} + \Sigma^{(3)}_{\lambda\alpha} - \Sigma^{PHD}_{\lambda\alpha}(\omega) +$$

$$+\sum_{n}\Big(\sum_{m\ell m'\ell'\ell''} \frac{V_{\lambda\ell',m\ell}(C^{n}_{m\ell}+C^{n}_{\ell m})(C^{n}_{m'\ell''}+C^{n}_{\ell''m'})V_{m'\ell'',\alpha\ell'}}{\omega-\omega^{o}_{\ell'}+\omega^{ex}_{n}-i\eta} +$$

$$+\sum_{m\ell m'm''\ell'} \frac{V_{\lambda m',m\ell}(C^{n}_{m\ell}+C^{n}_{\ell m})(C^{n}_{m''\ell'}+C^{n}_{\ell'm''})V_{m''\ell',\alpha m'}}{\omega-\omega^{o}_{m'}-\omega^{ex}_{n}+i\eta}\Big), \tag{5.43}$$

where $\Sigma^{PHD}_{\lambda\alpha}(\omega)$ has the same structure as $\Sigma^{(2)}_{\lambda\alpha}(\omega)$ except the denominators are shifted to be:

$$\Sigma^{PHD}_{\lambda\alpha}(\omega):(\omega-\omega^{o}_{m'}-\omega^{o}_{m}+\omega^{o}_{\ell}+V_{mm,\ell\ell}-V_{m\ell,m\ell}+V_{m'm',\ell\ell}-V_{m'\ell,m'\ell}+i\eta)$$

and

$$(\omega+\omega^{o}_{m}-\omega^{o}_{\ell}-\omega^{o}_{\ell'}-V_{mm,\ell\ell}+V_{m\ell,m\ell}-V_{mm,\ell'\ell'}+V_{m\ell,m\ell'}-i\eta) \tag{5.44}$$

Thus $\Sigma^{SC}_{\lambda\alpha}(\omega)$ contains contributions from the screened exchange to all orders(both diagonal and off-diagonal particle-hole interaction terms), but the cross terms are between diagonal particle-particle and hole-hole ladders and diagonal particle-hole ladders only.

VI. CONCLUSIONS

We haved showed some very simple methods to correct the excitation energies starting with Hartree-Fock solutions in sections II, III and IV and obtain very good results for a large number of electronic systems. Then in section V we showed how the excitation structure of a many-body system can be separated into hole renormalization (correlation and relaxation) effects, particle renormalization effects, and particle-hole interactions which justifies the approximations of sections II, III and IV. The numerical results can be found in the references given.

REFERENCES

1. T.C. Collins, A.B. Kunz and P.W. Deutsch, Phys. Rev. A10, 1034 (1974).
2. A.B. Kunz, Phys. Rev. B6, 606 (1972).
3. A.B. Kunz, D.J. Michish and T.C. Collins, Phys. Rev. Lett. 31, 756 (1973).
4. M.W. Ribarsky, Phys. Rev. A12, 1739 (1975).
5. W.H. Adams, J. Chem. Phys. 34, 89 (1961).
6. T.L. Gilbert, in "Molecular Orbitals in Chemistry, Physics and Biology," ed. by P.O. Löwdin and P. Pullman (Acad. Press, N.Y., 1964).
7. M. Inoue, C.K. Mahutte and S. Wang, Phys. Rev. C2, 539 (1970).
8. Y. Toyozawa, Progr. Theoret. Phys. (Kyoto) 12, 421 (1954).

COUPLED CLUSTER MANY ELECTRON THEORY AND THE COHESIVE ENERGY OF LARGE SYSTEMS CONSISTING OF WEAKLY INTERACTING SUBSYSTEMS

J. ČÍŽEK [+]

Lehrstuhl für Theoretische Chemie
Friedrich Alexander Universität Erlangen - Nürnberg
852 Erlangen, BRD

1. INTRODUCTION

The first goal of this lecture in the summer school is to present the coupled cluster many electron theory formulated with the use of the Feynman-like diagrams. The second goal is to show that this theory is applicable to the calculation of ground state energy of large systems consisting of weakly interacting subsystems.

This paper is considered as a supplement to the series of papers [1-7] and consequently only basic aspects of the theory will be repeated. First let us briefly review the papers [1-7]. In the first two papers of this series the general form of equations for cluster components was presented in diagrammatical form and further these equations were explicitly written for the approximation in which only biexcited clusters were considered. This approximation is called "coupled-pair many electron theory" (CPMET). In the third paper the equations of CPMET were rederived with the use of the traditional quantum mechanical methods. The main purpose of this paper was to show the simplicity of the diagrammatical method. In the paper [4] the explicit form of equations for connected singly, doubly and triply excited clusters was presented. This approximation is called extended coupled-pair many electron theory; (ECPMET). It was shown, that in contrast to the case of tetraexcited states, the disconnected triexcited clusters are negligible relative

---

[+] A.P. Sloan Fellow 1974-78.
Permanent Address: Department of Applied Mathematics and Department of Chemistry, University of Waterloo, Ontario, Canada, N2L 3GL - Guelph-Waterloo Center for Graduate Work in Chemistry, Ontario, Canada.

J.-M. André et al. (eds.), Quantum Theory of Polymers, 103-116.

to the connected ones. In the same paper the molecule BH_3 was studied with the use of ECPMET. The paper [5] is a sort of review paper in which the relationship of the CPMET to some other theories is studied and in which some new results in the application of CPMET and ECPMET are presented. The paper [6] is related to other papers only from a methodological point of view; the subject of the paper is the time-independent diagrammatical approach to perturbation theory of fermion systems. In the paper [7] the graphical techniques of spin algebras are combined with diagrammatical approach in order to get the spin-adapted form of the coupled cluster theory.

After this short review we shall mention some other series of papers which are relevant to the coupled cluster many electron theory. We shall first quote papers in which cluster expansion was introduced. In the context of nuclear physics the cluster expansion was put forward in the series of papers [8 - 11]. On the atomic and molecular side the cluster expansion was used in the papers [12 - 18]. Reformulation of the earlier version of the theory and application to nuclear physics can be found in the series of papers [19 - 24]. The possible application of coupled cluster theory for solid state theory is discussed in paper [25]. Further let us point out that the first actual application of coupled pair theory for electron gas was presented in paper [26]. Finally let us mention that one chapter in the book [27] is devoted to coupled-pair many electron theory.

II. MOTIVATION FOR THE EXPONENTIAL ANSATZ

We shall be looking for the wave function describing the ground state of the closed shell system in the form

$$|\tilde{\underline{\Psi}}\rangle = e^{\hat{T}}|\Phi\rangle. \tag{1}$$

As mentioned in the discussion in paper [1] our motivation for the study of this Ansatz was coming from two different sources.

On the atomic and molecular side it was shown by Sinanoglu that a rather simplified theory based on this Ansatz is leading to very interesting practical results. Further Primas [18] has shown that this Ansatz provides the correct answer to the problem which has been referred to in recent years as the size consistency problem.

On the other hand we realized that the ladder theory of nuclear matter and the ring diagram theory of electron gas can be united with some atomic and molecular theories by developing a systematic method for the calculation of matrix elements of the operator $\hat{T}$. Let us mention that in this time the resulting equations of these earlier attempts were either not explicit enough to allow their practical exploitation or, as in the electronic structure studies, did not consider properly the intercluster coupling terms [2]. This approximative treatment of inter-coupling terms led to the theory which was not invariant with respect to the unitary transformation among occupied orbitals as well as to that among unoccupied orbitals. This was another reason for the developing of the systematic method for the calculation of matrix elements of $\hat{T}$, because these transformations and consequent possible localization of the orbitals could lead to many useful approximative schemes, as we shall see later on in this paper.

In the framework of second quantization there are basically two approaches which can be used for the derivation of coupled cluster theory. The first approach is a purely algebraic one and in the second approach the technique of diagrams is used.

We feel that in the derivation of the above mentioned equations we face a number of problems of combinatorial character. It is well known that in combinatorial problems the graphical reformulation of the problem is often very helpful. This is indeed the case in our problem and we feel that by introduction of diagrams we achieve a considerable simplification of all necessary manipulation and in addition the structure of the theory is more transparent. Furthermore the diagrammatical formulation permitted us to establish the relation of this theory to the many body perturbation theory in Goldstone form. Finally the diagrammatic method provides a very simple rule how to eliminate spin variables for the case of spin independent forces which is the situation in atomic and molecular physics. Later is was shown that the diagrammatical approach is very convenient in order to get the spin adapted form of the theory [7].

The items listed above are the main reasons why we presented the theory in a diagrammatic form. Perhaps we should stress that when using diagrammatic methods it is necessary to have very precise rules stating the correspondence between the diagram and pertinent scalar factor or operator. This is not always the case in some papers and careless use of diagrams can lead to serious errors.

The algebraic approach was sketched in early papers [8 - 10] and it was revived and simplified in more recent papers [19 - 24]. The algebraic approach was adapted by Harris [20] for the derivation of equations of coupled pair many electron theory in spinorbital form. The author calls the algebraic approach straightforward albeit tedious. This is certainly true for the case considered by Harris and it is possible to expect that such type of approach will be even more tedious for more accurate forms of the coupled cluster theory as for example for the extended coupled-pair theory.

It has been shown [28] that algebraic approach can be considerably simplified when the algebraically formulated Wick theorem is used. In this algebraic formulation of Wick theorem the concept of Pfaffian is a very essential one.

Nevertheless we feel that the diagrammatical approach is

preferable even with respect to this simplified algebraic approach, both from the point of view of complexity of operations as well as from the point of view of the handling of the spin.

III. BASIC EQUATIONS

Let us consider a system having a non-degenerate ground state consisting of fixed atomic nuclei and of 2n electrons. Then, neglecting relativistic and magnetic effects, the Hamiltonian H of our problem is given by the following equations:

$$\hat{H} = \hat{Z} + \hat{V} \tag{2}$$

$$\hat{Z} = \sum_i \hat{z}_i, \quad \hat{V} = \sum_{i<j} \hat{v}_{i,j}, \tag{3}$$

where $\hat{z}_i$ is a one-particle operator corresponding to the sum of the kinetic and nuclear field energies and $\hat{v}_{i,j}$ is a two-particle operator of the interelectronic Coulomb repulsion. Our problem is to find the ground-state eigenvector and the corresponding eigenvalue E of the Hamiltonian (1):

$$\hat{H}|\Psi\rangle = E|\Psi\rangle \ . \tag{4}$$

The exact wave function will be written in the form of a cluster expansion, which has been discussed in the preceeding chapter:

$$\Psi = e^{\hat{T}}|\Phi\rangle, \tag{5}$$

where $|\Phi\rangle$ represents the ground-state wave function in the one-particle approximation. Therefore, the operator $\hat{T}$ is a logarithm of the so-called "wave operator".

We shall use the normalization

$$\langle \Phi | \Psi \rangle = 1. \tag{6}$$

Using this normalization we can write for the energy

$$E = \langle \Phi | \hat{H} e^{\hat{T}} | \Phi \rangle \tag{7}$$

and the equation (4) can be rewritten in the form:

$$(\hat{H} - \langle\Phi|H|\Phi\rangle)e^{\hat{T}}|\Phi\rangle = (\langle\Phi|\hat{H}|e^{\hat{T}}\Phi\rangle - \langle\Phi|H|\Phi\rangle)e^{\hat{T}}|\Phi\rangle . \tag{8}$$

This is an implicit form of the system of non-linear equations for the components of the operator $\hat{T}$ and the expression

$$\langle\Phi|\hat{H}|e^{T}\Phi\rangle - \langle\Phi|\hat{H}|\Phi\rangle \tag{9}$$

represents the correlation energy, provided that $|\Phi\rangle$ is the HF determinant.

Our next task will be to rewrite this system in an explicit form. As indicated in the proceeding chapter we shall use

1) The formalism of creation and annihilation operators.
2) Particle-hole formalism.
3) Wick's theorem.
4) Diagram techniques.

In order to define creation and annihilation operators we introduce a complete set of orthonormal spin-orbitals.

$$|A\rangle = |a\rangle|\alpha\rangle \tag{10}$$

where $|a\rangle$ and $|\alpha\rangle$ designate the space and the spin parts of the spin-orbital $|A\rangle$, respectively. Generally, capital letters will be used to designate spin-orbitals, lower case letters will be associated with orbitals, while Greek letters will be used for spin functions. In addition to this general system of spin-orbitals we shall use spin-orbitals whose space parts are eigenfunctions of the operator

$$(\hat{z} + \hat{u})|a\rangle = \omega_a|a\rangle , \tag{11}$$

where $\hat{u}$ is an arbitrary spin-independent one-particle Hermitian operator. Let us note explicitly that Hartree-Fock spin-orbitals fall within this class.

The creation and annihilation operators defined in the system of spin orbitals (10) will be designated

$$\hat{X}_A^+ \quad \text{and} \quad \hat{X}_A$$

respectively.

These operators satisfy the following anticommutation relations:

$$\hat{X}_A^+\hat{X}_B^+ + \hat{X}_B^+\hat{X}_A^+ = 0 ,$$

$$\hat{X}_A\hat{X}_B + \hat{X}_B\hat{X}_A = 0\,, \tag{12}$$
$$\hat{X}_A^+\hat{X}_B + \hat{X}_B\hat{X}_A^+ = \langle A|B\rangle\,.$$

The operators $\hat{Z}$ and $\hat{V}$ may then be expressed through the operators $\hat{X}_A^+$ and $\hat{X}_A^+$ in the following forms:

$$\hat{Z} = \sum_{C_1,B_1} \langle C_1|\hat{z}|B_1\rangle \hat{X}_{C_1}^+\hat{X}_{B_1} \tag{13}$$

$$\hat{V} = \frac{1}{2c} \sum_{C_1,B_1,C_2,B_2} \langle C_1C_2|\hat{v}|B_1B_2\rangle_c \hat{X}_{C_1}^+\hat{X}_{C_2}^+\hat{X}_{B_2}\hat{X}_{B_1} \tag{14}$$

$(c=1,2)$

where

$$\langle C_1C_2|\hat{v}|B_1B_2\rangle_2 = \langle C_1C_2|\hat{v}|B_1B_2\rangle_1 - \langle C_1C_2|\hat{v}|B_2B_1\rangle_1 \tag{15}$$

A non-antisymmetrized matrix element is usually denoted $\langle C_1C_2|\hat{v}|B_1B_2\rangle_1 = \langle C_1C_2|\hat{v}|B_1B_2\rangle$. The matrix elements $\langle C_1|\hat{z}|B_1\rangle$ and $\langle C_1C_2|\hat{v}|B_1B_2\rangle_1$ are defined in accordance with Dirac's notation.

Due to the spin-independency of our Hamiltonian the following relations are valid:

$$\begin{aligned} \langle C_1C_2|\hat{v}|B_1B_2\rangle_1 &= \langle c_1c_2|\hat{v}|b_1b_2\rangle\langle\gamma_1|\beta_1\rangle\langle\gamma_2|\beta_2\rangle \\ \langle C_1|\hat{z}|B_1\rangle &= \langle c_1|\hat{z}\; b_1\rangle\langle\gamma_1|\beta_1\rangle \end{aligned} \tag{16}$$

In (16) the notation introduced in (10) is used.

The operator $\hat{T}$ introduced above may be expressed through the operators $\hat{X}_A^+$ and $\hat{X}_A$ in the following form

$$\hat{T} = \sum_{j=1}^{2n} \hat{T}_j\,, \tag{17}$$

where

$$\hat{T}_j = \left(\frac{1}{j!}\right)^c \sum_{\substack{A''_j \ldots A''_1 \\ A'_j \ldots A'_1}} \langle A''_j \ldots A''_1 | \hat{i} | A'_j \ldots A'_1 \rangle_c \prod_{i=1}^{j} \hat{X}^+_{A''} \hat{X}_{A'} \quad c=1,2 \tag{18}$$

represents an operator of the j-fold excited state. In (18) the occupied and virtual spinorbitals are denoted by singly and doubly primed capitals, respectively. Later on we shall use the similar system of denotation for orbitals.

To establish a relationship between antisymmetrized and non-antisymmetrized matrix elements of the opertor T let us define a permutation P;

$$P = \begin{pmatrix} j & \ldots & 1 \\ k_j & \ldots & k_1 \end{pmatrix}$$

Then we get

$$\langle A''_j \ldots A''_1 | \hat{i} | A'_j \ldots A'_1 \rangle_2 = \sum_P (-1)^P \langle A''_j \ldots A''_1 | \hat{i} | A'_{k_j} \ldots A'_{k_1} \rangle_1 \tag{19}$$

In the following considerations it will be useful to regard the ground state $|\Phi\rangle$ as a "new vacuum state" as well as to interpret the excitation of a particle as the creation of a hole in a non-excited spin-orbital followed by the creation of a particle in an excited spin-orbital.

We introduce both normal product N and contractions with respect to this new "vacuum state"[1,2,6]. It can be shown [1,2,6] that the contractions may be expressed as

$$\begin{aligned} \overbrace{X_A X_B} &= 0 \\ \overbrace{X^+_A X_B} &= h(A)\langle A|B\rangle \\ \overbrace{X_A X^+_B} &= p(A)\langle A|B\rangle \\ \overbrace{X^+_A X^+_B} &= 0 \end{aligned} \tag{20}$$

where we used the functions p(A) and h(A), defined on the set of spin orbital indices as

$$\begin{aligned} p(A') &= 0, \qquad p(A'') = 1, \\ h(A') &= 1, \qquad h(A'') = 0. \end{aligned} \tag{21}$$

We shall be using both Wick theorem and the generalized Wick theorem [1,2,6].

First we shall use the Wick theorem in order to write operators $\hat{Z}$ and $\hat{V}$ in "normal" form, [1,2,6] :

$$\hat{Z} = \sum_{C_1,B_1} \langle C_1|\hat{z}|B_1\rangle N[\hat{X}^+_{C_1}\hat{X}_{B_1}] + \langle\Phi|\hat{Z}|\Phi\rangle \tag{22}$$

$$\begin{aligned}\hat{V} = \frac{1}{2^c} \sum_{C_1,B_1,C_2,B_2} \langle C_1C_2|\hat{v}|B_1B_2\rangle_c N[\hat{X}^+_{C_1}\hat{X}^+_{C_2}\hat{X}_{B_2}\hat{X}_{B_1}] \\ + \sum_{C_1,B_1} \langle C_1|\hat{g}|B_1\rangle N[\hat{X}^+_{C_1}\hat{X}_{B_1}] + \langle\Phi|\hat{V}|\Phi\rangle \\ (c=1,2)\end{aligned} \tag{23}$$

where

$$\langle C_1|\hat{g}|B_1\rangle = \sum_{A'} \langle C_1A'|\hat{v}|B_1A'\rangle_a \tag{24}$$

For this matrix element we can further write

$$\langle C_1|\hat{g}|B_1\rangle = \langle c_1|\hat{g}|b_1\rangle\langle\gamma_1|\beta_1\rangle \tag{25}$$

where

$$\langle c_1|\hat{g}|b_1\rangle = \sum_{a'} (2\langle c_1a'|\hat{v}|b_1a'\rangle - \langle c_1a'|\hat{v}|a'b_1\rangle) \tag{26}$$

Again, the one- or two-particle operators, the mean values of which we shall calculate later on, may be expressed in the same way as the operators $\hat{Z}$ and $\hat{V}$. Using relations (22)-(26) the total Hamiltonian may be written as follows:

$$\begin{aligned}\hat{H} - \langle\Phi|\hat{H}|\Phi\rangle = \sum_{C_1,B_1} \langle C_1|f|B_1\rangle N[\hat{X}^+_{C_1}\hat{X}_{B_1}] \\ + \frac{1}{2^c} \sum_{C_1,B_1;C_2,B_2} \langle C_1C_2|\hat{v}|B_1B_2\rangle_c N[\hat{X}_{C_1}\hat{X}_{C_2}\hat{X}_{B_2}\hat{X}_{B_1}] \\ (c=1,2)\end{aligned} \tag{27}$$

Finally let us emphasize that the transcription of the operators (13) and (14) into the "normal" form will result in a significant simplification of further considerations.

Our final goal in this chapter is to write the equations (8) and (9) in the explicit form. For this purpose we shall use the generalized Wick theorem and the technique of skeletons and diagrams. For the lack of space we have to refer the reader to the papers [1,2] for the exact definitions and intermediate results. The concept of the skeleton is purely topological while the diagram represents an operator or c number. Roughly speaking we get the skeleton if we strip the diagram of lables denoting spinorbitals. The basic idea of the procedure is that we express the operator $(\hat{H}-\langle\Phi|\hat{H}|\Phi\rangle)$ by means of so-called H diagrams, the operator $\hat{T}$ by means of the so-called T diagrams and finally the operator $e^{\hat{T}}$ by means of M diagrams. A M diagram is simply a set of T diagrams. The corresponding skeletons are called H, T and M skeletons, respectively. Then in order to express

$$(\hat{H}-\langle\Phi|\hat{H}|\Phi\rangle)e^{\hat{T}}|\Phi\rangle$$

we shall use the generalized Wick theorem. When using the generalized Wick theorem we express each contraction by joining of two lines of H and M skeletons. Skeletons which we get by performing several (zero included) contractions between the H and M skeleton we shall call R skeletons. Then we distinguish connected and disconnected skeletons. This distinction permits us to factorize the left hand side of (8) into $e^{\hat{T}}$ and the operator which is expressed through diagrams corresponding to connected R skeletons. The final result is

$$\sum_{r\in\Delta_j^c} w_R^c(r)\sum_{\sigma}\hat{D}_R^c(r,\sigma)|\Phi\rangle = \tag{28}$$

$$= (E-\langle\Phi|H|\Phi\rangle)|\Phi\rangle \qquad j=0,1,2,3\ldots \quad c=1,2$$

Before explaining the denotation let us stress that equation (28) is completely equivalent to the Schrödinger equation (5). In (28) the symbol r is representing the R skeleton, symbol σ is representing the spinorbitals which are labeling the R skeleton and operator $\hat{D}_R(r,\sigma)$ is representing the operator corresponding to the given R diagram. The symbol $w_R^c(r)$ is the numerical factor associated with the topology of the R skeleton. Before explaining the symbol Δ^c we should define the quasi-equivalency of skeletons. Let us introduce the concept of degenerate skeletons. By a degenerate skeleton we understand the skeleton which we get from an ordinary skeleton by replacing all non oriented lines by vertices [2,7]. Then two skeletons are quasi-equivalent when their degenerate counterparts are topologically identical. Now we can define Δ_j^c: Δ_j^1 is a set of topologically distinct connected R skeletons having j open paths. The set Δ_j^1 may be divided into several disjunct sets each containing topologically quasi-equivalent connected R skeletons. Further we shall designate by Δ_j^2 a set of connected R skeletons formed in such a way that from each subset of the quasi-equivalent connected R skeletons mentioned above we select one representative skeleton.

Let us emphasize that in the case c=1 we deal with the non-antisymmetrized version of the theory while for c=2 we deal with the antisymmetric version of the theory. The first version requires drawing of more diagrams than the second one but will permit simple elimination of the spin for the case of spin-independent forces. For the discussion of this version see paper [4]. The second version is providing a convenient first step for a spin adapted theory [7]. In addition the second version is handy in nuclear physics where the internucleon forces are spin dependent.

Finally let us stress that up to this point our considerations were completely general. Now we assume that $|\Phi\rangle$ is a HF determinant and $\hat{T}$ is approximated by $\hat{T}_2$. In this case we shall require that the equations (28) are fulfilled for j=0 and j=1. The explicit form of the equations for the case c=1 is given both in [1] and [2]. The explicit form of the equations for the case c=2 is presented both in [2] and [7]. There we present the case c=2 in modified form. First let us introduce a simplified denotation

$$\hat{t}_2 \equiv t\ ,\quad \hat{f} \equiv f\ ,\quad \hat{v} \equiv v\ ,$$

$$D_1' \equiv 1\ ,\quad D_2' \equiv 3\ ,\quad D_3' \equiv 5\ ,\quad D_4' \equiv 7\ ,$$

$$D_1'' \equiv 2\ ,\quad D_2'' \equiv 4\ ,\quad D_3'' \equiv 6\ ,\quad D_4'' \equiv 8\ .$$

For j=0 we have

$$E-\langle\Phi|\hat{H}|\Phi\rangle = \tag{29}$$

$$= \frac{1}{4}\sum_{\substack{1,3\\2,4}}\langle 13|v|24\rangle_2\langle 24|t|13\rangle_2$$

For j=2 after projection of expression (28) on the state

$$\hat{X}_2^+\,\hat{X}_1\,\hat{X}_4^+\,\hat{X}_3^+|\Phi\rangle$$

we get from equation (68) in [2] and also from equation (15) in [7]:

$$\sum_6\{\langle 2|f|6\rangle\langle 64|t|13\rangle_2$$
$$+\langle 4|f|6\rangle\langle 26|t|13\rangle_2$$
$$-\sum_5\{\langle 5|f|1\rangle\langle 24|t|53\rangle_2$$
$$+\langle 5|f|3\rangle\langle 24|t|15\rangle_2$$
$$+\langle 24|v|13\rangle_2$$
$$+\sum_{5,6}\{\langle 25|v|16\rangle_2\langle 46|t|35\rangle_2$$
$$+\langle 45|v|36\rangle_2\langle 26|t|15\rangle_2$$
$$-\langle 45|v|16\rangle_2\langle 26|t|35\rangle_2$$
$$-\langle 25|v|36\rangle_2\langle 46|t|15\rangle_2\}$$

$$+\frac{1}{2}\sum_{6,8}\langle 24|v|68\rangle_2\langle 68|t|13\rangle_2$$

$$+\frac{1}{2}\sum_{5,7}\langle 57|v|13\rangle_2\langle 24|t|57\rangle_2$$

$$+\sum_{5,6,7,8}\Big\{\langle 57|v|68\rangle_2 \times$$

$$\times\Big[\langle 26|t|15\rangle_2\langle 48|t|37\rangle_2$$

$$-\langle 46|t|15\rangle_2\langle 28|t|37\rangle_2$$

$$-\frac{1}{2}\langle 68|t|15\rangle_2\langle 24|t|37\rangle_2$$

$$+\frac{1}{2}\langle 68|t|35\rangle_2\langle 24|t|17\rangle_2$$

$$-\frac{1}{2}\langle 26|t|57\rangle_2\langle 48|t|13\rangle_2$$

$$+\frac{1}{2}\langle 46|t|57\rangle_2\langle 28|t|13\rangle_2$$

$$\frac{1}{4}\langle 24|t|57\rangle_2\langle 68|t|13\rangle_2\Big]\Big\} = 0$$

The spin independence of the operators $\hat{f}$ and $\hat{v}$ can be used for coupled pair many electron theory in two different ways. The first possibility is to consider the case c=1 and use the procedure described in [1] where to each closed loop a factor of 2 is assigned and each spinorbital is replaced by an orbital. The second possibility is to develop "Spin Adapted Cluster Many Electron Theory" [7]. In higher approximations than CPMET this approach is of outmost importance for clear understanding and transparent formulation of the pertinent system of nonlinear equations. Even for CPMET this approach is useful for practical application of the theory.

IV. PROPOSAL FOR THE APPLICATION OF CPMET

Let us consider a linear chain consisting of weakly interacting identical closed shell molecules.

--- (M) --- (M) --- (M) --- (M) ---

First let us deal with an isolated molecule. We assume that the molecule is of such type that correlation effects in ground state are well described by CPMET.

Now let us try to propose on the basis of the above assumptions the procedure for the calculation of the ground state energy for the whole infinite chain. The first step in this procedure is the solution of the HF problem for the infinite chain. The second step is the localisation of the HF occupied orbitals. We have assumed that we deal with weakly interacting closed shell molecules and therefore we can expect that in this case the localized HF orbitals are qualitatively similar to the HF orbitals of the isolated molecule. The third step is the choice of convenient localized virtual orbitals, which are orthogonal to all occupied HF orbitals in each site.

Having in mind this description of an infinite chain it is reasonable to assume that CPMET would work for the whole chain. The reasons are the following: First, because we start from the HF picture we can neglect due to the Brillouin theorem singly excited clusters. Further we have assumed weak interaction between the molecules. This means that two "independent biexcitations" in two "different sites" should be well described by the term:

$$\frac{\hat{T}_2^2}{2}|\Phi\rangle$$

and therefore we do not face any "size consistency problem".

It remains to describe the procedure how to solve the equations of CPMET. The proposed procedure is somewhat related to the method described in a series of papers by Klein et al [29-33]. First let us define "length" of the t_2 element. Each t_2 element is characterized by two occupied orbitals and by two unoccupied orbitals. The four orbitals are localized on four sites, some of them might be identical, in the extreme case all four orbitals are localized on one site. By the "length" of t_2 element we understand the number of links between the extreme left and extreme right site.

In the zero approximation we consider only t_2 elements of size zero, in other words we consider t_2 elements having all four orbitals localized in the same site. The corresponding system of nonlinear equations is therefore of rather small size and easily solvable.

In the first approximation we consider only t_2 matrix elements of length zero and one, in other words we consider all t_2 matrix elements having all four orbitals localized either on one site or on two neighbouring sites. The system of non-linear equations is getting more complicated, but we can facilitate the solution if we take for the first iteration the values of "zero length t_2 elements" from zero order approximation.

In the n'th approximation we shall take into account all t_2 matrix elements of length zero, one, two ... n-1, n. We can facilitate the solution of the system of non-linear equations by a similar manipulation as described above.

We have assumed only weak interaction between the molecules and therefore we can expect that the energy per unit will converge before n would be very high. The most simple system of this type is the chain of ethylene molecules.

$$
\begin{array}{cccc}
H_2C & H_2C & H_2C & H_2C \\
\| \cdots & \| \cdots & \| \cdots & \| \\
H_2C & H_2C & H_2C & H_2C
\end{array}
$$

This system is now under study in the Institute of Theoretical Chemistry in Erlangen.

ACKNOWLEDGMENTS

The stay of the author in Erlangen as Richard Merton Visiting Professor was made possible through the stipend from the Deutsche Forschungsgemeinschaft which is hereby gratefully acknowledged. This work has been supported by a National Research Council of Canada Grant in - Aid - of Research, which is also gratefully acknowledged.

I would like to express my sincere gratitude to Professor J. Ladik as well as to all other members of his Institute for their kind hospitality and stimulating atmosphere during my stay in Erlangen.

I am also very much oblighed to Professor J. Paldus for many helpful discussions on the subject of this paper.

Finally I thank Miss S. Patzak for the careful typing of the manuscript.

REFERENCES

1. J. Čížek, J. Chem. Phys. 45, 4256 (1966).
2. J. Čížek, Adv. Chem. Phys. 14, 35 (1969).
3. J. Čížek and J. Paldus, Int. J. Quant. Chem. 5, 359 (1971).
4. J. Paldus, J. Čížek and J. Shavitt, Phys. Rev. A5, 50 (1972).
5. J. Paldus and J. Čížek, in Energy, Structure and Reactivity, D.W. Smith and W.B. Mac Rae (Wiley, N.Y., 1973) p. 198.
6. J. Paldus and J. Čížek, Adv. Quant. Chem. 9, 105 (1975).
7. J. Paldus: Spin Adapted Coupled Cluster Many-Electron Theory, J. Chem. Phys. (in press).
8. F. Coester, Nucl. Phys. 7, 421 (1958).
9. F. Coester and H. Kümmel, Nucl. Phys. 17, 477 (1960).
10. H. Kümmel, in Lectures on the Many Body Problem, E.R. Caianiello, Ed. (Academic Press, N.Y. 1962) p. 265.
11. J. da Providencia, Nucl. Phys. 61, 87 (1965).
12. O. Sinanoglu, J. Chem. Phys. 36, 706 (1962).
13. O. Sinanoglu, J. Chem. Phys. 36, 3198 (1962).
14. O. Sinanoglu, Adv. Chem. Phys. 6, 315 (1964).
15. O. Sinanoglu, Adv. Chem. Phys. 14, 237 (1969).
16. H.J. Silverstone and O. Sinanoglu, J. Chem. Phys. 44, 1899 (1966).
17. H.J. Silverstone and O. Sinanoglu, J. Chem. Phys. 44, 3608 (1966).
18. H. Primas, in Modern Quantum Chemistry Vol. II, O. Sinanoglu, Ed. (Academic Press, N.Y. 1965), p. 45.
19. H. Kümmel, Nucl. Phys. A176, 205 (1971).

20. H. Kümmel and K.H. Lührmann, Nucl. Phys. A191, 525 (1972).
21. K.H. Lührmann and H. Kümmel, Nucl. Phys. A194, 225 (1972).
22. J.G. Zabolitzky, Nucl. Phys. A228, 272 (1974).
23. J.G. Zabolitzky, Nucl. Phys. A228, 285 (1974).
24. H. Kümmel and J.G. Zabolitzky, Phys. Rev. C7, 547 (1973).
25. F.E. Harris, in Electrons in Finite and Infinite Structures, Proceedings of the NATO ASI in Gent, Belgium 1976, P. Pharizean, Ed. (Plenum Press, N.Y.) in press.
26. D.L. Freeman, Coupled-Pair Many-Electron Theory applied to the Electron Gas: Inclusion of Ring and Exchange Effects, Phys. Rev. B, in press.
27. A.C. Harley, Electron Correlation in Small Molecules, (Academic Press, New York, 1976).
28. J. Čížek and J. Paldus, to be published.
29. D.J. Klein, J. Chem. Phys. 64, 4868 (1976).
30. D.J. Klein, Mol. Phys. 31, 783 (1976).
31. D.J. Klein, Mol. Phys. 31, 811 (1976).
32. D.J. Klein and M.A. Garcia-Bach, J. Chem. Phys. 64, 4873 (1976).
33. D.J. Klein and M.A. Garcia-Bach, Mol. Phys. 31, 797 (1976).

FOURIER REPRESENTATION METHODS IN ELECTRONIC STRUCTURE STUDIES OF CRYSTALS AND POLYMERS

Frank E. HARRIS

Department of Physics
University of Utah
Salt Lake City, Utah (U.S.A.)

I. INTRODUCTION

A complete theoretical approach to electronic structures of infinitely extended periodic systems must include two features. That which has tended to receive more attention is the effect of neighboring groups on the charge distribution in the vicinity of each atom of the system. The other feature, often not explicitly discussed, is the long-range effect of the inifite amount of charge (of both signs) of which the system is composed.

The infinite range of the Coulomb force makes the electrostatic potential sensitively dependent upon the size, shape and detailed specification of the system, and even the convergence of the Madelung-type summations for the potential may depend upon the system specification. We therefore open this series of lectures with an examination of the convergence problems associated with charge arrays.

Both the summation of long-range contributions to the potential and the evaluation of quantum mechanical electronic interactions are facilitated by the use of Fourier representation methods. We therefore proceed next to a discussion of the main characteristics of Fourier representation theory, identifying the features which make its use so advantageous. We then describe the application of these techniques to make calculations in the Hartree-Fock approximation for systems with three-dimensional or one-dimensional periodicity.

J.-M. André et al. (eds.), Quantum Theory of Polymers, 117-135.

To obtain substantial agreement with experiment it is necessary to go beyond the Hartree-Fock approximation by estimating or calculating the correlation energy. In a final section, we therefore examine various ways in which the correlation energy might be studied, and report a few results which have been obtained thus for.

II. MADELUNG SUMS

Lattice sums over charge arrays (Madelung sums)are notable for their slow convergence. Even where there are no formal convergence difficulties, the numbers of terms needed for high accuracy may be prohibitive if a straightforwerd term-by-term evaluation is carried out. For example, consider the simple system consisting of an infinite equally spaced linear array of point charges of alternating sign, for which the potential at the position of a positive charge, V+, is given by

$$V+ = 2 \sum_{n=1}^{\infty} \frac{(-1)^n}{na} = -\frac{2}{a} \ln 2 \qquad (1)$$

where a is the distance between neighboring charges. If evaluated directly, 10^4 terms are needed to yield four significant figures in the sum. However, one would usually at least combine the contributions into neutral groupings (here unit cells containing one charge of each sign); then about 50 cells in either direction will be needed for four figure accuracy. The situation is far worse if derivatives of the potential are to be considered. The electric field at a positive charge, $\mathcal{E}+$, should vanish; the intra-cell contribution ($1/a^2$) is only cancelled to four figures if 100 cells in each direction are included.

The summation of eq.(1) also has the formal characteristic of conditional convergence. From a physical point of view this presents no problem as it is clear that what is intended is the limit as the number of charges is extended indefinitely in both directions (so that the numbers of negative and positive charges remain equal, except for a possible extra charge at each end of the line). But in systems with two or three-dimensional periodicity, the situation may be more complicated. Since the potentials of charges, dipoles and quadrupoles vary respectively with distance as r^{-1}, r^{-2} and r^{-3}, three dimensional arrays of charges or such multipoles may diverge ($\int r^{-p} d\vec{r}$ behaves as $\int r^{-p+2} dr$, and is divergent at large r for $p \leqslant 3$). In two dimensions, arrays of charges or dipoles can yield divergent results. In actually realizable situations, the charges or multipoles are arranged in a way causing the potentials to remain finite; the conditional convergence reflects this fact and indicates that the summations must be carried out in a manner consistent with the assumed physical situation.

Finally we call attention to the fact that for a three-dimensional periodic system, the potential will also depend upon the value of the trace of the second moment tensor (as well as upon its other components which are reflected in the quadrupole moment)[1]. The trace contribution is not size, shape or position dependent but must be considered in connection with calculations of energies involved in addition or removal of an electron.

Many periodic systems consist of crystallographic unit cells possessing non vanishing low order moments. In such cases, it is nevertheless possible to obtain convergent Madelung sums by defining a "repeating unit" other than the crystallographic unit cell. For example, a linear array of charges of alternating sign at separation a can be obtained by repetition of the charge sequence $+\frac{1}{2}$, -1, $+\frac{1}{2}$ at intervals 2a; this repeating unit has zero dipole moment. Any macroscopic sample can be regarded as composed of repeating units with vanishing low-order moments with charges, or fractions of charges near the boundaries of the sample. The potential inside such a sample will differ from that calculated for the repeating-unit array by the easily calculable effect of the added boundary charge.

The various methods which may be used to carry out Madelung summations carry implications as to the nature of the sample to which they apply. For example, a direct calculation by summation over cells or shells will apply to a sample whose boundary specification must match that of the cell or shell units. In using such a method, it will be easily noticed whether convergence to a unique result is being obtained; failure to obtain such a result can be traced to an inappropriate definition of the implied repeating unit.

In line with the above observation, we note that the integral representation techniques for accelerating convergence of lattice sums (cf. that of Ewald) imply particular sample specifications. This statement is formally obvious from the fact that these techniques usually convert the summations to absolutely convergent forms. Closer investigation reveals that the acceleration techniques involve interchanges of summations and integrations under conditions which are only satisfied by a particular sample specification. In the case of the Ewald method in three dimensions, the implied requirement is that the sample have a vanishing net charge, dipole moment, and second moment tensor (all components)[2]. The same conditions are found to be implied by the Fourier representation technique to be described below.
The classical Madelung summation problem has also been discussed

by the present author in the notes from an earlier NATO-supported summer institute(3).

III. FOURIER REPRESENTATION THEORY

We start this section by giving some standard formulas involving Fourier transforms, using them to introduce the notations and as a basis for a discussion of the advantages the Fourier representation formalism can provide. A function $f(\vec{r})$, where $\vec{r}$ is a three-dimensional vector, has a transform denoted $f^T(\vec{q})$, where $\vec{q}$ is the transform variable (which may be regarded as in a three-dimensional space reciprocal to the $\vec{r}$ space), according to the formula

$$f^T(\vec{q}) = \int e^{i\vec{q}.\vec{r}} f(\vec{r})d\vec{r} \tag{2}$$

In eq.(2), the integration is over an infinite range of $\vec{r}$. Equation (2) can be inverted to yield the <u>inversion formula</u>

$$f(\vec{r}) = \frac{1}{(2\pi)^3} \int e^{-i\vec{q}.\vec{r}} f^T(\vec{q})\, d\vec{q} \tag{3}$$

An important consequence of eq.(2) is that it causes the effect of the origin of $\vec{r}$ to occur only in the exponential $\exp(i\vec{q}.\vec{r})$; this has the effect that a shift in origin only multiplies a transform by a phase factor :

$$[f(\vec{r}-\vec{R})]^T(\vec{q}) = e^{i\vec{q}.\vec{R}} f^T(\vec{q}) \tag{4}$$

Some important functions have simple transforms. For example,

$$\left(\frac{1}{r}\right)^T = \frac{4\pi}{q^2} \tag{5}$$

$$(e^{-\zeta r})^T = \frac{8\pi^{1/2}\zeta^{5/2}}{(q^2+\zeta^2)^2} \tag{6}$$

$$(e^{-ar^2})^T = \left(\frac{\pi}{a}\right)^{3/2} e^{-q^2/4a} \tag{7}$$

We see that the 1s Slater-type orbital (STO) has a simple transform. STO's of higher quantum numbers also have tractable transforms.

This fact facilitates the use of STO's in calculations of periodic systems.
By writing a function as the inverse of its transform, we may express it as an integral called its Fourier representation . An important example is

$$\frac{1}{r} = \frac{1}{(2\pi)^3} \int d\vec{q}\ e^{-i\vec{q}.\vec{r}} \left(\frac{4\pi}{q^2}\right) \tag{8}$$

Fourier transforms have the property that convolution of two functions in $\vec{r}$ space corresponds to multiplication of their transforms in $\vec{q}$ space. In particular for two functions f and g,

$$\left[\int d\vec{r}'\ f(\vec{r}')\ g(\vec{r}-\vec{r}') \right]^T = f^T(\vec{q})\ g^T(\vec{q}) \tag{9}$$

In combination with the inversion theorem, eq.(9) yields several vary useful formulas including

$$\int d\vec{r}_1 d\vec{r}_2 f(\vec{r}_1)\ r_{12}^{-1}\ \ g(\vec{r}_2-\vec{R}) = \frac{1}{2\pi^2} \int \frac{d\vec{q}}{q^2}\ f^T(\vec{q}) g^T(-\vec{q}) e^{-i\vec{q}.\vec{R}} \tag{10}$$

$$[f(\vec{r})g(\vec{r}-\vec{R})]^T\ (\vec{q}) = \frac{1}{(2\pi)^3} \int d\vec{p}\ f^T(\vec{p})\ g^T(\vec{q}-\vec{p})\ e^{i(\vec{q}-\vec{p}).\vec{R}} \tag{11}$$

Equation (10) is the reduction of the electron repulsion integral introduced by Bonham , Peacher and Cox [4]; it converts the very complicated geometric structure of the integral into a more easily handled exponential form and removes the irrational form implied by $1/r_{12}$.

A particularly useful transform and representation are those of the Dirac delta function $\delta(\vec{r})$, for which

$$[\delta(\vec{r})]^T = 1 \tag{12}$$

and hence

$$\delta(\vec{r}) = \frac{1}{(2\pi)^3} \int e^{-i\vec{q}.\vec{r}}\ \ d\vec{q} \tag{13}$$

If the integral in eq.(13) is replaced by a summation over points of a lattice, the analogous formula can be written

$$\frac{V_o}{(2\pi)^3} \sum_{\mu} e^{-i\vec{k}\vec{r}_\mu} = \sum_{\mu} \delta(\vec{k}-\vec{q}_\mu) \tag{14}$$

where the $\vec{r}_\mu$ are the points of a lattice of unit-cell volume V_o, and the $\vec{q}_\mu$ are the points of the lattice reciprocal to the $\vec{r}_\mu$. (In the case of a cubic $\vec{r}_\mu$ lattice of cell dimension _a_, $V_o = a^3$ and the reciprocal lattice is also cubic, of cell dimension $2\pi/a$). Equation (14) is sometimes called a lattice orthogonality theorem and is discussed in a number of texts (5).

If we multiply both sides of eq.(14) by a transform $f^T(\vec{k})$ and integrate in $\vec{k}$ over a full three-dimensional space, the result can be reduced to the form

$$\sum_{\mu} f(\vec{r}_\mu) = \frac{1}{V_o} \sum_{\mu} f^T(\vec{q}_\mu) \tag{15}$$

showing that a sum of _f_ over the $\vec{r}_\mu$ lattice is equivalent to a sum of its transform f^T over the reciprocal lattice $\vec{q}_\mu$. This result is known as the <u>Poisson summation formula</u>. Since functions which are diffuse in a given space are more concentrated locally in its reciprocal space, the Poisson formula provides a means for accelerating the convergence of lattice sums.

The usefulness of the Fourier representation technique for handling the long-range aspects of the Coulomb force can be illustrated by considering a classical lattice sum. For a simple cubic lattice of cell dimension _a_ with a unit positive charge at each lattice point and a unit negative charge at a distance $\vec{s}$ from each lattice point, the potential at lattice point $\vec{r}_\mu$ is given by the conditionally convergent sum

$$V(\vec{r}_\mu) = \sum_{\nu\neq\mu} \frac{1}{|\vec{r}_\nu - \vec{r}_\mu|} - \sum_{\nu} \frac{1}{|\vec{r}_\nu + \vec{s} - \vec{r}_\mu|} \tag{16}$$

Using now eq.(8), we write

$$V(\vec{r}_\mu) = \frac{1}{2\pi^2} \sum_{\nu\neq\mu} \int \frac{d\vec{q}}{q^2} e^{-i\vec{q}.(\vec{r}_\nu - \vec{r}_\mu)} - \frac{1}{2\pi^2} \sum_{\nu} \int \frac{d\vec{q}}{q^2} e^{-i\vec{q}.(\vec{r}_\nu + \vec{s} - \vec{r}_\mu)} \tag{17}$$

Interchanging the order of summation and integration, then adding (and subtracting again) a term $\nu = \mu$ to complete the first sum, and finally using eq.(14), we have

$$V(\vec{r}_\mu) = \frac{1}{2\pi^2}\int \frac{d\vec{q}}{q^2}\left[\left(\frac{2\pi}{a}\right)^3 \sum_\mu \delta(\vec{q}-\vec{q}_\mu) - 1 - \left(\frac{2\pi}{a}\right)^3 \sum_\mu \delta(\vec{q}-\vec{q}_\mu)e^{-i\vec{q}.\vec{s}}\right]$$

$$= \frac{1}{2\pi^2}\left[\left(\frac{2\pi}{a}\right)^3 \sum_\mu \frac{1}{q^2} - \int \frac{d\vec{q}}{q^2}\right] - \frac{1}{2\pi^2}\left(\frac{2\pi}{a}\right)^3 \sum_\mu \frac{e^{-i\vec{q}_\mu.\vec{s}}}{q_\mu^2} \tag{18}$$

The infinite $q_\mu=0$ contributions to the two sums are opposite in sign and are dropped, after which the individually divergent sum and integral (enclosed in large brackets) combine to give a convergent result which, for the lattice now under discussion, has the value $-(8.913633...)(2\pi/a)$. Thus,

$$V(\vec{r}_\mu) = \frac{-8.913633}{\pi a} - \frac{4\pi}{a} \sum_{\substack{\mu \\ q_\mu \neq 0}} \frac{e^{-i\vec{q}_\mu.\vec{s}}}{q_\mu^2} \tag{19}$$

The propriety of the mathematical steps leading to eq.(19) can be confirmed for a sample all of whose moments vanish through order 2. We see that the final result consists of a "lattice structure" term dependent on the scale (and symmetry) of the lattice plus a term describing effects due to the placement of charge within each cell at a distance $\vec{s}$ from the cell origin. If $\vec{s}$ is averaged over the cell, the second term vanishes, showing that the first term can also be interpreted as the potential due to a charge lattice embedded in a uniform compensating background; the second term therefore describes contributions to the potential caused by deviation from the uniform-background situation.

In systems with two-dimensional periodicity, a development much like that just illustrated is also available. However, for systems with periodicity in only one dimension, some modifications in the procedure are necessary, owing to the fact that a uniform compensating line of charge yields a potential which is divergent for points on the line. A different grouping of terms is therefore necessary. On convenient possibility amounts to expressing the potential as the Fourier representation of the deviation in potential from the limiting case of equally spaced charges of alternating sign [6].

IV. HARTREE-FOCK APPROXIMATION

We now proceed to see how the Fourier representation technique facilitates the evaluation of the detailed interactions among nuclei and electrons. We consider ab initio calculations in which no approximations are introduced to simplify the treatment of exchange, using the exact non-relativistic zero-order (electrostatic) electronic Hamiltonian. Illustrating for a simple lattice of hydrogen atoms, the Hamiltonian is, in atomic units (energy in hartrees , distance in bohrs)

$$H = \sum_i - \frac{1}{2} \nabla_i^2 + \sum_{i<j} h(\vec{r}_i, \vec{r}_j) \tag{20}$$

where subscripts i,j refer to electrons, $\vec{r}_i$ is the coordinate vector of electron i, and

$$h(\vec{r}_1, \vec{r}_2) = \frac{1}{r_{12}} - \frac{1}{N} \sum_\mu \frac{1}{|\vec{r}_1 - \vec{R}_\mu|} - \frac{1}{N} \sum_\mu \frac{1}{|\vec{r}_2 - \vec{R}_\mu|}$$

$$+ \frac{1}{2N^2} \sum_{\mu \neq \nu} \frac{1}{R_{\mu\nu}} \tag{21}$$

In eq.(21), $\vec{R}_\mu$ is a point at which a stationary proton is situated, the system contains N protons and N electrons, and h is accurate to order 1/N.

The Hartree-Fock solution is assumed to be an antisymmetrized product of orbitals of Bloch type with the symmetry of the lattice; these orbitals are constructed from Bloch basis states. Letting $|\vec{k}\rangle$ be the Hartree-Fock Bloch orbital of Bloch wave vector $\vec{k}$, and $|\vec{k}_\alpha\rangle$ a basis Bloch state (also of Bloch wave vector $\vec{k}$), we have

$$|\vec{k}\rangle = \sum_\alpha C_{k\alpha} |\vec{k}_\alpha\rangle \tag{22}$$

There are several possibilities for the choice of the $|\vec{k}_\alpha\rangle$, of which we illustrate three :

(1) LCAO or tight-binding :

$$|\vec{k}_\alpha\rangle = \sum_\alpha e^{i\vec{k}.\vec{R}_\mu} \phi_\alpha(\vec{r} - \vec{R}_\mu) \tag{23}$$

(2) Modulated plane wave (MPW) :

$$|\vec{k}_\alpha\rangle = e^{i\vec{k}.\vec{r}} \sum_\mu \phi_\alpha(\vec{r}-\vec{R}_\mu) \tag{24}$$

(3) Plane wave :

$$|\vec{k}_{\vec{K}}\rangle = e^{i(\vec{k}-\vec{K}).\vec{r}} \tag{25}$$

In these formulas, ϕ_α is an atomic orbital, so eqs.(23) and (24) describe lattice sums of such orbitals centered at the points R_μ. In eq.(25), $\vec{K}$ is a reciprocal-lattice vector which plays the same role as α does in eqs. (22)-(24).

The Hartree-Fock energy is the expectation value of the Hamiltonian for a wave function built from the doubly occupied orbitals of lowest energy, which occupy the region in $\vec{K}$ bounded by what is known as the Fermi surface. In the present units, the $\vec{k}$-space volume of the occupied orbitals is $\frac{1}{2}\left(\frac{a}{2\pi}\right)^3$ for a system containing one electron per lattice cell. For the illustrated system,

$$\langle H\rangle = 2N\left(\frac{a}{2\pi}\right)^3 \int \frac{\langle\vec{k}|-\frac{1}{2}\nabla^2|\vec{k}\rangle}{\langle\vec{k}|\vec{k}\rangle}\, d\vec{k} + 2N^2\left(\frac{a}{2\pi}\right)^6 \int \frac{\langle\vec{k}\vec{k}'|h|\vec{k}\vec{k}'\rangle - \frac{1}{2}\langle\vec{k}\vec{k}'|h|\vec{k}'\vec{k}\rangle}{\langle\vec{k}|\vec{k}\rangle\langle\vec{k}'|\vec{k}'\rangle}\, d\vec{k}\, d\vec{k}' \tag{26}$$

where the integrals are over the volume within the Fermi surface. The Hartree-Fock wavefunction is that resulting when $|\vec{k}\rangle$ and the Fermi surface are varied to achieve a minimum in $\langle H\rangle$. Techniques for accomplishing this variation have been described elsewhere [7].

The matrix elements appearing in eq.(26) can be evaluated with the aid of Fourier representation methods. This approach will not only render them tractable, but will also provide a resolution of the conditional convergence problems which must be present in the quantum-mechanical as well as in the classical treatment. To verify the convergence problem, note that $\langle\vec{k}\vec{k}'|h|\vec{k}\vec{k}'\rangle$ consists of electron-electron, electron-nuclear, and nuclear-nuclear contributions whose long-range behavior parallels that found classically. For example, the term $\langle\vec{k}\vec{k}'|r_{12}^{-1}|\vec{k}\vec{k}'\rangle$ can be shown to depend upon the sample size as $N^{2/3}\langle\vec{k}|\vec{k}\rangle\langle\vec{k}'|\vec{k}'\rangle$; when multiplied by N^2 in eq.(26) this term clearly diverges relative to N. The divergent part of $\langle\vec{k}\vec{k}'|r_{12}^{-1}|\vec{k}\vec{k}'\rangle$ of course cancels against that of an electron-nuclear interaction term; the cancellation is obtained by dropping the $\vec{q}_\mu=0$ reciprocal-space contributions and by taking as the representation of $\Sigma_{\nu\neq\mu} R_{\mu\nu}^{-1}$ the limit given as the first term of eq.(19).

Proceeding in more detail, consider next the evaluation of $\langle\vec{k}\vec{k}'|r_{12}^{-1}|\vec{k}\vec{k}'\rangle$ for a single function $|\vec{k}\rangle$ of the form given in eq. (24). We choose this form because it is that which leads to the simplest calculations while illustrating the key features of the calculational method. Calculations based on eq.(23) are also practical. In actuality, the integral $\langle\vec{k}\vec{k}'|r_{12}^{-1}|\vec{k}\vec{k}'\rangle$ will be formed as a linear combination of $\langle\vec{k}_\alpha\vec{k}_\beta'|r_{12}^{-1}|\vec{k}_\gamma\vec{k}_\delta'\rangle$, each calculated by the method illustrated. For the illustrative case,

$$\langle\vec{k}\vec{k}'|r_{12}^{-1}|\vec{k}\vec{k}'\rangle = \sum_{\mu\nu\lambda\sigma}\int \frac{d\vec{r}_1 d\vec{r}_2}{r_{12}} \phi^*(\vec{r}_1-\vec{R}_\mu)\phi^*(\vec{r}_2-\vec{R}_\nu)\phi(\vec{r}_1-\vec{R}_\lambda)\phi(\vec{r}_2-\vec{R}_\sigma)$$

$$= \sum_{\mu\nu\lambda\sigma} \frac{1}{2\pi^2}\int \frac{d\vec{q}}{q^2}[\phi^*(\vec{r}-\vec{R}_\mu)\phi(\vec{r}-\vec{R}_\lambda)]^T(\vec{q})[\phi^*(\vec{r}-\vec{R}_\nu)\phi(\vec{r}-\vec{R}_\sigma)]^T(-\vec{q}), \tag{27}$$

where we have used eq.(10) with $\vec{R}=0$. Applying eq.(4) and moving the summations inward as much as possible, eq.(27) may be rewritten

$$\langle\vec{k}\vec{k}'|r_{12}^{-1}|\vec{k}\vec{k}'\rangle = \frac{1}{2\pi^2}\int\frac{d\vec{q}}{q^2}\sum_{\mu\nu} e^{-i\vec{q}.(\vec{R}_\mu-\vec{R}_\nu)}\,\Phi(\vec{q})\;\Phi(-\vec{q}) \tag{28}$$

with

$$\Phi(\vec{q}) = \sum_\lambda [\phi^*(\vec{r})\;\phi(\vec{r}-\vec{R}_\lambda)]^T(\vec{q})$$
$$= \sum_\lambda \langle\phi(\vec{r})\;|e^{-i\vec{q}.\vec{r}}|\phi(\vec{r}-\vec{R}_\lambda)\rangle \tag{29}$$

Application of eq.(14) replaces the sum of exponentials by a delta function sum, yielding (after exclusion of the point $\vec{q}_\mu=0$)

$$\langle\vec{k}\vec{k}'|r_{12}^{-1}|\vec{k}\vec{k}'\rangle = \frac{4\pi}{a^3}\sum_{\substack{\mu\\ q_\mu\neq 0}}\frac{1}{q_\mu^2}\,\Phi(\vec{q}_\mu)\;\Phi(-\vec{q}_\mu) \tag{30}$$

Further analyses similar to that leading to eq.(30) lead for the simple cubic lattice to

$$\langle\vec{k}\vec{k}'|h|\vec{k}\vec{k}'\rangle = \frac{4\pi}{a^3}\sum_{\substack{\mu\\ q_\mu\neq 0}}\frac{1}{q_\mu^2}[\Phi(\vec{q}_\mu)\Phi(-\vec{q}_\mu)-\Phi(\vec{q}_\mu)\Phi(0)-\Phi(0)\Phi(-\vec{q}_\mu)]$$
$$-\frac{8.913633}{\pi a}\;\Phi(0)\Phi(0) \tag{31}$$

It can be shown that eq.(31) is correct for a sample with zero moments through order 2.

Turning next to the exchange matrix element $<\vec{k}\vec{k}'|h|\vec{k}'\vec{k}>$, we obtain non-negligible contributions only for the r_{12}^{-1} part of h, for which an analysis yields

$$<\vec{k}\vec{k}'|r_{12}^{-1}|\vec{k}'\vec{k}> = \frac{4\pi}{a^3} \sum_{\mu} \frac{1}{|\vec{q}_\mu+\vec{k}'-\vec{k}|^2} \Phi(\vec{q}_\mu)\Phi(-\vec{q}_\mu) \qquad (32)$$

Now, however the derivation does not call for the suppression of the term $\vec{q}_\mu=0$, and because $\vec{k}'-\vec{k}$ is generally non zero no pathological behavior results. In fact, the term with $\vec{q}_\mu=0$ corresponds to the exchange energy of a uniform electron gas, while the nonzero $\vec{q}_\mu$ terms describe effects due to nonuniformity.

Finally, we note from eq.(29) and the definition of $<\vec{k}|\vec{k}>$ that

$$<\vec{k}|\vec{k}> = N\Phi(0) \qquad (33)$$

independently of the value of $\vec{k}$, and that

$$<\vec{k}|-\frac{1}{2}\nabla^2|\vec{k}> = \frac{1}{2}k^2<\vec{k}|\vec{k}> + N\sum_{\mu}<\phi(\vec{r})|(-\frac{1}{2}\nabla^2|\phi(\vec{r}-\vec{R}_\mu)> \qquad (34)$$

Except for the last term of eq.(34), all the matrix elements reduce to expressions involving $\Phi(\vec{q})$, the quantity defined in eq.(29). This quantity, in term, may be reduced via eqs.(11) and (14) to

$$\Phi(\vec{q}) = \frac{1}{a^3}\sum_{\mu}\phi^T(\vec{q}-\vec{p}_\mu)\,\phi^T(\vec{p}_\mu) \qquad (35)$$

where the $\vec{p}_\mu$ are summed over the reciprocal lattice. The evaluation of $\Phi(\vec{q})$ is one of the lengthier parts of an actual numerical calculation and it may therefore be desirable to use methods to accelerate convergence of the sum in eq.(35). A procedure for so doing is currently in press (8).
More detail on these matrix element evaluations has been given elsewhere (2,7).

Three further observations regarding the formalism are relevant. First, we note that the calculations pose no severe difficulties for any atomic functions ϕ whose transforms are known. The acceptable ϕ therefore include STO's, in vivid contrast to the situation for polyatomic molecules, where STO integrals remain notoriously difficult and cumbersome to evaluate. Second, we see that the nonuniformity part of the exchange matrix element (the terms of eq.(32) with $\vec{q}_\mu \neq 0$) corresponds closely with the nonuniformity part of the Coulomb electron-electron matrix element (the summation of eq.(30)). Except for small $\vec{q}_\mu$ (i.e. $\vec{q}_\mu$ within one lattice point of $\vec{q}_\mu = 0$) the difference between $\vec{q}_\mu$ and $|\vec{q}_\mu + \vec{k}' - \vec{k}|$ cannot be large, and these nonuniformity contributions must be of similar significance. We conclude that nonuniformity is of comparable effect on the Coulomb and exchange energies.

Our third observation is that we have achieved a decomposition into uniform-limit and non-uniformity contributions to both the kinetic and potential energy. To the extent that a real system has a nearly uniform electron distribution this will also constitute a partitioning into larger and smaller contributions. The "large" contributions consist of the uniform-distribution kinetic energy (the free-electron result $\frac{1}{2} k^2$), the uniform-lattice coulomb energy ($-8.913633/\pi a$), and the free-electron exchange energy (obtainable from the $\vec{q}_\mu = 0$ term of eq.(32)). "Small" contributions are the remainder of the kinetic energy and the $\Phi(\pm\vec{q}_\mu)\Phi(0)$ terms of eq.(31). The remaining terms of eqs.(31) and (32) may be classified as "very small", as each contains two "small" factors. This organization of the various energy contributions may be used to help identify the ranges of validity of the uniform-distribution and more general one-electron models for extended systems. Only the "very small" terms act to break down the one-electron model.

Much of the foregoing analysis is restricted to systems with three-dimensional periodicity by the use of the three-dimensional lattice orthogonality theorem. However, in one-dimensionally periodic systems an analogous procedure can be applied for the periodic dimension, and in simple cases complete energy expressions can be evaluated. A more complete discussion will be reported elsewhere(9).

V. HARTREE-FOCK RESULTS

The above described methods have been applied to several systems possessing three-dimensional periodicity, and to one linear system (the infinite linear chain of equally-spaced hydrogen atoms). The results indicate in general the practicality of the methods, and also provide some insight as to the nature of the wavefunctions and the quality of the local exchange approximation.

We look first at the atomic hydrogen crystal, where results have been obtained in simple cubic, body-centered cubic (bcc) and face-centered cubic (fcc) geometries (10,11). Quite good results are already obtained from an LCAO or MPW basis built from a single

1s STO per atom, with the STO screening parameter optimized in the crystal calculation. Extension to included additional STO basis functions of A_{1g} symmetry (s,g,...) has little effect on the energy and at this level of flexibility the wavefunction and energy seem well converged. The equilibrium interatom spacings are qualitively in accord with estimates made long ago (12,13) and the fcc and bcc structures are calculated to be nearly identical in energy. The Fermi surfaces were found to be nearly spherical, confirming the view of this system as highly metallic.

To determine the effect of the choice of atomic basis orbitals on the results, calculations were also made for "cut-off STO's", in which the atomic function drops discontinuously to zero at the Wigner-Seitz cell boundaries. By taking several such functions the slope dicontinuities accompanying this basis can be minimized; it was found that three cut-off STO's per atom gave results of a quality comparable to those achievable with two conventional STO's. Calculations were also made with muffin-tin orbitals (STO's within a sphere, constant outside it). The muffin-tin energies differed from those of the more accurate calculations by several per cent.

The effect of the restriction to A_{1g} atomic orbitals was investigated by Oddershede _et al_ (14) and by Ramaker and Kumar (15). By using a plane wave expansion or MPW functions with $\vec{k}$ values outside the first Brillouin zone, these investigators were able to estimate an energetic contribution of about 0.020 hartree for other-symmetry orbitals. Major changes in the wavefunctions and Fermi surface were not observed, and the simpler calculations appear satisfactory for a qualitative discussion of the properties of the Hartree-Fock approximation for the atomic hydrogen crystal.

Examining first the wavefunction, we find a degree of inhomogeneity relatively similar to that of the free hydrogen atom. Figure 1 compares the free-atom and Hartree-Fock wavefunctions (the latter at $\vec{k}=0$). It is clear that a plane-wave approximation would be significantly in error; in fact, it would yield an error in the energy of about 15%.

The spatial inhomogeneity also raises questions as to the adequacy of local exchange approximation based on properties of the uniform electron gas. Such local exchange approximations have been advanced by Slater (16), Gaspar (17), and Kohn and Sham (18), and can be written in the common form

$$V_{ex}(\vec{r}) = -\frac{3\alpha}{2\pi}\left(3\pi^2\,\rho(\vec{r})\right)^{1/3} \qquad (36)$$

where $\alpha=1$ yields the original Slater result and $\alpha=2/3$ corresponds to the work of Kohn and Sham.

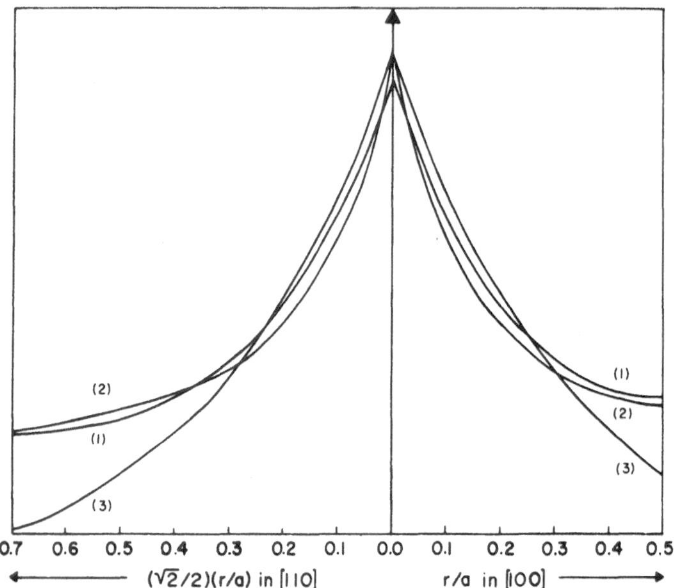

Figure 1.: Atomic-hydrogen crystal orbitals ($\vec{k}=0$) and free-atom wavefunction; crystal simple cubic with lattice spacing a=2.75 bohr.
(1) crystal orbital built from one 1s STO;
(2) crystal orbital built from three "cut-off" STO's;
(3) free-atom 1s orbital

Given the Hartree-Fock wavefunctions and energies, it is possible to find the local potential $V_{ex}(\vec{r},\vec{k})$ which (for each $\vec{k}$ value) can reproduce $|\vec{k}>$ and the corresponding energy. Local potentials determined in this way are compared with the Slater and Kohn-Sham potentials in Figure 2.

Figure 2 shows that the <u>shape</u> of the accurate local potential depends very weakly on k, but that its placement on the energy scale is quite sensitive. We may also note that the Kohn-Sham potential has rather nearly the correct shape over the large part of the unit cell far from the nucleus. We conclude that it should be possible to find local potentials which could yield relatively accurate wavefunctions, and that the Kohn-Sham potential is not too bad in this respect. However, as is well known, it is not appropriate to use a local-potential model as an accurate guide to orbital energies.

We look next at the energy calculated in the Hartree-Fock approximation.
The Hartree-Fock energies of the cubic hydrogen crystals were calculated in the two-STO basis at approximately -0.468 hartrees/atom.

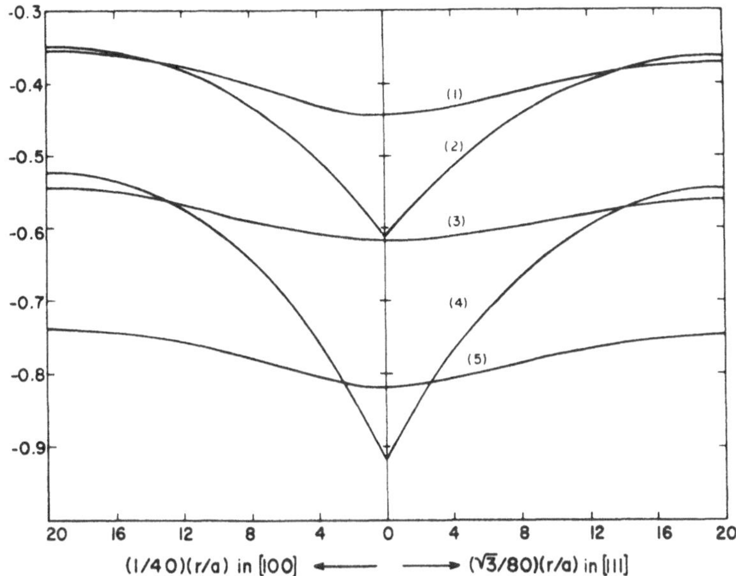

Figure 2.: Accurate and approximate local exchange potentials (in hartrees) for a bcc atomic hydrogen crystal of lattice spacing a = 3.3 bohr.
(1) accurate potential for $\vec{k}$ on Fermi surface;
(2) Kohn-Sham approximate potential;
(3) accurate potential averaged over occupied $\vec{k}$;
(4) Slater approximate potential;
(5) accurate potential for $\vec{k} = 0$.

After allowing for the effect of higher angular terms (-0.020 hartree) and correlation (estimated at -0.027 hartree[16]), the experimental energy may be predicted to lie no lower than -0.515 hartree/atom, only slightly below that of the free hydrogen atom (-0.500 hartree) and well above that of the hydrogen molecule (-0.587 hartree/atom). It may be concluded that, at least at zero pressure, the molecular solid will be more stable than the metallic crystal at low temperature.

Two additional sets of calculations contribute further information relative to the stability and structure of atomic hydrogen lattices. We discuss first our approximate Hartree-Fock treatment of the one-dimensional system consisting of an infinite line of equally spaced hydrogen atoms [9]. Finite cluster or crystal-orbital calculations relevant to the present discussion of this system also have been carried out by "Kislow" et al [19] (the name is actually "Liskow"), Kertesz, Koller and Azman [20], Berggren and Martino [21], Calais [22] and André [23].
These calculations are all consistent with a Hartree-Fock energy of about -0.527 hartree/atom, and an equilibrium internuclear

separation larger than that in the H_2 molecule . The calculations by Kertesz et al also indicate that the equally spaced chain will be unstable relative to formation of H_2 molecules at the same overall (linear) density.

The linear-chain energy is lower than that of the cubic hydrogen crystal, suggesting the possible stability of a filamentary structure for the three-dimensional atomic crystal. Such a structure had previously been proposed by Brovman et al [24]. Our calculations on space-filling arrays of filaments [25] suggest that such a structure would be more stable than fcc at atomic volumes greater than about 22.3 $bohr^3$.

The remaining relevant set of calculations examines the relative stability of atomic and molecular hydrogen crystals. Ramaker, Kumar and Harris[26] made approximate Hartree-Fock calculations for a simple cubic lattice containing two hydrogen atoms per unit cell, one at the cell origin and the other on the principal diagonal. If the second atom is at the center of this diagonal, the bcc atomic hydrogen lattice results; other positions of the second atom correspond to molecule formation. These calculations indicate the molecular phase to be more stable than either the cubic or filamentary structures down to a volume of about 17 $bohr^3$/atom. We conclude that there is no range in which the filamentary structure is the most stable. The relative energies of the cubic and molecular structures indicate a first-order transition at a pressure of about 2.5 Mbar.

Tuning briefly to the Li crystal, we have a greater opportunity to compare calculations [27,28] with experiment. As expected, the Hartree-Fock approximation yields bands which are too broad to correspond to soft X-ray emission spectra. In Li the bottom of the band is fairly well placed, but the top is considerably too high. Use of the Hartree-Fock wavefunctions to estimate intensities does not remove the long-standing discrepancies between theory and experiment. The other significant point of comparison with experiment is the momentum distribution of the valence electrons, as given by Compton X-ray scattering. The Hartree-Fock wavefunction provides a significant improvement over earlier calculations based on plane wave states, and defines the magnitude of the effect of electron correlation on the Compton profiles.

VI. ELECTRON CORRELATION

Extensive systems present special problems for the treatment of electron correlation. The standard configuration-interaction (CI) methods of atomic and molecular calculation cannot be effective because configurations involving macroscopic degrees of excitation are needed to obtain an extensive contribution to the correlation energy.

The most widely used approach to correlation, other than CI, has been many-body perturbation theory (MBPT) (29). MBPT has the advantage that the perturbation series consists only of terms which scale appropriately with system size, and in addition has the feature that many of the terms can be associated with easily visualized aspects of the interparticle interactions. However, as is well known, the orders of MBPT contributions do not necessarily correlate with their significance ordering, and in some problems (e.g. the uniform electron gas) many of the individual terms are infinite. While formal methods have developed for the removal and control of divergences, the overall situation remains far from unambiguous. There is still really no clearcut algorithm for obtaining the MBPT series in a significance ordering. Nevertheless, for want of a well developed alternative, MBPT remains a popular approach to the correlation problem.

In addition to the inherent difficulties of MBPT, there also remains the question of the choice of zero-order problem from which to apply a perturbation. Plane waves have been popular as zero-order states because they are easy to use and well defined. However, for systems containing real nuclei the more inhomogeneous Hartree-Fock states provide a starting point with a smaller perturbation, and should be expected to yield a more rapidly convergent expansion. The first demonstration that the use of Hartree-Fock states was practical was given by Monkhorst and Oddershede (30) who carried out MBPT in the random phase approximation for the atomic hydrogen crystal. They not only showed such calculations to be practical, but also found that the convergence was enough improved to make extremely advisable the use of inhomogeneous zero-order states.

In electronic systems a promising alternative to traditional MBPT is the coupled cluster expansion, introduced in nuclear physico by Coester and Kümmel (31) and first used in electronic structure theory by Cizek (32). This expansion is equivalent to a condensed form of MBPT, and leads to a natural hierarchy of approximations defined by the maximum number of particles whose unfactorizable interaction is to be considered. The coupled cluster method is discussed in detail elsewhere in this volume.

The first coupled-cluster calculation ever made in an extended system has just been reported. Freeman (33) has considered the uniform electron gas keeping clusters of size 2. He found it possible to include exchange contributions beyond the random phase approximation, and showed that it was possible to find stable iterative schemes for solving the nonlinear coupled cluster equations. The most exciting aspect of Freeman's calculation is that it demontrates the practicality of the method in the context of a non-trivial system. It is to be hoped that this very powerful method will see rapid exploitation in the years to come.

ACKNOWLEDGEMENTS

This work was supported in part by U.S. National Science Foundation Grant CHE-7501284. The author gratefully acknowledges his colleagues who have contributed much to his understanding and to the progress of the work, especially Drs. H.J. Monkhorst, L. Kumar, and J. Delhalle. Special thanks are due Melle P. Lonnoy for converting illegible scrawls into flawless typescript.

REFERENCES

1. R.N. Euwema and G.T. Surratt, J. Phys. Chem. Solids, 36, 67 (1975).
2. F.E. Harris, *Theoretical Chemistry, Advances and Perspectives* (Academic Press, New York, 1975), vol.1, pp. 147-218.
3. F.E. Harris, in *Electronic Structures of Polymers and Molecular Crystals*, J.M. André, J. Ladik and J. Delhalle, eds. (Plenum Publishing Co., New York, 1975), pp. 453-477.
4. R.A. Bonham, J.L. Peacher and H.L. Cox, J. Chem. Phys., 40, 3083 (1964).
5. See, for example, J. Callaway, *Quantum Theory of the Solid State* (Academic Press, New York, 1974), pp. 352ff.
6. F.E. Harris, J. Chem. Phys., 56, 4422 (1972).
7. See, for example, F.E. Harris, in *Computational Methods in Theory*, P.M. Marcus, J.F. Janak, and A.R. Williams, eds., (Plenum Publishing Corp., New York, 1971), pp. 517-541.
8. F.E Harris, J. Math. Phys. (to appear ca. December 1977).
9. J. Delhalle and F.E. Harris, to be published.
10. F.E. Harris and H.J. Monkhorst, Phys.Rev.Lett., 23, 1026 (1969).
11. F.E. Harris, L. Kumar and H.J. Monkhorst, Phys. Rev. B, 7, 2850 (1973).
12. E. Wigner and H.B. Huntirgton, J. Chem. Phys., 3, 764 (1935).
13. R. Kronig, J. De Boer and J. Korringa, Physica, 12, 245 (1946).
14. J. Oddershede, L. Kumar and H.J. Monkhorst Int. J. Quantum Chem., 85, 447 (1974).
15. D.E. Ramaker and L. Kumar, private communication (1974).
16. J.C. Slater, Phys. Rev., 81, 385 (1951).
17. R. Gaspar, Acta Phys., 3, 263 (1954).
18. W. Kohn and L.J. Sham, Phys. Rev., 140, A1133 (1965).
19. D.H. Kislow, J.M. McKelvey, C.F. Bender and H.F. Schaefer,

Phys. Rev. Lett., 32, 933 (1974).
20. M. Kertesz, J. Koller and A. Azman, Theoret. Chim. Acta, 41, 89 (1976).
21. K.F. Berggren and F. Martino, Phys. Rev., 184, 484 (1969).
22. J.L. Calais, Ark. Fysik, 29, 511 (1965).
23. J.M. André, J. Chem. Phys., 50, 1536 (1969).
24. E.G. Brovman, Yu.Kagan and A. Kholas, Zh. Eksp. Teor. Fiz., 61, 2429 (1971) (Soviet Physics-JETP 34, 1300 (1972)).
25. F.E. Harris and J. Delhalle, to be published.
26. D.E. Ramaker, L. Kumar and F.E Harris, Phys. Rev. Lett., 34, 812 (1975).
27. L. Kumar, H.J. Monkhorst and F.E. Harris, Phys. Rev., B9, 4084 (1974).
28. L. Kumar and H.J. Monkhorst, J. Phys. F. 4, 1135 (1974).
29. J. Goldstone, Proc. Roy. Soc., A 239, 267 (1957).
30. H.J. Monkhorst and J. Oddershede, Phys. Rev. Lett., 30, 797 (1973).
31. F. Coester and H. Kümmel, Nucl. Phys., 17, 477 (1960).
32. J. Cizek, J. Chem. Phys., 45, 4256 (1966).
33. D.L. Freeman, Phys. Rev., B 15, 5512 (1977).

LINKED-CLUSTER PERTURBATION THEORY FOR CLOSED AND OPEN-SHELL SYSTEMS: DERIVATION OF EFFECTIVE π-ELECTRON HAMILTONIANS*

B. H. Brandow

Theoretical Division, Los Alamos Scientific Laboratory
University of California, Los Alamos, New Mexico
USA

ABSTRACT. The Brueckner-Goldstone form of linked-cluster perturbation theory is derived, together with its open-shell analog, by an elementary time-independent approach. This serves to focus attention on the physical interpretation of the results. The open-shell expansion is used to provide a straightforward justification for the effective π-electron Hamiltonians of planar organic molecules.

1. INTRODUCTION

Linked-cluster perturbation theory has long been recognized as a powerful tool for the accurate calculation of electronic correlation energies in chemical systems, beginning with the pioneering work of Kelly [1] in 1963. The range of chemical applications of this formalism has since been gradually expanding, from light closed-shell atoms (e.g., beryllium and neon) to heavier atoms and to small molecules such as water and methane, nearly always with the restriction to those states which can, in first approximation, be represented by a closed-shell or single-determinant wavefunction. Quite recently, however, there have been some chemical applications [2] of a degenerate or open-shell version of this formalism [3-5]. This generalization allows one to deal with systems whose ground states require more than one determinant in order to obtain the proper quantum numbers, for the total spin etc. This also allows excited states to be treated on the same footing as ground states, so that the correlation energies of excited states can also be calculated accurately [2].

*This work was carried out under the auspices of the U.S.E.R.D.A.

J.-M. André et al. (eds.), Quantum Theory of Polymers, 137-167.

The full power and generality of this open-shell formalism has, however, not yet been generally recognized by the community of theoretical chemists. On the one hand, this is now formally applicable to *any* electronic system, even those (such as bulk samples of magnetic insulator materials [5,6], as well as superconductors) which must be considered to have a macroscopically large number of valence or open-shell particles. On the other hand, the energy results are expressed in the form of an effective Hamiltonian for the valence particles. There are, of course, many examples throughout physics and chemistry of the use of such effective Hamiltonians, in which one generally employs some semi-empirical parameters. The open-shell perturbation formalism provides a very general means for deriving or formally justifying the use of such effective Hamiltonians, and it also generates formal recipes for the *ab initio* calculation of their parameters. There is also a corresponding linked-cluster perturbation expansion for the expectation values of any operator (for the ground and low excited states), as well as for the associated transition matrix elements. This result can be used, for example, to calculate electromagnetic transition elements.

The purpose of these lectures is to provide a relatively easy introduction to the linked-cluster perturbation formalism, emphasizing its basic mathematical structure and physical interpretation, for both the original closed-shell form and the more recent open-shell generalizations. Only time-independent and temperature-independent results will be discussed, and, in contrast to much of the previous literature, the derivations will not invoke time- or temperature-dependence at any stage. Since a full account is quite impossible in the presently allotted space and time, we recommend that anyone seriously interested in this subject should also see a recent and more expanded version of the present material [5], where many further references are given. (Connections with the time-dependent derivation of Goldstone are also discussed there.) We should also mention that the degenerate perturbation formalism is by no means unique, for either many-body or "ordinary" (non-many-body) quantum systems. The many alternative formulations and derivations now available have been discussed in two review articles [7,8], which also provide historical background and some critical comparisons between the various methods.

We turn, now, to the relevance of this formalism for polymer systems. Although *ab initio* (LCAO-SCF) calculations have been carried out for some simple polymers, this procedure is necessarily of limited use because (1) correlation corrections are known to be quite significant, and (2) accurate SCF calculations are often prohibitively difficult, especially for the complex polymers of biological interest. On the other hand, the use of effective π-electron Hamiltonians has been very successful for many organic molecules, both for excited-state [9] and ground-state [10] calculations. This approach has been extended to hetero-atom monomers of biological interest [11], and also to the corresponding complex polymers [12]. Since such effective Hamiltonians ("all valence electron" as well as π-electron Hamiltonians) will undoubtedly play a major role in future developments, we shall discuss the use of the present open-shell formalism to derive effective π-electron Hamiltonians.

2. DEGENERATE BRILLOUIN-WIGNER PERTURBATION THEORY

There are basically just two types of perturbation theory, namely, the Brillouin-Wigner (BW) and the Rayleigh-Schroedinger (RS) forms. We shall begin with the BW form because this is algebraically simplest, and also because this provides a convenient point of reference for all of the later developments.

Consider a very general quantum system, not necessarily a "many body" system. Starting with the usual formulas,

$$H = H_o + V, \tag{2.1}$$

$$\Psi = \sum_i a_i \Phi_i, \tag{2.2}$$

$$H\Psi = E\Psi, \tag{2.3}$$

$$H_o \Phi_i = E_i \Phi_i, \tag{2.4}$$

we immediately obtain

$$(E - E_i)a_i = \langle \Phi_i | V | \Psi \rangle. \tag{2.5}$$

We then select a certain number d of the Φ_i's to span a degenerate subspace or "model subspace" D. The corresponding E_i's may be only quasi-degenerate; exact degeneracy is not required. With this choice of D we associate a degenerate projection operator P, a degenerate or "model" wavefunction

$$\Psi_D = P\Psi = \sum_{i\varepsilon D} a_i \Phi_i, \tag{2.6}$$

and a resolvent or Green's function

$$G = \sum_{i \notin D} \frac{|\Phi_i\rangle\langle\Phi_i|}{E - E_i} = \frac{Q}{E - H_o}, \tag{2.7}$$

where P + Q = I. Then (2.2) and (2.5) can be combined in the form

$$\Psi = \Psi_D + \mathcal{G}V\Psi$$

$$= \sum_{n=0}^{\infty} (\mathcal{G}V)^n \Psi_D . \tag{2.8}$$

It is convenient to define a wave operator or "model operator" Ω such that

$$\Psi = \Omega\Psi_D, \tag{2.9}$$

together with the condition $\Omega P = \Omega$, and also to define a reaction matrix, effective interaction, or "model interaction"

$$\mathcal{V} = V\Omega . \tag{2.10}$$

Substitution in (2.8) yields

$$\Omega = P + \mathcal{G}V\Omega$$

$$= \sum_{n=0}^{\infty} (\mathcal{G}V)^n P, \tag{2.11}$$

$$\mathcal{V} = VP + V\mathcal{G}\mathcal{V}$$

$$= V \sum_{n=0}^{\infty} (\mathcal{G}V)^n P. \tag{2.12}$$

Now (2.5) can be rewritten as

$$(E - E_i)a_i = \langle \Phi_i | V\Omega | \Psi_D \rangle$$

$$= \sum_{j \varepsilon D} \langle \Phi_i | \mathcal{V} | \Phi_j \rangle a_j . \tag{2.13}$$

This is valid for <u>all</u> amplitudes a_i, whether in D or not. In particular this applies to the a_i's in Ψ_D, and these d equations can be expressed in the form of a d-dimensional secular equation

$$[H_o + P\mathcal{V}(E) - EI]A = 0, \tag{2.14}$$

where A is a column vector composed of the amplitudes a_i in (2.6). We have now obtained a formally exact version of degenerate perturbation theory [13,14]. This is expressed in the convenient form of a "model Hamiltonian", about which we shall say much more in Section 5. It is sufficient for now to recognize that all of the "nondegenerate" or "virtual" basis states Φ_i are concealed within $\mathcal{V}$.

Iteration of the first lines of Eqs. (2.8), (2.11), and (2.12) leads to perturbation expansions of the Brillouin-Wigner form, as displayed above. The two identifying characteristics of this BW form should be noted: (1) This has the very simple formal structure of a geometric series in the operator $\mathcal{G}V$. (2) The desired (and presumably unknown) energy E appears in all of the energy denominators.

3. LINKED-CLUSTER EXPANSION FOR NONDEGENERATE SYSTEMS

We first consider systems whose unperturbed ground state is nondegenerate. In this case $d = 1$, $P = |\Phi_o\rangle\langle\Phi_o|$, and one immediately obtains

$$\begin{aligned} \Delta E = E - E_o &= \langle\Phi_o|V|\Psi\rangle = \langle\Phi_o|\mathcal{V}|\Phi_o\rangle \\ &= \langle\Phi_o|V + V\mathcal{G}V + V\mathcal{G}V\mathcal{G}V + \ldots|\Phi_o\rangle. \end{aligned} \tag{3.1}$$

This is the standard BW expansion for the interaction energy. It should be noted that the intermediate normalization convention

$$\langle\Phi_o|\Psi\rangle = \langle\Phi_o|\Omega|\Phi_o\rangle = \langle\Phi_o|\Phi_o\rangle = 1 \tag{3.2}$$

is implicit in this result. This also follows from (2.8).

3.1. Diagrammatic Notation

In applying the expansion (3.1) to a many-body system, one quickly finds that the higher-order terms involve a bewildering variety of different types of intermediate-configuration sequences (sequences of Φ_i's), and that it is necessary to keep track of certain of their distinguishing features. To make any sense of this complex situation, it is quite essential to have some sort of efficient shorthand notation. This problem has been neatly solved by the diagrams introduced by Goldstone [15], some low-order examples of which are shown in Fig. 1. Here diagrams (a) and (c) represent the "direct" first-order and second-order terms from (3.1), whereas (b) and (d) show the corresponding exchange terms. The combination (a) + (b) represents the potential energy of the Hartree-Fock approximation. The connection with the various terms $\langle V\rangle$, $\langle V\mathcal{G}V\rangle$, etc. in (3.1) follows from the rule that the successive horizontal levels, proceeding from bottom to top, correspond to the successive V's and $\mathcal{G}$'s in the individual terms

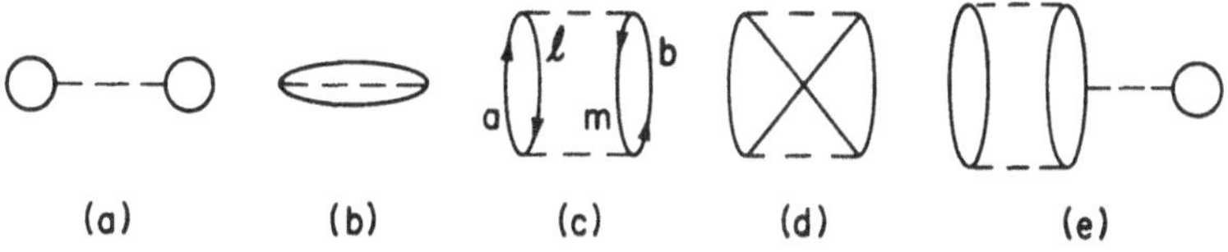

Fig. 1. Some low-order Goldstone diagrams.

of (3.1), proceeding from right to left. Thus in diagrams (c) and (d), the bottom (top) dashed lines represent matrix elements of the two-body parts of the right-hand (left-hand) interactions V in $\langle\Phi_o|VGV|\Phi_o\rangle$. This correspondence is usually described by saying that "time runs upwards" in the Goldstone diagrams.

The two downgoing lines in diagram (c) represent "hole" orbitals, namely, orbitals that were occupied in Φ_o but are vacant in the particular intermediate Φ_i represented here. The two upgoing lines represent "particle" orbitals, that is, orbitals occupied in this Φ_i but not in Φ_o. Line segments that begin and end at the same horizontal level, as seen in diagrams (a), (b), and (e), represent orbitals in Φ_o ("passive hole orbitals") that remain occupied before and after the associated interaction V. The solid lines always form closed loops, in which the arrow directions proceed in a continuous manner, which simply means that the total number of particles is conserved at every V interaction. It is actually a great convenience, especially for large systems, that the hole orbitals which do not interact are not shown at all. In view of this use of Φ_o as a reference state, Φ_o is often called the "vacuum state" in analogy to the diagram convention of quantum electrodynamics.

Each solid line segment in Fig. 1 should be thought of as carrying the label of an orbital from the basis generated by the one-body operator H_o. These labels are eventually summed over, the hole (particle) labels running over all orbitals within (outside of) Φ_o. These orbital summations (together with a sum over topologically distinct diagrams) are equivalent to the Φ_i summations from (2.7).

One of the nice features of these diagrams is that the appropriate energy denominators [from (2.7)] can be determined immediately by inspection. The general rule for any intermediate-state level is

$$(E_o + \Delta E - H_o) = \Delta E + \Sigma(\text{all downgoing line energies}) - \Sigma(\text{all upgoing line energies}). \tag{3.3}$$

Thus in diagram (c), for example,

$$e(\text{diagram c}) = \Delta E + (\varepsilon_\ell + \varepsilon_m) - (\varepsilon_a + \varepsilon_b), \tag{3.4}$$

where the ε_i's are the orbital eigenvalues obtained from H_o. There are correspondingly simple and completely mechanical rules, based on topological features of the diagrams, for keeping track of the various factors of ½, and similarly for the minus signs associated with the exchange matrix elements. By means of these rules the algebraic counterpart of any diagram is easily determined, thus one ordinarily proceeds by first drawing all of the topologically distinct diagrams (for a given order), and then translating these into formulas. For a particularly convenient form of these rules, see Appendix B of [4]. Simple derivations for some of these rules may be found in [16]. In the higher orders, however, it may become difficult to correctly identify all of the topologically distinct diagrams; any confusion at this point can now be eliminated by means of a computer program [17].

3.2. The Unlinked-Cluster Problem

Let us now consider a macroscopic sample of some quantum fluid (liquid or gas) such as the electron gas or nuclear matter, where the basis orbitals ϕ_ℓ, Φ_a, etc. are just plane waves. The "direct" diagrams from the first three orders in (3.1) are shown in Fig. 2, together with a few of the many fourth-order diagrams. For a fixed bulk density $\rho = N/\text{volume}$, it is physically obvious that ΔE must be proportional to the total number of particles N. Unfortunately, however, this feature is not at all obvious from the individual terms of the BW expansion.

From elementary counting arguments (see Appendix C of [4]), one finds that each linked (topologically connected) piece of a diagram contributes an overall factor of N, arising from the various summations and orbital normalization factors within the matrix elements. On the other hand, each of the ΔE terms within the energy denominators (3.3) contributes a factor of N^{-1}. Thus diagrams (a) and (b) of Fig. 1 are $\mathcal{O}(N)$, (c) and (d) are $\mathcal{O}(1)$, (e) is $\mathcal{O}(N^{-1})$, whereas the unlinked third-order diagram in Fig. 2 is $\mathcal{O}(1)$. It turns out that the first-order diagrams (a) and (b), corresponding to $\langle\Phi_o|V|\Phi_o\rangle$, are the <u>only</u> BW terms whose contribution is $\mathcal{O}(N)$; all other BW diagrams are $\mathcal{O}(1)$ or smaller [$\mathcal{O}(N^{-1})$, $\mathcal{O}(N^{-2})$, etc.]. The smallness of these individual contributions is compensated, of

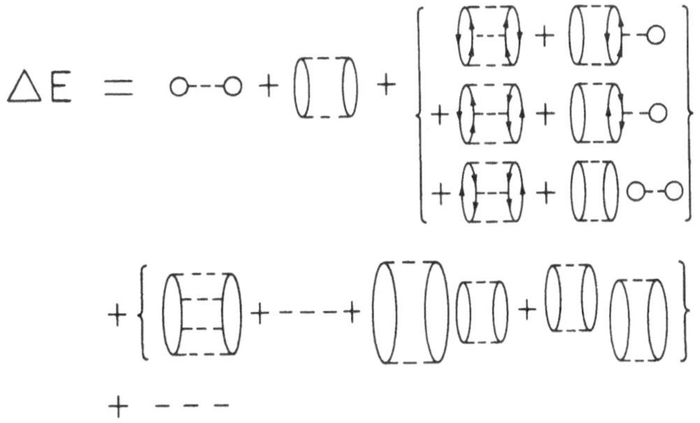

Fig. 2. The first few orders of "direct" diagrams from the BW energy expansion (3.1).

course, by the huge number of unlinked diagrams that are generated by the higher-order terms in (3.1). The complete BW expansion is thus still formally correct, although it is utterly impractical for large N. On the other hand, if one were to go over to the Rayleigh-Schroedinger (RS) form of perturbation theory in a naive manner, by simply ignoring the ΔE's in all of the energy denominators, one would find terms proportional to arbitrarily high positive powers of N (N^2, N^3, etc.), due to the unlinked diagrams. This is equally unsatisfactory.

It was Brueckner [18,19] who first drew attention to this problem. He proceeded to show, however, that when the RS expansion is worked out with proper care, all of the unlinked-diagram terms mutually cancel and one is simply left with linked (fully-connected) diagram terms, all of which scale properly with N. This is the famous linked-cluster result. Brueckner demonstrated this cancellation explicitly in low orders (eventually up to sixth order in V[19]), but he did not have a general proof covering all of the higher-order terms. The first general proof was presented by Goldstone [15], and many other derivations have appeared since (see [4], Introduction, for references). It is only fair, however, to add that Hugenholtz [20] obtained this result independently, at almost the same time as Goldstone. We shall now outline what we consider to be the simplest derivation of all [3,4].

3.3. Cancellation of Unlinked Diagrams

One way the BW expansion (3.1) can be converted into the RS expansion is by the following procedure. The first step is to formally expand ΔE out of all the energy denominators of (3.1), each BW denominator being replaced by a geometric series involving RS-type denominators:

$$\frac{1}{E_o + \Delta E - H_o} = \frac{1}{E_o - H_o} + \frac{1}{E_o - H_o}(-\Delta E)\frac{1}{E_o - H_o} + \frac{1}{E_o - H_o}(-\Delta E)\frac{1}{E_o - H_o}(-\Delta E)\frac{1}{E_o - H_o} + \cdots \quad (3.5)$$

It follows from (3.3) that these RS denominators are determined entirely by the sets of upgoing and downgoing orbital eigenvalues at the corresponding intermediate-state levels. This expansion (3.5) can be represented diagrammatically as in Fig. 3, where each solid horizontal bar represents an "insertion" (a fictitious diagonal interaction) of numerical value ($-\Delta E$). We now have an expansion of ΔE in terms of diagrams, most but not all of which themselves contain ΔE. This expansion may thus be substituted back into itself to eliminate the ΔE's from the right-hand side of (3.5) and Fig. 3, as illustrated in Fig. 4. At this point we observe that the first term on the right-hand side of Fig. 4 is

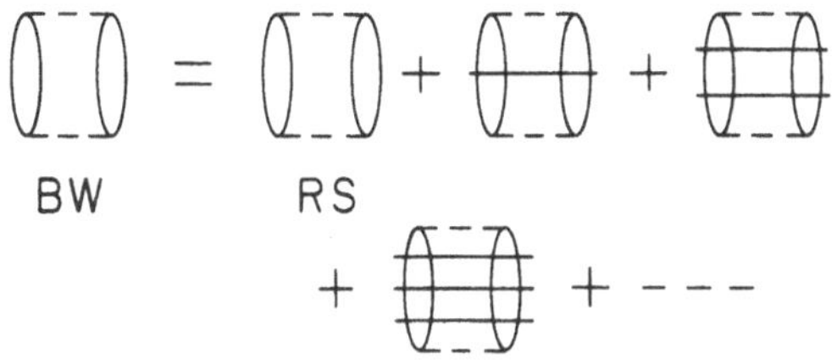

Fig. 3. Diagrammatic representation of the denominator expansion (3.5).

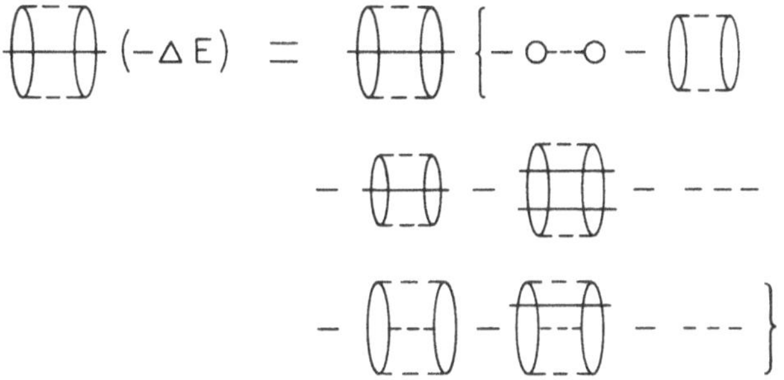

Fig. 4. Insertion of the double expansion for ΔE, due to (3.1) and (3.5), into one of the diagrams of Fig. 3.

just equal and opposite to the leading (third-order) unlinked term seen in Fig. 2, so these terms cancel identically. It can easily be shown that this type of cancellation is completely general -- all of the unlinked terms from the original BW series are cancelled by terms arising from the $(-\Delta E)$'s generated by (3.5), and vice versa. The only surviving terms are the fully linked diagrams with no $(-\Delta E)$ insertions, such as the first term to the right in Fig. 3. This is just the Brueckner-Goldstone result. Before the general proof can be completed, however, one must apply two other important concepts which we shall now discuss.

3.4. The Factorization Theorem

To go beyond the simple example of cancellation just described, one must employ the factorization theorem. Consider the RS-type denominators [from the first term of the series (3.5)] for the two different unlinked fourth-order diagrams shown in Fig. 2. Let us assume the same set of orbital labels (unsummed) for both diagrams, so that these diagrams differ only in the relative "time" orders of their lower interactions. Let e_L and e_R be the RS denominators appropriate for the left-hand and right-hand linked parts, each part considered separately. Then the complete energy denominator products for these two diagrams are

$$\frac{1}{e_L}\left(\frac{1}{e_L + e_R}\right)\frac{1}{e_L} , \quad \frac{1}{e_L}\left(\frac{1}{e_L + e_R}\right)\frac{1}{e_R} ,$$

respectively. We now observe that the sum of these two products is simply $(e_L e_R e_L)^{-1}$, which corresponds to the negative of the second diagram on the right-hand side of Fig. 4. This is summarized in Fig. 5.

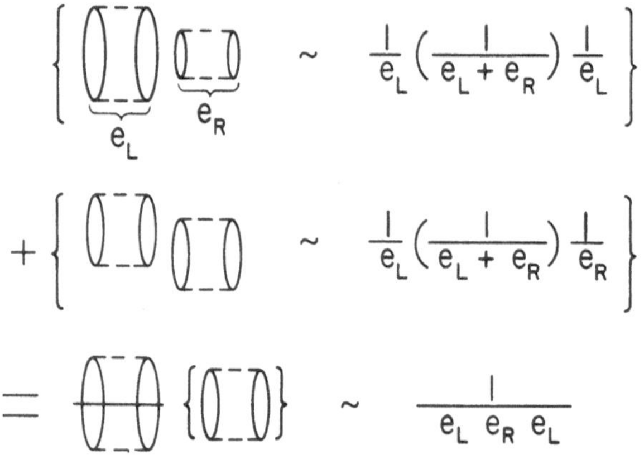

Fig. 5. Illustration of the factorization theorem. The first two diagrams contain "off shell" insertions, while the last has an "on shell" insertion.

Let us now rephrase what has just been done. The denominators of the last-mentioned diagram are all of the most elementary RS form, since the second-order "insertion" is evaluated "on the energy shell". In other words, the denominator (e_R) of the insertion part is not influenced by the local excitation energy (e_L) of the rest of the diagram (the "skeleton") to which it belongs. We have just seen that this simplified RS diagram is equivalent to a sum of diagrams in which the insertion part is "off the energy shell", meaning that at least one of its denominators now contains an excitation energy (e_L) from the rest of the diagram, in addition to its own excitation energy e_R. In the general case, the corresponding sum of diagrams is characterized as follows. The top of the inserted part is placed at the same intermediate-state level as the ($-\Delta E$) insertion (horizontal bar in Fig. 3) which it represents. The remainder of the insertion is then allowed to assume all possible relative "time" orderings with respect to the lower part of the original (skeleton) diagram, such that the original time ordering within the inserted part is preserved.

The factorization theorem is a purely algebraic identity which shows that the form of the prescription just given is valid for diagrams of arbitrary complexity. That is, each of the summations just described, of diagrams with off-energy-shell denominators, leads to the corresponding on-shell diagram which is needed to complete the cancellation. This theorem can be proved by induction [20, §7; 21], building upon the example just given.

3.5. Exclusion-Violating Diagrams

The other important ingredient of the general proof is the matter of "ignoring the exclusion principle in intermediate states", an idea dating back to the first paper on the use of diagrams in quantum electrodynamics [22]. The orbital summations for the various "insertions" in Fig. 4 must obviously be independent of the orbitals in the "skeleton" diagram on the left-hand side of this figure, since these insertions arise from the original expansion (3.1). This contrasts with the corresponding diagrams in Fig. 2, where, according to the exclusion principle for Slater determinants, the orbital labels at each horizontal level must all be distinct. It would thus appear that the cancellation of unlinked diagrams cannot be complete. Goldstone showed how to resolve this problem by a judicious addition and subtraction of "exclusion-violating" diagrams. An example is shown in Fig. 6, where the normally occupied orbital m has been emptied twice. Diagram 6(a) is a physically unallowed contribution to one of the fourth-order diagrams of Fig. 2, whereas diagram 6(b) has, according to the standard diagrammatic rules, a precisely equal and opposite numerical value. Thus no harm is done by adding both of these diagrams to the terms originally in Fig. 2. Thanks to diagram 6(a) and its generalizations, we now obtain a precise cancellation of all the unlinked diagrams, but only at the cost of including the extra diagrams 6(b), etc., among the set of surviving fully linked diagrams. These "extra" linked diagrams are all $\mathcal{O}(N)$ for macroscopic systems, as desired.

One of the marvels of the Goldstone formalism is that the required compensation diagrams, Fig. 6(b) etc., can all be

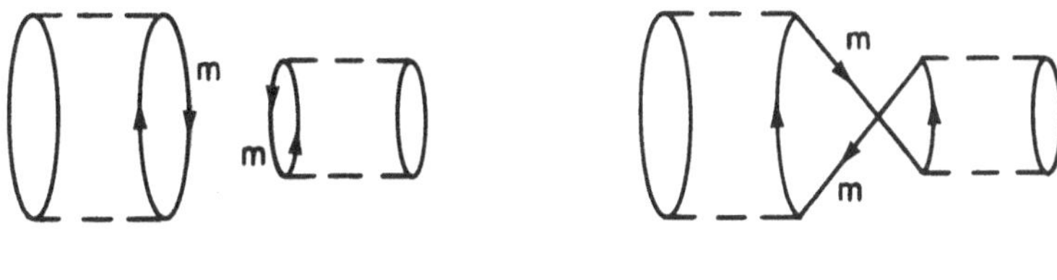

Fig. 6. Effect of ignoring the exclusion principle in intermediate states: (a) a nonphysical unlinked diagram; (b) an "exclusion violating" linked diagram which exactly compensates for (a).

generated quite automatically, by the simple recipe that (1) linked diagrams of all possible distinct topologies must be included, and (2) all of the orbital-label summations must be carried out independently. This is what is meant by "ignoring the exclusion principle in intermediate states". The formal meaning of this addition and subtraction is that one has abandoned the original Slater-determinant description of the intermediate configurations Φ_i, replacing this by a "second-quantized" description. The consistent use of fermion creation and annihilation operators automatically generates all of the necessary linked exclusion-violating diagrams, and also accounts for the remarkable simplicity of the diagram rules. A nice introduction to the formalism of second-quantization (the occupation-number representation) has recently been presented by Manne [16]; see also Ambegaokar [23], or Brown [24].

3.6. Concluding Remarks

We are now ready for the general proof of cancellation. This is really quite simple (see [3] or [4]), but since it adds no further insight, this will not be discussed here. The important concepts have all been explained, namely, (1) the use of diagrams, (2) the RS form of perturbation theory, (3) the factorization theorem, and (4) ignoring the exclusion principle in intermediate states. It should be mentioned that there are also corresponding linked-cluster results for the complete wavefunction Ψ, for its norm $\langle\Psi|\Psi\rangle$ [which is > 1, in view of (3.2)], and for the expectation value $\langle \mathcal{O} \rangle$ of a general operator $\mathcal{O}$. These expansions are all discussed in [4] and [5].

In the specific diagrammatic examples given above, we have ignored the possible presence of an auxiliary one-body potential (such as the Hartree-Fock potential) which will, in most applications, have been included within H_0, and must then also appear (accompanied by a minus sign) in the perturbation term V. The presence of such a one-body interaction within V is very straightforward to handle formally, and does not alter the preceding developments in any significant way.

4. PHYSICAL INTERPRETATION AND GENERAL PHILOSOPHY OF APPLICATIONS

Since the original motivation for this formalism was to deal with systems having a macroscopic number of particles, a quantum chemist will naturally ask what advantages, if any, this might offer for relatively few-electron systems such as atoms and small molecules. The answer involves the physical interpretation of this formalism [4], which we now discuss.

For a macroscopic system one believes intuitively that particles should not interact to any significant extent when they are farther apart than some appropriate correlation length. This notion leads to the argument, familiar in statistical mechanics, that a macroscopic system can be divided into many comparatively small subsystems, each of which is still large enough to be considered approximately independent. That is, the "boundary" contributions are still small compared to the bulk contributions. Let us pursue this idea by considering a fictitious "cellular" system, where the various subsystems are rigorously isolated from each other by means of physical barriers. This guarantees complete independence for the various subsystems s, whereby the energy is now rigorously just the sum of the subsystem energies,

$$E_o = \sum_s E_{os} \sim N, \ \Delta E = \sum_s \Delta E_s \sim N. \tag{4.1}$$

The wavefunction has a product form,

$$\Psi = \prod_s \Psi_s , \tag{4.2}$$

and thus the (normalized) overlap of Ψ onto Φ_o is exponentially small,

$$\langle\Phi_o|\Psi\rangle = \prod_s \langle\Phi_{os}|\Psi_s\rangle \sim e^{-\alpha N} . \tag{4.3}$$

[In terms of the intermediate normalization convention (3.2), however, one finds that $\langle\Psi|\Psi\rangle \sim e^{+2\alpha N}$.] But this near-vanishing overlap is nothing to be concerned about, since the individual overlaps $\langle\Phi_{os}|\Psi_s\rangle$ need not be small, and all expectation values are simply additive,

$$\frac{\langle\Psi|\mathcal{O}|\Psi\rangle}{\langle\Psi|\Psi\rangle} = \sum_s \frac{\langle\Psi_s|\mathcal{O}_s|\Psi_s\rangle}{\langle\Psi_s|\Psi_s\rangle} \sim N. \tag{4.4}$$

When one considers the BW expansion for this cellular system, one sees immediately that it is physically inappropriate to have all of the "other" $\Delta E_{s'}$'s (for $s'\neq s$) occurring within the energy denominators of subsystem s. Going over to the RS form of perturbation theory is clearly a sensible step here, since this removes <u>all</u> of the E_s's from all of the denominators. Moreover, the factorization theorem also follows very simply from the notion of physically independent subsystems, as Hugenholtz [20, Appendix] has pointed out. It is therefore quite reasonable to expect the RS perturbation theory to give a simple additive result for the energy, with no cross terms involving more than one subsystem s.

The present linked-cluster derivation is obviously very closely related to this elementary picture of physically independent subsystems. The comparison shows that each linked-cluster diagram (even before the orbital summation) is behaving formally as if it represents an independent subsystem. Upon reflection, however, this result may appear to be too strong, since the wavefunctions corresponding to the various cluster terms are actually spatially interpenetrating. (In realistic applications, the "hole" or normally-occupied orbitals usually extend throughout the entire volume of the system.) The clusters must therefore interact, both dynamically (through V) and statistically (via the exclusion principle). This is indeed true. But the Brueckner-Goldstone formalism has the nice feature that these expected "cluster-cluster" interactions are always represented by higher-order linked clusters, whereby the "original" clusters may legitimately be thought of as acting quite independently. Diagram 6(b), for example, can be regarded as a case of two simple clusters interacting statistically.

The practical value of this formalism lies in the fact that one can usually find a judicious way of grouping the various cluster terms together (i.e., of partially summing the perturbation series), which corresponds physically to decomposing the many-body system into an appropriate set of subsystems. By this we mean that most of the correlation energy is accounted for within these clusters, the omitted (higher-order) clusters having (hopefully!) only a small effect. Since the computational effort generally increases very rapidly with the size of the system (or subsystem), such a "cluster decomposition" can be of great value even for small systems such as the lighter atoms. For larger systems this type of approach is quite indispensable. The optimal procedure will naturally depend on the physical characteristics of the system. In our opinion, the most significant feature of the present approach, as compared to rival many-body formalisms, is the very great flexibility which it allows in tailoring the approximation scheme to the characteristics of the system at hand. It seems to us that this formalism is the most "open and unbiased" about the many possibilities available -- the possibilities for ordering and partially summing the infinite series. Examples of cluster schemes suitable for various systems are discussed in many places, but we shall not pursue this matter here. We do wish to mention, however, that the nonperturbative coupled-cluster formalism of Coester, Kümmel, Cížek, and Paldus appears to be quite well suited to the particular needs of electronic systems, as demonstrated in [25]. That formalism is actually quite closely related to the present one [6].

5. OPEN-SHELL SYSTEMS -- A GENERAL FORMALISM FOR EFFECTIVE HAMILTONIANS AND EFFECTIVE OPERATORS

We shall now discuss the extension of the linked-cluster perturbation formalism to open-shell systems. Besides offering a practical method for dealing with such systems, this provides a very important bonus -- a general formalism for deriving effective Hamiltonians and other effective operators. These are operators that, when suitably used within a "model subspace" (degenerate or quasi-degenerate), will correctly reproduce the energies and other matrix elements of some subset of the exact eigenstates.

The simplest way to introduce the concept of an effective or model Hamiltonian is by the method of Löwdin [14,26]. Here one partitions the Hilbert space into two disjoint subspaces, P and Q, such that the Schroedinger equation becomes a 2 × 2 block matrix equation,

$$\begin{pmatrix} H_{PP} & H_{PQ} \\ H_{QP} & H_{QQ} \end{pmatrix} \begin{pmatrix} \Psi_P \\ \Psi_Q \end{pmatrix} = E \begin{pmatrix} \Psi_P \\ \Psi_Q \end{pmatrix} . \qquad (5.1)$$

This can be regarded formally as two linear equations in two unknowns. The variable Ψ_Q can then be easily eliminated to produce the "projected" Schroedinger equation

$$[H_{PP} + H_{PQ}(E - H_{QQ})^{-1} H_{QP}] \Psi_P = E \Psi_P . \qquad (5.2)$$

This procedure, generally known as "partitioning of the Hamiltonian", is obviously independent of perturbation theory. In practice, however, the concept of a model Hamiltonian is usually applied to a many-body system, and thus one needs a perturbative approach in order to cast the results into a suitable linked-cluster form. Our starting point will therefore be the degenerate BW perturbation result (2.14). [The latter can, of course, be readily derived from (5.2).] One might suppose that other many-body techniques, such as Green's functions or the Coester-Kümmel-Čižek method, could also be used for this purpose. But there are some formal difficulties peculiar to degenerate systems. To our knowledge, these obstacles have been fully surmounted only within the perturbative context. The historical background and some critical comparisons with the various alternative formulations and derivations of degenerate perturbation theory are reviewed elsewhere [7,8].

5.1. Diagrammatic Conventions

As in the closed-shell case, the first step towards a useful linked-cluster result is the introduction of suitable diagrammatic conventions. The present model subspace D[see (2.6)] is defined in an obvious manner: the "model" Φ_i's are all those (N + n)-body Slater determinants in which a given set of <u>core orbital</u> shells are fully occupied by N <u>core particles</u>, while the remaining n valence particles are all distributed among a given set of valence orbitals. (Of course there are corresponding results for systems with valence holes.) For the moment we shall assume exact degeneracy, as if all of the valence orbitals belong to a single subshell, but this restriction will be removed later on.

The goal is to calculate the set of all matrix elements

$$V_{ij} = \langle \Phi_i | V | \Phi_j \rangle \tag{5.3}$$

for i, j within the model subspace. To specify one of the model Φ_i's, it is clearly sufficient to know which n of the valence orbitals are occupied, since, by definition, the N core orbitals are always occupied in these model states. This suggests that the various perturbation terms in (5.3) should be represented by means of n "incoming" valence lines at the bottom, to specify Φ_j, and n "outgoing" valence lines at the top, for Φ_i. The N core particles will be represented by the same "vacuum convention" used previously, that is, we shall only indicate the deviations from their initial closed-shell configuration.

With these conventions we now apply the BW expansion (2.12) to some particular matrix element V_{ij}. The resulting diagrams may be classified into three general categories, examples of which are shown in Fig. 7 for the case n = 3. A valence diagram is shown in (a), in which all of the interactions are connected (directly or indirectly) to one or more of the external valence lines. A core diagram is shown in (b), where all of the valence lines are completely passive. Diagrams (c) and (d) are "mixed" terms containing both core and valence interaction processes. Note that at each intermediate level at least one of the particles

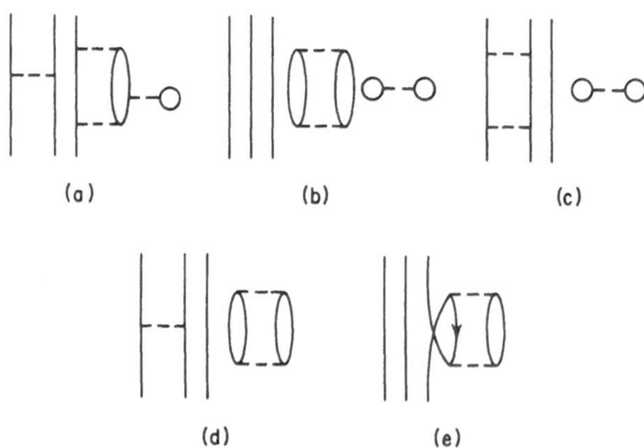

Fig. 7. Diagrams from the BW expansion (2.12) for V_{ij}: (a) and (e) are valence diagrams, (b) is a core diagram, while (c) and (d) are of mixed character.

must be excited to a higher orbital than in the initial configuration Φ_j, because none of the allowed intermediate states Φ_ℓ [from (2.7)] are degenerate with the model states.

We continue to ignore exclusion in intermediate states, in order to make the set of core diagrams precisely the same as if the valence particles did not exist. This necessitates the introduction of some core-valence interference terms, analogous to Fig. 6(b), one of which is shown in Fig. 7(e). This represents a "blocking" of one of the assumed core-correlation processes, due to the presence of a valence particle in one of the orbitals that was presumed (for the purpose of defining these core processes) to be vacant. All such core-valence interference terms are classified as valence diagrams, similar to example (a).

5.2. Separation of Core and Valence Contributions

We decompose the total energy eigenvalue into core and valence terms, and then further separate each of these into unperturbed and interaction contributions,

$$E = E_c + E_v$$

$$= E_{oc} + \Delta E_c + E_{ov} + \Delta E_v \ . \tag{5.4}$$

Since in (2.14) the H_o acts only within the model subspace, one finds

$$H_o \, A = (E_{oc} + H_{ov}) \, A = (E_{oc} + E_{ov}) \, A \ . \tag{5.5}$$

We now define ΔE_c to be just the interaction energy for the previous non-degenerate case, in which all of the valence particles are physically absent. This ΔE_c is clearly given by the sum of all core diagrams, Fig. 7(b) etc., from which it follows that ΔE_v is defined by all of the remaining diagrams.

The next step is a ΔE expansion and diagram cancellation analysis similar to that outlined in §3.3. Although ΔE_c is now expanded from all of the present BW energy denominators, ΔE_v is only expanded from some of these denominators. (See [4] for details.) This leads to the following results: (1) ΔE_c is reduced to just the sum of linked core diagrams, i.e., the Brueckner-Goldstone result; (2) all diagrams of mixed character, such as (c) and (d) above, are now cancelled identically. We also note that the set of noninteracting valence lines in (b) is equivalent to the unit operator I for the model subspace. These results mean that the model interaction operator simplifies considerably,

$$P\mathcal{V}(E) \rightarrow I\Delta E_c + \mathcal{V}_v \, (E_v), \tag{5.6}$$

where $\mathcal{V}_v$ is given by just the valence diagrams, (a) and (e) of Fig. 7 etc. The net result is that (2.14) now reduces to

$$[H_{ov} + \mathcal{V}_v(E_v) - I\, E_v]\, A = 0, \tag{5.7}$$

an expression in which all explicit reference to "core" quantities has disappeared. One should note, however, that ΔE_v still appears within all the energy denominators of $\mathcal{V}_v$. This important result was first obtained by Bloch and Horowitz [13].

We could have used (5.5) to reduce this result further, to

$$[\mathcal{V}_v(E_v) - I\, \Delta E_v]\, A = 0, \tag{5.8}$$

but we chose not to, for the following reason. For many applications one needs results which are not restricted to exact degeneracy. Our strategy is to carry through the entire analysis assuming exact degeneracy, and then extend the results to the quasi-degenerate case by adding to V a diagonal degeneracy-breaking interaction,

$$\begin{aligned} V_{db} &= \sum_{i\varepsilon D} |\Phi_i\rangle\, \Delta E_i \langle\Phi_i| \\ &= \sum_v a_v^\dagger\, a_v\, \Delta\varepsilon_v \end{aligned} \tag{5.9}$$

where the second line expresses the model-determinant energy shifts ΔE_i in terms of shifts in the valence orbital eigenvalues. Thanks to their being diagonal, it is then an easy matter to formally sum out all of the V_{db} interactions. The result has the same general form as before, but with V_{db} now incorporated within H_{ov}.

We must also mention that the result (5.7) can and should be reduced still further before proceeding with the following developments, in order to simplify some of the core-particle excitations which occur within $\mathcal{V}_v$. This "reduction of the Bloch-Horowitz expansion" is a rather technical matter which is discussed in [5], and more fully in [4].

5.3. Degenerate Rayleigh-Schroedinger Perturbation Theory

In spite of its appearance the result (5.7) is not yet suitable for many-body systems, because (1) the diagrams of $\mathcal{V}_v$ are not all fully linked, and (2) the desired interaction energy ΔE_v still appears within all of the denominators. For example, diagram 7(a) must be regarded as an effective three-body interaction, contrary to the intuitive idea that this represents a two-body interaction together with a "self-energy" or effective one-body potential contribution. Besides making calculations more difficult (probably prohibitively so for $n > 3$), this feature considerably complicates the physical interpretation. Thus, even for $n = 2$, a fully linked form is necessary to distinguish cleanly between the one-body (effective potential) and the two-body (effective interaction) aspects of $\mathcal{V}_v$. And when n is large, the individual perturbation terms tend to have unphysical magnitudes because of the ΔE_v's in their denominators. These difficulties are a reflection of the fact that although the core part is now treated by the fully linked RS formalism, the valence part of the system is still being treated by a BW type of expansion. It should now be clear that one needs a Rayleigh-Schroedinger type of expansion for $\mathcal{V}_v$.

Let us return to the original degenerate secular equation (2.14), for the moment ignoring all many-body features and assuming exact degeneracy. The BW expansion for $P\mathcal{V}(E)$ can be converted to an RS type of expansion by formally expanding out the ΔE dependence,

$$P\mathcal{V}(E_o + \Delta E) = \sum_{r=0}^{\infty} \mathcal{V}^{(r)}[-\Delta E]^r, \tag{5.10}$$

where, according to the Taylor formula,

$$\mathcal{V}^{(r)} = P\, \frac{(-1)^r}{r!}\, \frac{d^r \mathcal{V}(E_o)}{d\, E_o{}^r} \;. \tag{5.11}$$

The minus signs serve to simplify the $\mathcal{V}^{(r)}$'s, since

$$\frac{(-1)^r}{r!} \left(\frac{d}{dE_o}\right)^r \frac{1}{E_o - H_o} = + \left(\frac{1}{E_o - H_o}\right)^{r+1} . \tag{5.12}$$

One can now make repeated use of (5.8), thanks to the exact degeneracy, to express (5.10) as

$$P\mathcal{V}(E_o + \Delta E) = \sum_{r=0}^{\infty} \mathcal{V}^{(r)}[-P\mathcal{V}(E_o + \Delta E)]^r \;. \tag{5.13}$$

Although ΔE still appears here, its first explicit occurrence is now in a term of fourth order in V, rather than second order as in the original $\mathcal{V}(E)$. The first explicit occurrence of ΔE can be pushed out further, to higher and higher orders in V, by applying this expansion repeatedly to the $[-P\mathcal{V}(E)]$ factors in the right-hand side of (5.13). The result of many applications of this procedure is to approach a limiting form, $\mathcal{W}$, in which ΔE no longer appears at all. This satisfies the equation

$$\mathcal{W} = \sum_{r=0}^{\infty} \mathcal{V}^{(r)}[-\mathcal{W}]^r \;. \tag{5.14}$$

This $\mathcal{W}$ constitutes the desired RS analog of $P\mathcal{V}$, such that the secular equation now becomes

$$[H_o + \mathcal{W} - I\,E]\,A = 0\,. \tag{5.15}$$

The explicit RS expansion may be obtained by first solving (5.14) recursively, to express $\mathcal{W}$ in terms of products of various $\mathcal{V}^{(r)}$'s, and then developing these $\mathcal{V}^{(r)}$'s in powers of V by means of (2.12) and (5.11). To third order, the result is

$$\begin{aligned} \mathcal{W} &= \mathcal{V}^{(0)} + \mathcal{V}^{(1)}[-\mathcal{V}^{(0)}] + \mathcal{V}^{(2)}[-\mathcal{V}^{(0)}]^2 \\ &\quad + \mathcal{V}^{(1)}[-\mathcal{V}^{(1)}][-\mathcal{V}^{(0)}] + \mathcal{O}\,[\mathcal{V}^{(r)}]^4 \\ &= PVP + PV\frac{Q}{e}VP + PV\frac{Q}{e}V\frac{Q}{e}VP \\ &\quad + PV\left(\frac{Q}{e}\right)^2 V(-P)\,VP + \mathcal{O}(V^4), \end{aligned} \tag{5.16}$$

where e stands for $E_o - H_o$.

5.4. The Folded-Diagram Expansion

We shall now apply this $\mathcal{V} \to \mathcal{W}$ development to the Bloch-Horowitz operator $\mathcal{V}_v(E_v)$ from (5.7), calling the resulting operator $\mathcal{W}_v$. The matrix multiplications and explicit minus signs in (5.14), (5.16) will be dealt with by means of "folded diagrams".

Consider a system with just two valence particles. The $\langle\Phi_i|\mathcal{V}_v^{(1)}[-\mathcal{V}_v^{(0)}]|\Phi_j\rangle$ terms can be represented by diagrams such as Fig. 8(a), where labels have been added to show the meaning of the various sections. The solid horizontal line in $\mathcal{V}^{(1)}$ stands for the repeated energy denominator, from (5.11) - (5.12), exactly as in the case of Figs. 3 and 4. The loop in the middle of this diagram stands for the projection operator P, together with the explicit minus sign from $[-P\mathcal{V}_v^{(0)}]$. Note that the associated matrix multiplication involves a sum over all intermediate states $\Phi_{i'}$, with i' in D. This can be accomplished here by summing each line segment within the loop over all of the valence orbitals. The loop therefore indicates these orbital summations, together with the explicit minus sign. These valence-orbital summations can all be done independently, since the resulting exclusion-violating terms will all cancel in the manner of Fig. 6.

We now "fold" diagram (a) in a zig-zag fashion, placing the folds at the bottom interaction of $\mathcal{V}_v^{(1)}$ and at the top of $\mathcal{V}_v^{(0)}$, so as to bring the top of $\mathcal{V}_v^{(0)}$ up to the same level as the horizontal bar in diagram (a). We also apply the factorization theorem, in the opposite direction from the way it was used in Fig. 5, to obtain finally the two folded diagrams (b) and (c). These diagrams have off-shell energy denominators, such that these denominators are now of just the same form as in the first two diagrams of Fig. 5. The matrix multiplications associated with the r > 1 terms in (5.13), as well as their "higher generation"

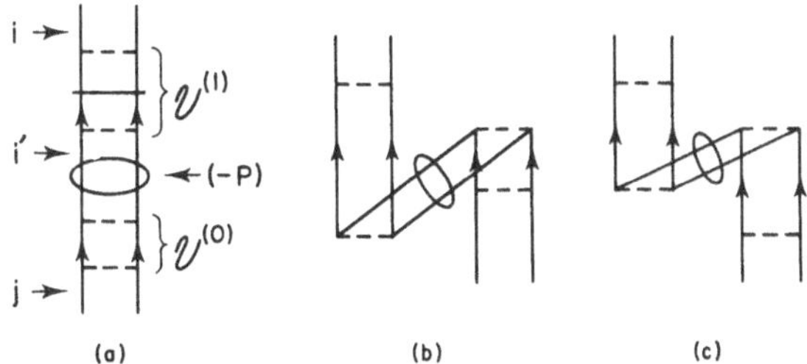

Fig. 8. Examples of folded diagrams: folded diagrams (b) and (c) are generated by the "unfolded" diagram (a).

analogs from the expansions of the $[-P\mathcal{V}_V(E_0 + \Delta E_V)]$ factors, are all to be treated in a similar manner, thus the perturbation series for $\mathcal{W}_V$ will include diagrams with arbitrarily many folds. The general "time ordering" topology of these multiply-folded diagrams is described in [5].

The practical value of these folded diagrams becomes clear when one is faced with unlinked valence diagrams such as (a) of Fig. 7. Consider the unlinked diagrams (a) and (b) of Fig. 9, for a system with four valence particles. In the present RS-type expansion, these two diagrams are cancelled identically by diagram (c). This becomes quite evident when diagram 9 (c) is manipulated in the manner shown in Fig. 8. This type of cancellation is found to be completely general [4], which means that <u>only fully linked valence diagrams appear within the expansion for</u> $\mathcal{W}_V$. To obtain the desired complete cancellation, however, it turns out to be essential (not merely a convenience) to "ignore the exclusion principle" while summing over the sets of valence-orbital lines enclosed by the various (-P) loops. In fact, one must consistently ignore the exclusion principle at <u>all</u> intermediate levels between the bottommost and topmost interactions of each diagram, where we are now referring to the "unfolded"

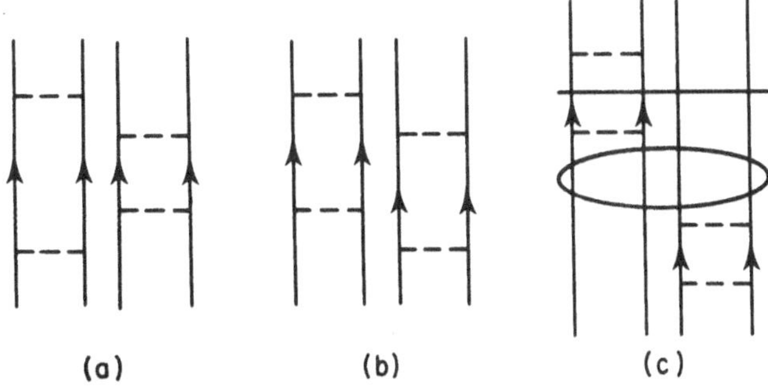

Fig. 9. An example of unlinked diagram cancellation: diagrams (a) and (b) are cancelled by (c).

forms [Fig. 8(a)] of these diagrams. In the resulting fully linked valence diagrams, any completely passive (noninteracting) valence lines may now be erased, since these lines no longer convey any useful information. This feature is very significant, since it paves the way for applications to systems with a large or even macroscopic number of particles.

This result does not mean, however, that all folded diagrams simply disappear by cancellation. Although the unlinked terms (folded or not) are all cancelled, one is still left with an infinite number of linked folded diagrams, such as (b) and (c) of Fig. 8, and these diagrams may contain arbitrarily many folds. These leftover folded diagrams are all analogous to the closed-shell diagrams in which some of the hole lines violate exclusion, as in Fig. 6(b). In fact, a one-to-one correspondence can be established by applying the present formalism to the case where the n valence particles completely fill the valence shell, so that the system is actually nondegenerate. At this stage, where the unlinked terms have all been eliminated, one may now simplify these linked folded diagrams by applying the factorization theorem in the "forward direction", i.e., in the manner of Fig. 5. Moreover, the algebraic structure of (5.14) provides much further opportunity for partial summation of these diagrams.

5.5. Hermiticity and Effective Operators

It must be recognized that the present effective interaction matrix $\mathcal{W}_v$ is not fully Hermitian. The eigenvalues of (5.15) are all real (by construction they reproduce certain of the exact eigenvalues), but the model eigenvectors A_α are generally not orthogonal. This can be understood from two different viewpoints. On the one hand, these A_α's represent projections of the exact eigenvectors onto the model subspace [see (2.6)], and projections of orthogonal vectors need not be orthogonal. On the other hand, the different eigenstates α generally have different eigenvalues E_α, whereas the original matrix operator $P\mathcal{V}(E)$ exhibits Hermiticity only when E is held fixed. It is therefore essential that $\mathcal{W}$ be nonhermitean, in order to faithfully reproduce the effect of $\mathcal{V}$. Diagrammatically, this lack of Hermiticity arises via (5.13), (5.14) and the convention of Figs. 8 and 9, where the "folded in" insertions are all connected to the bottoms of the skeleton parts [the explicit $\mathcal{V}^{(r)}$'s of (5.14)]. For most applications, however, one would like to have a Hermitean effective interaction. We shall now describe a relatively simple and convenient way to obtain a Hermitean analog of $\mathcal{W}_v$.

Assuming the intermediate normalization convention,

$$\langle\Psi_{D\alpha}|\Psi_\alpha\rangle = \langle\Psi_{D\alpha}|\Psi_{D\alpha}\rangle \equiv \langle A_\alpha|A_\alpha\rangle = 1, \tag{5.17}$$

the orthogonality condition for the exact eigenstates becomes

$$\begin{aligned}\langle\Psi_\alpha|\Psi_\beta\rangle &= \langle\Psi_{D\alpha}|\Omega_\alpha^\dagger\Omega_\beta|\Psi_{D\beta}\rangle \\ &= \langle\Psi_{D\alpha}|\Omega_\alpha^\dagger(P + Q)\,\Omega_\beta|\Psi_{D\beta}\rangle \\ &\equiv \langle A_\alpha|(I + \Theta)|A_\beta\rangle = N_\alpha\delta_{\alpha\beta}\ .\end{aligned} \tag{5.18}$$

The present Ω's carry subscripts because of their dependence on the eigenvalues E_α, E_β, via (2.7), (2.11). In obtaining the last line of (5.18), we have (1) used the RS expansion technique of §5.3, together with folded diagrams, to eliminate the explicit eigenvalue dependence of these Ω's, and (2) expressed $\Omega^\dagger\Omega$ as a matrix $(I + \Theta)$ acting within the model subspace, using the fact that $\Omega^\dagger P\Omega = P = I$ here, whereby $\Theta = \Omega^\dagger Q\Omega$. The last line of (5.18) suggests the introduction of a new set of model eigenvectors,

$$\hat{A}_\alpha = (I + \Theta)^{\frac{1}{2}}\, A_\alpha N_\alpha^{-\frac{1}{2}}, \tag{5.19}$$

which are now orthonormal by construction. The eigenvalue equation (5.15) can now be expressed in terms of these $\hat{A}$ vectors,

$$(H_{ov} + \mathcal{W}_v)\ (1 + \Theta)^{-\frac{1}{2}}\hat{A}_\alpha = E_v(I + \Theta)^{-\frac{1}{2}}\hat{A}_\alpha, \tag{5.20}$$

and thus

$$(I + \Theta)^{\frac{1}{2}}(H_{ov} + \mathcal{W}_v)(I + \Theta)^{-\frac{1}{2}}\hat{A}_\alpha \equiv (H_{ov} + \mathcal{K}_v)\ \hat{A}_\alpha = E_v\hat{A}_\alpha. \tag{5.21}$$

The new interaction matrix $\mathcal{K}_v$ is Hermitian by construction (it has real eigenvalues and orthogonal eigenvectors), but it is not manifestly Hermitean. In practice, this defect can be remedied by explicit Hermitization,

$$\mathcal{K}_v \to \frac{1}{2}[(I + \Theta)^{\frac{1}{2}}(H_{ov} + \mathcal{W}_v)(I + \Theta)^{-\frac{1}{2}} + \text{h.c.}]-H_{ov}\ . \tag{5.22}$$

For many-body systems one must deal with (5.21) or (5.22) by expanding the square-root operators into binomial series in Θ. [Note that truncation of (5.22) at any finite order in Θ will preserve Hermiticity.] The resulting series for $\mathcal{K}_v$ is fully linked [4], although this is not true for Θ by itself. For many purposes it is adequate to simply use

$$\mathcal{K}_v \approx \frac{1}{2}\ (\mathcal{W}_v + \mathcal{W}_v^\dagger), \tag{5.23}$$

since the error is $\mathcal{O}(V^4)$.

An added benefit from the use of $\mathcal{K}_v$ and the resulting $\hat{A}$ model vectors is that this leads to a convenient expression for the "effective" or "model" operators $M(O)$, whose effect within the model subspace is the same as that of the operator O acting on the corresponding exact eigenstates:

$$\langle \hat{A}_\alpha | M(O) | \hat{A}_\beta \rangle \equiv \frac{\langle \Psi_\alpha | O | \Psi_\beta \rangle}{(\langle \Psi_\alpha | \Psi_\alpha \rangle \langle \Psi_\beta | \Psi_\beta \rangle)^{\frac{1}{2}}}. \tag{5.24}$$

By using (5.19), one readily finds that

$$M(O) = (I + \Theta)^{-\frac{1}{2}} (\Omega^\dagger O \Omega)(I + \Theta)^{-\frac{1}{2}} . \tag{5.25}$$

In an open-shell system, the various matrix operators in (5.25) should all have valence subscripts v. The complete $M(O)$ matrix then contains an additional diagonal term $\langle O \rangle_c \delta_{\alpha\beta}$, where $\langle O \rangle_c$ is just the expectation-value contribution from the core particles, calculated as if all of the valence particles had been physically removed. The folded-diagram expansion for $M(O)$ has also been shown to be fully linked [4]. Besides providing general expectation values (for $\alpha = \beta$), this expansion should be useful in the calculation of electromagnetic transition amplitudes.

In concluding this section, we should point out that there are several other and apparently very different perturbation procedures which lead to the same operator K_v. These alternatives are reviewed in [7,8]. The present approach, based on (5.14) and (5.22), is by far the most convenient one for applications where it is desirable to carry out partial summations to infinite order. We should also mention that there is a formal (and sometimes practical) problem in determining just which d of the infinite number of exact eigenstates should be reasonably reproduced when using various approximation techniques to evaluate W_v or K_v. This has been the object of much research in nuclear physics, where it is known as the problem of "intruder states". The reader will find this discussed in several articles of a recent conference proceedings (see [8]), and in a forthcoming review article [27].

6. APPLICATION TO POLYMERS -- EFFECTIVE π-ELECTRON HAMILTONIANS

For any large system it is clear that the correlation effects must be handled by means of the cluster-decomposition idea (Section 4) which characterizes <u>all</u> forms of many-body theory. For organic molecules, however, there is abundant evidence (see Introduction) that correlation effects can be treated quite well by means of (a) empirical parameters within a simple type of π-electron Hamiltonian, and (b) explicit configuration mixing within the π-orbital subspace. (For non-planar molecules this treatment should really be extended to the σ electrons as well as the π's, as in the so-called "all valence electron" methods, since a rigorous distinction between π and σ orbitals is no longer possible. This problem is not considered here.) This is a very natural application for the present open-shell formalism, which, in fact, seems ideally suited for this purpose. After all, this formalism was originally developed for a very similar problem -- that of justifying the nuclear shell model [27].

Several of the early attempts to understand the success of π Hamiltonians in terms of many-body theory have been based on the cononical transformation approach, i.e., the Van Vleck-Kemble form of degenerate perturbation theory. (See Westhaus et al [28], and references therein.) It has been demonstrated that this is formally equivalent to the present K_V expansion [6, Appendix F; 7], assuming that the core-valence separation has been handled properly. The K_V form is, however, much more convenient [8] for high order treatments such as those involving partial summations to infinite order. The most extensive quantitative investigation to date, by Iwata and Freed [29], was based instead on a treatment equivalent to the degenerate BW formulation of §2 [see their Eqs. (5) - (10)], developed to second order in V. These calculations suffered greatly from the unlinked-cluster problems discussed in §5.3. From a strictly quantitative standpoint, the problem of the explicit E-dependence of all second-order terms appeared rather minor for their example of the ethylene molecule, which has only two "valence" electrons. But this nondiagrammatic BW formulation did involve major qualitative difficulties of interpretation -- the correct assignment of various correlation energy contributions to the individual α, β, and γ parameters was very seriously obscured. Specifically, the bulk of the correlation energy's state dependence was assigned to the σ core rather than to the various effective π matrix elements. To resolve this problem the authors then turned to a formalism of the present type [30]. They developed the degenerate RS expansion (K_V form) to third order by purely algebraic means, confirming the cancellation of unlinked third-order terms, and then expressed the results in terms of diagrams. Westhaus, on the other hand, has argued convincingly that his entire perturbation series must be linked [28, 31], even though his explicit results are limited to second order.

In contrast, the diagrammatic analysis outlined in §5 is quite complete. The results are available to all orders, for arbitrary numbers of core electrons, valence electrons, and/or valence holes. The resulting K_V diagrams can be simply drawn and used, following the rules in a mechanical fashion (as in [17], for example). The detailed physical interpretation of each diagram is usually obvious, at least in the lower orders, and every diagrammatic term is clearly assigned to a particular linked r-body matrix element of the effective Hamiltonian (with r = 1, 2, 3, etc.). One must recognize, however, that some infinite-order partial summations will probably be necessary for reasonably accurate and reliable results; this follows from experience with atomic ground-state correlations [1], for which the use of partial summations has been very successful. But since the general algebraic structure of K_v is fairly transparent [see (5.10)-(5.14), (5.19), and (5.22)], partial summations are really about as easy for K_v as for the closed-shell case. This is confirmed by the experience with K_v in nuclear physics [27]. The suggestion [30, 32] that this approach would be difficult or obscure for the open-shell case is therefore unwarranted.

6.1. The Unperturbed Hamiltonian

As a necessary preliminary for the present application, one must first choose an unperturbed Hamiltonian H_o which generates a suitable basis with localized π orbitals. This is a non-trivial matter, because the <u>canonical</u> Hartree-Fock orbitals, namely those which satisfy the usual eigenvalue form

$$F\,\psi_\alpha = \varepsilon_\alpha\,\psi_\alpha \tag{6.1}$$

of the HF equations, necessarily have a delocalized (Bloch function or molecular orbital) form. (This result is not quite as obvious as it may seem, but is quite generally true nevertheless [33].) Formally, however, the eigenfunctions ψ_α may be canonically transformed into a localized (Wannier-function) basis φ_n, where n is a site index, in which case the HF equations take on the <u>noncanonical</u> form

$$F\,\varphi_n = \Sigma_{n'}\varphi_{n'}\lambda_{n'n}\ , \tag{6.2}$$

$$\lambda_{n'n} = \langle\varphi_{n'}|F|\varphi_n\rangle \quad . \tag{6.3}$$

The equations (6.2) can now be transformed into a "pseudo-canonical" form, by the simple expedient of transferring the off-diagonal terms to the left-hand side. By adding suitable orbital-transfer operators one therefore obtains an H_o of the desired form,

$$H_o\varphi_{n\nu} = [F - \Sigma_{\ell\ell'\nu'}|\varphi_{\ell'\nu'}\rangle\lambda_{\ell'\ell\nu'}\langle\varphi_{\ell\nu'}|]\,\varphi_{n\nu} = \lambda_{nn\nu}\varphi_{n\nu}. \tag{6.4}$$

A "band" index $\nu(=\pi,\sigma)$ has been added here so that localized σ orbitals can be generated in the same manner as the localized π's, although this refinement is optional for the σ's.

Note that for the higher orbitals, which lack such off-diagonal λ terms, this H_o generates the usual canonical (delocalized) HF orbitals, thus leading to a complete basis. (In practice, however, it may well be advantageous to modify this H_o for the orbitals of the virtually-excited subspace [30, 34].) Direct calculation of the desired φ's by means of (6.2), (6.3) should be feasible, for example by adapting the usual LCAO-SCF technique to the procedure demonstrated in [6]. For sufficiently small and symmetrical systems, an obvious alternative is to simply calculate the corresponding canonical orbitals and then transform these ψ's into φ's, as in [29]. The open-shell nature of the π electrons leads, however, to some ambiguity in the choice of the Fock operator F. For maximum accuracy in the calculation of γ_{oo}, it would probably be best to handle this by means of an occupation average over all of the canonical orbitals ψ, such that the configuration at each carbon atom (in the φ representation) is 50% polar and 50% nonpolar. A less repulsive choice of F would probably give faster convergence for β. Some care will be required in any event to determine the "most physical" definition for β. A very similar H_o is discussed in [6], and a number of related localized-orbital proposals are reviewed in [35].

6.2. Diagrams and Results

The diagrams up to second order for the usual α, β, and γ parameters of the effective π-electron Hamiltonian are shown in Fig. 10. The n's are site (carbon atom) indices, with spin labels also implied where necessary. Each dashed line now represents an antisymmetrized (direct minus exchange) matrix element. We have attempted to deduce numerical values for these diagrams from the results of Iwata and Freed [29], recognizing that there is some ambiguity here. It appears that diagram (a) is about - 0.5 eV. Diagram (b) should be considerably larger and repulsive; this appears to be about + 1.8 eV, but has considerable uncertainty. The total α_n (including all higher-order terms) should represent the physical ionization potential of the planar CH_3 system (neutral methyl radical). The first-neighbor "resonance" integral $\beta_{n'n}$ appears to consist almost entirely of just the "bare" trans-

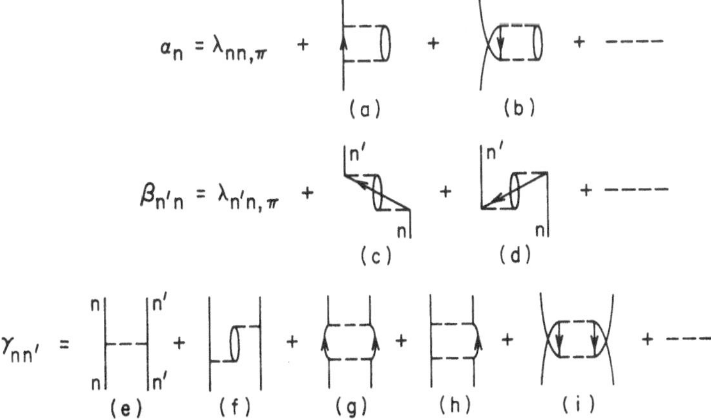

Fig. 10. Leading diagrams for the π-Hamiltonian parameters.

fer integral from (6.3), here about - 3.8 eV. The correlation (c) reduces β in absolute magnitude by only about 5%, while (d) was not well determined, but should be somewhat larger and of the same sign (positive). There may, however, also be a significant contribution to the effective β from "nonclassical" terms (see below).

For the $n' = n$ γ term, γ_{oo}, most of the correction to the "bare" Coulomb repulsion integral (e) is due to (f), which represents a polarization of the σ-electron core. This is not surprising, as the addition of a π electron should cause radial expansion of the three localized σ orbitals at the same site. (There should also be some transfer of σ-electron charge density towards the site from which the π charge originated, as in an optical transition to a more polar state, but such correlations must contribute instead to the screening of the more distant Coulomb parameters γ_{on}.) Diagram (e) is about +16 eV, and (f) about - 2.3 eV, for $n' = n$ (γ_{oo}). Diagram (g) shows a simple π-π correlation, while (h) represents a radial expansion of one of the π orbitals due to the repulsive effect of the other π electron. Diagram (g) amounts to - 0.3 eV, while (h) appears to be only - 0.03 eV. This remarkably small value for (h) is undoubtedly due to the initial choice of rather diffuse π orbitals, since the Iwata-Freed value (the $\pi_{vo} \rightarrow \pi_e$ entry in their Table IV) also includes a self-consistency correction associated with the choice of the "unperturbed" π orbitals. This is consistent with the larger (h) contribution (- 0.15 eV) to γ_{o1} (for $|n'-n| = 1$) where the "initial" π orbitals evidently now experience a significant radial contraction. Other γ_{o1} contributions which can be deduced are (e) $\approx$ + 8.5 eV, (f) $\approx$ - 0.17 eV, and (g) $\approx$ - 0.01 eV.

Diagram (b) is a "blocking" of a core (σ-electron) correlation process due to the presence of a single π electron, as explained in connection with Fig. 7(e). Since the σ-σ core correlations are attractive, diagram (b) must be repulsive. With the addition of a <u>second</u> π electron, however, the additional diagram (b) contribution now leads to a part of the "initial" σ-σ correlation being subtracted <u>twice</u>, so the latter processes must be added back again in order to maintain a correct counting of the "allowed" and "forbidden" core correlations. This is the meaning of the "double blocking" diagram (i). Unfortunately, its numerical value is not apparent from the tables in [29]; one may expect this to be attractive and much smaller than (b).

In the usual π-Hamiltonian model for ethylene, the difference between the excitation energies of the states V(a pure polar state) and T(a pure nonpolar state) is just γ_{oo}-γ_{o1}. This difference is observed to be 3.0 eV. With their non-diffuse π orbitals Iwata and Freed obtained 7.75 eV without correlation, and 4.85 eV with correlation, whereby their correlations have accounted for 60% of the initial discrepancy. As the authors have noted, the remaining discrepancy can very reasonably be attributed to the limited basis and the omitted higher-order perturbation terms. Bradford and Westhaus [36] have also reported some applications of their canonical transformation approach. (This is essentially the RS perturbation theory to second order, with shifted energy denominators, the latter corresponding to a simple type of partial summation.) It is interesting that they obtained about 50% of the correlation energy contribution to the V-state excitation energy of ethylene, and that they have attributed most of this discrepancy to their limited orbital basis. Further calculations are clearly desirable.

Some general comments should be added. (1) The empirically successful transferrability of the α, β, and γ parameters between molecules with similar local environments is immediately suggested by the form of the present localized π orbitals, which must closely resemble Löwdin-orthogonalized atomic orbitals. (2) We believe that the conventional neglect of many two-electron terms, usually described as the "zero differential overlap" approximation, is best justified in terms of (a) the individual smallness of these terms (these matrix elements typically being <0.1 eV each), and (b) the observation [29] that correlation may further reduce the absolute magnitudes for many of these terms. The alternative suggestion [30, 37] that the perturbation formalism should be set up and applied with nonorthogonal π orbitals does not appear to us to be useful. (3) Although the "π-σ separability" has been built into the structure of the formalism, and is thus not really a physical assumption, there <u>is</u> a related physical assumption in practice. Any consistent many-body formalism will predict effective many-body interaction terms (three-body terms, for example), whereas such terms are ignored in the usual empirical Hamiltonians. On the other hand, such "nonclassical" contributions to the effective Hamiltonian can (and should) be partially absorbed into the usual parameters, as discussed by Iwata and Freed [30, 38] for π Hamiltonians, and by the author [4, p. 799] for the Landau theory of liquid ^{3}He.

One final comment may spare the reader some confusion. The diagrams shown in Fig. 10 for α and β clearly correspond to the leading terms of the expansion for the mass operator or self-energy operator of the causal one-body Green's function, evaluated at the "quasi-particle pole" energy. In higher orders, however, it becomes apparent that the one-body terms of the present expansion are generating a "reducible" form of this operator, rather than the more familiar "irreducible" form obtained from time-dependent perturbation theory. The latter form is generally more useful in practice, by virtue of its being "more highly summed". The connection between these two forms has been demonstrated in detail in [39]. Other connections between the K_v expansion and previous formal results are reviewed in [5].

REFERENCES

1. H. P. Kelly, Phys. Rev. 131, 684(1963) and 136, B896(1964); H. P. Kelly, Adv. Chem. Phys. 14, 129(1969).
2. U. Kaldor, Phys. Rev. Lett. 31, 1338(1973); U. Kaldor, J. Chem. Phys. 63, 2199(1975); P. S. Stern and U. Kaldor, J. Chem. Phys. 64, 2002(1976).
3. B. H. Brandow, Proc. Int. Sch. Phys. "Enrico Fermi" 36, 496 (1966).
4. B. H. Brandow, Rev. Mod. Phys. 39, 771(1967).
5. B. H. Brandow, Adv. Quantum Chem. 10(in press).
6. B. H. Brandow, Adv. Phys. (in press).
7. D. J. Klein, J. Chem. Phys. 61, 786(1974).
8. B. H. Brandow, in Effective Interactions and Operators in Nuclei, B. R. Barrett, ed. (Springer-Verlag, Berlin and New York, 1975).
9. R. G. Parr, The Quantum Theory of Molecular Electronic Structure, (Benjamin, New York, 1964).
10. M. J. S. Dewar, The Molecular Orbital Theory of Organic Chemistry, (McGraw-Hill, New York, 1969).
11. H. Berthod, C. Giessner-Prettre, and A. Pullman, Theoret. Chim. Acta. 5, 53(1966).
12. J. Ladik, D. K. Rai, and K. Appel, J. Mol. Spectry. 27, 72 (1968).
13. C. Bloch and J. Horowitz, Nucl. Phys. 8, 91(1958).
14. P.-O. Löwdin, J. Math. Phys. 3, 969(1962).
15. J. Goldstone, Proc. Roy. Soc. (London) A239, 267(1957).
16. R. Manne, Int. J. Quantum Chem. Symp. (in press).
17. U. Kaldor, J. Comput. Phys. 20, 432(1976).
18. K. A. Brueckner, Phys. Rev. 100, 36(1955).
19. K. A. Brueckner, in The Many-Body Problem, C. de Witt, ed. (Dunod, Paris, 1959).
20. N. M. Hugenholtz, Physica 23, 481(1957).
21. L. M. Frantz and R. L. Mills, Nucl. Phys. 15, 16(1960).
22. R. P. Feynman, Phys. Rev. 76, 749(1949).
23. V. Ambegaokar, in Superconductivity, R. D. Parks, ed. (Dekker, New York, 1969), Vol. 2, p. 1359.
24. G. E. Brown, Many-Body Problems (North-Holland Publ., Amsterdam, 1972), p. 5.
25. J. Paudus, J. Čižek, and I. Shavitt, Phys. Rev. A 5, 50(1972); D. L. Freeman, Phys. Rev. A or B (in press).
26. P.-O. Löwdin, J. Chem. Phys. 19, 1396(1951).
27. P. J. Ellis and E. Osnes, Rev. Mod. Phys. (in press).
28. P. Westhaus, E. G. Bradford, and D. Hall, J. Chem. Phys. 62, 1607(1975). See also S. Fischer, Int. J. Quantum Chem. Symp. 3, 651(1970).
29. S. Iwata and K. F. Freed, J. Chem. Phys. 61, 1500(1974).

30. S. Iwata and K. F. Freed, J. Chem. Phys. 65, 1071(1976).
31. P. Westhaus, Int. J. Quantum Chem. Symp. 7, 463(1973).
32. K. F. Freed, J. Chem. Phys. 60, 1765(1974).
33. B. H. Brandow, Phys. Rev. B 12, 3464(1975).
34. H. P. Kelly, Phys. Rev. 136, B 896(1964); H. J. Silverstone and M.-L. Yin, J. Chem. Phys. 49, 2026(1968); S. Huzinaga and C. Arnau, Phys. Rev. A 1, 1285(1970); J. H. Miller and H. P. Kelly, Phys. Rev. A 3, 578(1971). See also Ref. 29.
35. T. L. Gilbert, in Molecular Orbitals in Chemistry, Physics, and Biology, P.-O. Löwdin and B. Pullman, eds. (Academic Press, New York, 1964), p. 405; T. L. Gilbert and A. B. Kunz, Phys. Rev. B 10, 3706(1974).
36. E. G. Bradford and P. Westhaus, J. Chem. Phys. 64, 4276 (1976).
37. V. Kvasnicka, Phys. Rev. A 12, 1159(1975).
38. S. Iwata and K. F. Freed, Chem. Phys. Lett. 38, 425(1976).
39. B. H. Brandow, Ann. Phys. (N.Y.) 64, 21(1971), see Appendix C.

METHODS FOR THE TREATMENT OF DISORDERED SYSTEMS WITH POSSIBLE APPLICATIONS TO APERIODIC POLYMERS

F. Martino*

Department of Physics
The City College of CUNY
New York, New York 10031

I. INTRODUCTION

It is a truism that the real world is never exactly described by the kind of simple and regular models that solid state physicists have capitalized on over the years. Almost all real materials are disordered - even laboratory samples of "perfect" crystals contain impurities, vacancies, etc. The traditional techniques of solid state physics chose to ignore this truism because of the tremendously effective tools translational symmetry provides - the Bloch theorem, a conserved k vector, and the other benefits provided by group theory. These traditional techniques, applied to systems in which imperfection played a minor or nonexistent role in macroscopic behavior, have provided brilliant successes over the years. During the last twenty years, however, there has been an enormous increase in interest in materials whose disordered nature was not only significant to their macroscopic behavior, but often was the principle reason for startingly new and/or useful kinds of behavior. A partial list would include alloys, mixed crystals, liquid crystals, doped semiconductors, and of course amorphous semiconductors. Finally, of course, we must cap this list with the most interesting and startling of all materials, biological matter. This increase in interest has coincided with progress in both theoretical and experimental investigations of disordered systems. On the experimental side the greatest success has been achieved in the areas of amorphous semiconductors and related new sorts of electronic devices, as well as in random alloys. Another area of recent success is that of computer "experiments" on large cluster disordered systems, which with the advent of very large and fast computers, have reached a degree of sophistication which provides information of

J.-M. André et al. (eds.), Quantum Theory of Polymers, 169-184.

great value to both experimentalists and theoreticians. At the very least they provide useful touchstones with which theoretical approximations can be compared. Finally the theory of the behavior of elementary excitations in disordered systems has developed increasingly rapidly, starting from various forms of simple effective medium theories developed over twenty-five years ago through the development of the so-called coherent potential approximation by a number of different workers into a highly effective and diverse body of mathematical tools.

Indeed the field is so large that we should begin with a series of apologies for the many aspects of the problem this paper will not discuss. Chief among these, perhaps the most interesting of all, and certainly one of the most obscure and difficult, stems from a question raised by P.W. Anderson concerning the nature of the wave functions of an electron in a disordered array. The question may be stated in a number of ways: under what circumstances will the electron be spatially restricted, rather than extended throughout the sample, or when will the electron fail to quantum mechanically tunnel out of its original location after an infinite amount of time, or when will an electron in a partially filled band require a non-vanishing activation energy to propagate? For reasons mentioned in the discussion of section IV we must ignore this problem. We shall here limit ourselves to discussions of densities of states (energy spectra) and other such single-particle properties of elementary excitations as may be investigated by means of an averaged single-particle Green's function $\langle G \rangle$. The averages are taken over disordered microscopic configurations the sum of which correspond to the actual macroscopic sample. There are many different sorts of disorder in this sense, and we shall catalogue these below. We shall concentrate our attention, however, on those sorts of disorder which are deemed to be most appropriate to the discussion of aperiodic polymers - including but not exclusively periodic polymers with a random distribution of impurities. Consistent with this approach we have also chosen to ignore the enormous progress made in computer modeling of disordered systems, as being less appropriate to aperiodic polymers. A review of these developments is found in reference 3.

As a final general disclaimer we should state that this paper in no way purports to be an exhaustive general review of the field. There are a number of excellent reviews of various aspects of the problems presented here, containing large bibliographies, and we have chosen to reference these rather than attempt to assign detailed historical credit for the developments (references 1-6). Our purpose is a more modest, or even tentative one: to bring to the attention various techniques for the treatment of disorder, which have been recently applied with some success to systems with relatively simple basic units, and to suggest the possibility of applying these same techniques to more complex units. This attempt should be taken in the same spirit as the recent applications to polymers of traditional techniques of solid state physics for periodic systems.

Section II contains a sketchy review of Green's function formalism sufficient for the formalisms of section III, which deals with various approximations at the heart of the theory of configuration averaged Green's functions in disordered systems. Finally section IV contains a summary and brief discussion.

II. REVIEW OF GREEN'S FUNCTION FORMALISMS

We shall find it convenient to define our Hamiltonian either in a spatial (site) representation or its Fourier transform (wave vector) representation. The site representation is most convenient for characterizing the various forms of disorder we may choose to deal with. We take the one particle model Hamiltonian to be

$$H=\sum_{i\mu} \varepsilon_i^{\mu} a_{i\mu}^{+} a_{i\mu} + \sum_{i\neq j} \sum_{\mu\upsilon} t_{ij}^{\mu\upsilon} a_{i\mu}^{+} a_{j\nu} \tag{1}$$

where $a_{I\mu}^{+}$ $(a_{i\mu})$ creates (annihilates) an elementary excitation in the state μ at the site i with an energy ε_i^{μ}, and $t_{ij}^{\mu\upsilon}$ is the transfer energy between the state μ at i and υ at j Workers in the field, for the sake of convenience, have divided the types of disorder into various classes. With the understanding that this division is somewhat arbitrary, and that most real disordered systems will be some combination of the various types, we list them: In the case where the sites, i, j, form some sort of periodic array, but the local potentials and energies vary, the disorder is termed "substitutional" or "compositional". Compositional disorder may be "site-diagonal" (only the ε_i^{μ} are random) or "off-diagonal" (the system is statistically homogeneous, all $\varepsilon_i^{\mu}\equiv\varepsilon^{\mu}$, but the $t_{ij}^{\mu\upsilon}$, are disordered), or, in general, both. In the case of glasses or amorphous systems, the disorder is termed "positional" or "structural". This latter case can be divided into single species liquids or amorphous solids ($\varepsilon_i^{\mu}\equiv\varepsilon^{\mu}$; $t_{ij}^{\mu\upsilon}=t^{\mu\upsilon}(R_{ij})$), liquid alloys, and the most complex forms of disorder, in which the ε_i^{μ}, $t_{ij}^{\mu\upsilon}$ and the site positions are all random.

The wave vector representation of the Hamiltonian is found by Fourier analyzing equation (1).

$$H = \sum_{k\mu} E^{\mu}(\underline{k})\, a_{\mu}^{+}(\underline{k}) a_{\mu}(\underline{k}) + \sum \sum_{\mu\upsilon \underline{kk}'} T^{\mu\upsilon}(\underline{k},\underline{k}')\, a_{\mu}^{+}(\underline{k}) a_{\nu}(\underline{k}') \tag{2}$$

In a perfect crystalline array the translational symmetry implies that the wave-vector $\underline{k}$ is conserved and any excitation may be labeled by a $\underline{k}$ in the first Brillouin zone and a band label μ.

Thus the existence of a non-zero $T^{\mu\upsilon}(\underline{k},\underline{k}')$ in equation (2) may be taken to imply imperfections, or disorder, in the system.

If we wish to describe disordered defects in a perfect crystal in the spatial representation (equation (1) then we may define deviations from perfect values, $\varepsilon_i^{\mu}(o)$, $t_{ij}^{\mu\upsilon}(o)$ by

$$\Delta_{ij}^{\mu\upsilon} = (\varepsilon_i^{\mu} - \varepsilon_i^{\mu}(o))\delta_{\mu\upsilon}\,\delta_{ij} + (t_{ij}^{\mu\upsilon} - t_{ij}^{\mu\upsilon}(o)). \tag{3}$$

In the one-band model we may neglect the superscripts in equation (3). If the defect is very localized and has very low concentration, then the second term in equation (3) may be ignored. Simple examples will be given below.

Should we wish to calculate the elementary excitations of even a periodic array we must solve the secular equation

$$|(E - \varepsilon_i^{\mu})\,\delta_{\mu\upsilon} - t^{\mu\upsilon}(\underline{k})| = o \tag{4}$$

where $t^{\mu\upsilon}(\underline{k}) = \sum_j t_{ij}^{\mu\upsilon} \exp\{i\,\underline{k}\cdot(\underline{R}_i - \underline{R}_j)$

In a disordered system such a solution is in general far too difficult, and in any case would give information not accessible to the experimentalist, since $\underline{k}$ is not a good quantum number. The Green's function method, on the other hand, presents a fairly direct connection between experimentally measured quantities (for example electrical conductivity) and correlation functions given in terms of Green's functions. This directness is sometimes achieved at the price of loss of physical understanding of the approximations made, but in any case Green's functions are a major tool in the understanding of disordered systems. We give only a very brief and sketchy outline here. Thorough treatments and bibliographies may be found in, for example, references 7-10.

We define an operator $\Psi^+(\underline{x})$ $(\psi(\underline{x}))$ which creates (annihilates) a particle at a point $\underline{x}$ in space. Then the particle density operator is given by

$$\rho(\underline{x}) = \int d^3x'\,\psi^+(\underline{x}')\,\delta(\underline{x} - \underline{x}')\,\psi(\underline{x}') \tag{5}$$

Using the time-dependent operator in the Heisenberg representation $\psi(\underline{x}t)$ we can examine the probability amplitude that a particle created at time t at a point $\underline{x}$ has moved to another point $\underline{x}'$ at a later time t'>t. The average expectation value for this process may be defined as

$$\langle\psi(\underline{x}'\,t')\psi^+(\underline{x}\,t)\rangle = \mathrm{Tr}\,\alpha\psi(\underline{x}'t')\psi^+(\underline{x}\,t) \tag{6}$$

where the proper statistical operator is denoted by α. We also define a chronological operator T such that

$$T(\psi(\underline{x}'t')\,\psi^+(\underline{x}t)) = \begin{cases} \psi(\underline{x}'t')\,\psi^+(\underline{x}t) & \text{if } t' > t \\ -\psi^+(\underline{x}t)\,\psi(\underline{x}'t') & \text{if } t > t' \end{cases} \tag{7}$$

(noting that for Fermions the first line of 7 is always sufficient). Then we define a one-particle Green's function by

$$G\,(x't';xt) = -i\,\langle T(\psi\,(\underline{x}'t')\,\psi^+\,(\underline{x}t))\rangle \tag{8}$$

If the system has spatial invariance and is homogeneous in time then we may use relative coordinates and write

$$G\,(\underline{x}t) = -i\,\langle T(\,\psi(\underline{x}t)\,\psi^+(00)\rangle \tag{9}$$

this definition for G can be shown to be equivalent to that of a Green's function for the Schrödinger equation. That is, for a non-interacting electron gas,

$$(i\hbar\,\frac{\partial}{\partial t} + \frac{\hbar}{2m}\,\nabla^2)\;G(\underline{x}t) = \delta(\underline{x})\;\delta(t) \tag{10}$$

The two particle Green's function is analogously defined

$$G(\underline{x}_1't_1'\;\underline{x}_2't_2';\;\underline{x}_1t_1\;\underline{x}_2t_2) = \langle T(\psi(\underline{x}_1't_1')\psi(\underline{x}_2't_2')\psi^+(\underline{x}_1t_1)\psi^+(\underline{x}_2t_2)\rangle \tag{11}$$

We will also require the single and double Fourier transforms $G(\underline{k}t)$ and $G(\underline{k}\omega)$ where

$$G(\underline{k}t) = \int d^3\,x\,\exp(-i\underline{k}\cdot\underline{x})\;G(\underline{x}t) \tag{12}$$

$$G(\underline{k}\omega) = \int\,dt\,\exp(i\omega t)\;(G(\underline{k}t) \tag{13}$$

As a simple instructive example we consider the case of a non-interacting Fermion gas in the ground state. Expanding $\psi(\underline{x}\;t)$ in the free particle eigenstates we have

$$\psi(\underline{x}t) = \sum_{\underline{k}'} a_{\underline{k}'}\,e^{-i\omega_{k'}\,t}\;e^{i\underline{k}'\cdot\underline{x}} \tag{14}$$

where $\omega_k = k^2/2m$. Using equation (9), and taking the ground state to be $|0\rangle$ we find

$$G(\underline{x}t) = -i\langle 0|T(\sum_{\underline{k}'\underline{k}''} a_{\underline{k}'}\,a^+_{\underline{k}''}\,e^{-i\omega_{k'}t}\;e^{i\underline{k}'\cdot\underline{x}})\,|0\rangle \tag{15}$$

We then may use equation (15) in equation (12) to find

$$G(\underline{k}t) = -i\langle 0|T(\sum_{\underline{k}'\underline{k}''} a_{\underline{k}'}\,a^+_{\underline{k}''}\,e^{-i\omega_{k'}t}\int d^3x\;e^{i(\underline{k}'-\underline{k})\cdot\underline{x}})\,|0\rangle \tag{16}$$

$$= -i\langle 0|T(\sum_{\underline{k}'} a_{\underline{k}'}\,a^+_{\underline{k}'}\,e^{-i\omega_{k'}t}\quad\delta(\underline{k}'-\underline{k})\;|0\rangle$$

$$= \begin{cases} -i\,(1-n_k)\,\exp\,(-i\omega_k t) & t > o \\ i\,n_k\,\exp\,(-i\omega_k\,t) & t < o \end{cases} \tag{17}$$

where the number density of the state k is given by $n_k = a_k^+ a_k = 0,1$. Equation (17) leads further to

$$G(\underline{k}\omega) = \lim_{\delta(k)\to o} 1/(\omega-\omega_k + i\delta(k)) \tag{18}$$

where $\delta_k >(<) o$ for $k>(<)$ the Fermi level, k_f.

We have mentioned above the connection between Green's functions and experimentally measurable quantities. Among the most fundamental of such properties are the spectral densities, or densities of states. The relationship with the Green's functions follows from equation (18). If we define the spectral density $\rho(\underline{k}\ \omega) = \delta\ (\omega-\omega_k)$ the property of the limit in equation (18), $\lim_{\varepsilon\to+o} 1/(x+i\varepsilon)= \mathcal{P}(1/x)-i\pi\ \delta(x)$, implies the relationship

$$\rho(\underline{k}\omega) = -\ (1/\pi)\ \text{Im}\ G(k\omega) \tag{19}$$

and the density of states per unit energy is given by ($\hbar = 1$ everywhere)

$$\rho(\omega)=-\frac{1}{\pi}\ \text{Im}\ \sum_k G(\underline{k}\omega) \tag{20}$$

Another useful example is that of electrical conductivity (and other such transport properties). These properties are related to correlation functions involving four a operators, and in turn to two-particle Green's functions of the sort given in equation (11). For example, the Kubo formula for electrical conductivity is given by

$$\sigma_{\mu\upsilon}(\omega)=\lim_{\eta\to+0} \frac{1}{v}\int_0^\infty dt\ \exp(-i\ \omega t)\ \exp(-\eta t)\int_0^\beta d\lambda\ \langle J_\mu(0)J_\upsilon(t+i\lambda)\rangle \tag{21}$$

where $\underline{J}$ is the current density operator and $\beta = (k_BT)^{-1}$.
For the Hamiltonian in equation (1) (site representation) the z component of J is given by

$$J_z = 2e \sum_{ij} (\underline{R}_i - \underline{R}_j)_z\ t_{ij}\ a_i^+ a_j \tag{22}$$

where we have used the one-band model for simplicity. Thus we must deal with correlation functions of the form $\sum_{keij}\langle a_k^+(t)a_e(t)a_i^+(0) a_j(0)\rangle$ which are related to the two-particle Green's functions.

For (harmonic) Hamiltonians of the form in equations (1) or (2), however, the two-particle Green's functions can always be exactly written as sums of products of two one-particle functions. This can be seen by first considering the equations of motion

$$\frac{i\partial a_i}{\partial t} = [a_i(t), H] = \varepsilon_i a_i + \sum_j t_{ij} a_j \tag{23}$$

which in turn (defining the matrix element)

$$G_{ij} = -i \left\langle T(a_i(t) a_j^+(t)) \right\rangle$$

leads to

$$\frac{i\partial G_{ij}}{\partial t} = 2\pi\ \delta(t) \left\langle [a_i a_j^+]_+ \right\rangle + \varepsilon_i G_{ij} + \sum_l t_{il} G_{lj} \tag{24}$$

Fourier transforming equation (24) leads to a matrix equation

$$\underline{G}\ (\omega) = (\omega \underline{1} - \underline{H})^{-1} \tag{25}$$

which can always be diagonalized in principle. Note that in the case of a disordered system it is prohibitively difficult in general actually to do so, but since quantities like the density of states depend on the trace over $\underline{G}$, these can be evaluated on some other more convenient basis. On the basis in which $\underline{G}$ is diagonal, however, $G_{ij}(\omega) = \delta_{ij}/(\omega - \varepsilon_i)$ and our two-body function may be written

$$\begin{aligned} G_{ijkl} &= \left\langle T(a_i(t) a_j(t) a_k^+(o) a_l^+\ (o)) \right\rangle \\ &= \left\langle a_i^+(t) a_i(o) \right\rangle G_{ij}(t) + \left\langle a_j^+\ (o) a_j(t) \right\rangle G_{ii}\ (t) \end{aligned} \tag{26}$$

As a final illustration of the connection of Green's functions with physical properties we show that the ground-state energy is determined by the one-particle Green's function. From equation (17) we find (for the ground state, $|o\rangle$)

$$n_k = \left\langle o\ | a^+(\underline{k})\ a\ (\underline{k}) |\ o \right\rangle = -i\ G\ (\underline{k},\ t = -o). \tag{27}$$

Using the Hamiltonian of equation (2) we see that

$$\begin{aligned} a^+(\underline{k})\ [H, a(\underline{k})] &= a^+(\underline{k}) \sum_{k'} (E(\underline{k}')(a^+(\underline{k}')[a(\underline{k}'), a(\underline{k})]_+ - \delta(\underline{k}-\underline{k}') a(\underline{k}')) \\ &+ \sum_{k''} T(\underline{k}', \underline{k}'')(a^+(\underline{k}')[a(\underline{k}''),\ a(\underline{k})]_+ - \delta(\underline{k} - \underline{k}')\ a(\underline{k}''))) \end{aligned} \tag{28}$$

thus we find

$$\sum_k a^+(\underline{k})\ [H,\ a_k] = -\ H \tag{29}$$

so that the ground-state energy may be written

$$E(o) = \left\langle o\ |H|\ o \right\rangle = -\sum_k \left\langle o\ | a^+(\underline{k})\ [H, a(\underline{k})]\ | o \right\rangle \tag{30}$$

If the Hamiltonian also includes two-body interactions, then the equivalent of equation (30) (in a diagonalized basis) is given by

$$E\ (0) = 1/2 \sum_k (E(k) - \langle 0|\ a^+(\underline{k})\ [H,\ a(k)]\ |0\rangle\ ). \quad (31)$$

From equation (27) the right-hand side of equation (30) is given by

$$\langle 0|a^+(\underline{k})\ [H,a(\underline{k})]\ |0\rangle = -\left[\frac{dG(kt)}{dt}\right]_{t=-0} = \frac{i}{2\pi} \int dEEG(\underline{k}E).$$

The last step above follows from the definition, equation (13). For up to two-body interactions, then

$$E(0) = \frac{1}{4\pi i} \sum_k \int dE(E(k) + E)\ G(\underline{k}E) \quad (32)$$

III. CONFIGURATION AVERAGED GREEN'S FUNCTIONS IN DISORDERED SYSTEMS

In the case of randomly disordered systems we must calculate some sort of averaged Green's function in order to obtain the physical properties mentioned in the previous section. We shall outline below the various degrees of approximation which have been attempted. These vary considerably in complexity and sophistication depending on the kind of disorder present and the degree of approximation acceptable for describing the elementary excitations. We may begin, however, by describing a form of aperiodicity which has been extensively and successfully tested in the literature, namely that of the single defect (for a review and bibliography see reference 11).

All standard techniques involve some sort of perturbation theory, in which one presumes knowledge of the Green's function of a perfect (periodic) system, G^o, and deviations from this system are described by terms like $\Delta^{\mu\nu}_{ij}$ in equation (3). Then the perturbation expansion about the perfect system (using matrix notation for convenience) is given by

$$\underline{G} = \underline{G}^o + \underline{G}^o\underline{\Delta}\underline{G}^o + \underline{G}^o\underline{\Delta}\underline{G}^o\underline{\Delta}\underline{G}^o + \cdots \quad (33)$$

there are various formal ways of expressing this expansion, among which are

$$\underline{G} = (\underline{1} - \underline{G}^o\underline{\Delta})^{-1}\ \underline{G}^o \quad (34)$$

which leads to the Dyson equation

$$\underline{G} = \underline{G}^o + \underline{G}^o\underline{\Delta}\ \underline{G}. \quad (35)$$

One may also define the $\underline{T}$ maxtrix

$$\underline{T} = \underline{\Delta}\ (\underline{1} - \underline{G}^o\ \underline{\Delta})^{-1} \quad (36)$$

such that

$$\underline{G} = \underline{G}^o + \underline{G}^o\ \underline{T}\ \underline{G}^o \tag{37}$$

For an isolated defect the number of non-zero matrix elements of $\underline{\Delta}$ is small, given by the extent of the defect. Thus the $\underline{T}$ matrix of equation (36) also has the same small number of non-zero elements. The simplest case is that of the single-site diagonal perturbation, for which $\underline{\Delta}$ is diagonal, with a single element, Δ.

Then the $\underline{T}$ matrix is also diagonal, with a single element, $T = \Delta/(1-\Delta G^o(0))$, where the defect is taken to be at the origin, and for a volume Ω,

$$G^{(o)}(0) = G(\underline{x} = 0, E) = \Omega^{-1}\Sigma_k (E-E(\underline{k}))^{-1}. \tag{38}$$

The excitation energies are at the poles of G, which in turn are given at small shifts from the poles of G^o except for a new pole in T, i.e., that given by $G^o(0) \equiv G(0,E) = 1/\Delta$. If Δ is much larger than the band-width, located at E_b, then $G(0) \simeq 1/(E-E_b)$ and $E \simeq E_b + \Delta$, which is clearly an isolated mode shifted out of the band at the defect energy. For small enough Δ the isolated mode disappears and we must solve

$$G_{ij} = G^o_{ij} + G^o_{io}\, G^o_{jo}\, \Delta\ /\ (1-\Delta\, G^o(0)). \tag{39}$$

As the defect covers larger spatial extent we must deal with larger matrices. For a two-site perturbation the $\underline{\Delta}$ matrix is given by

$$\underline{\Delta} = \begin{pmatrix} \delta_1 & \delta_{12} \\ \delta_{12} & \delta_2 \end{pmatrix} \tag{40}$$

and the poles of $\underline{T}$, from equation (36) are given by

$$1-(\delta_1+\delta_2)G^o(0)-2\delta_{12}G^o_{12}+(\delta_1\delta_2-\delta^2_{12})(G^o(0)^2 - G^{o2}_{12})=0. \tag{41}$$

A possible simplification of the model would be a system in which the transfer integrals t_{ij} of equation (1) are varying but the diagonal energies are approximately equal ($\varepsilon_i \approx \varepsilon_j$), in which case we have $\delta_1 \sim \delta_2 \sim \delta_{12}$ and the poles of T are given by

$$1-2\delta_{12}(G^o_{12}-G(0)) = 0. \tag{42}$$

Another possible simplification is one in which there are two different kinds of single-site diagonal perturbations. The poles are shifted from those given by equation (39) and we have

$$1-(\delta_1+\delta_2)G^o(0)+ \delta_1\delta_2\ (G^o(0)^2-G^{o2}_{12}) = 0. \tag{43}$$

Two identical single-site diagonal defects can also interfere, lifting the degeneracy in the isolated modes by the equation resulting from $\delta_1 = \delta_2$ in equation (43):

$$1 - \delta G^{(o)}(0) \pm \delta G^{(o)}_{12} = 0. \tag{44}$$

As the number of impurities, or deviations from an average periodic system, become large, exact solutions of the sort above rapidly become impossible. Isolated levels accumulate to form defect bands, and the bands of the perfect system become considerably changed. Under these circumstances approximate methods must be applied to solving equations like (33)-(37), and some sort of configuration averaged Green's function must be calculated. Of course, for large enough systems which are truly homogeneously random, any actual macroscopic experiment is sampling a configuration average of the system. For other systems which are not truly random, although aperiodic, we may choose to make the approximation of configuation averaging in any case.

For a homogeneously disordered system, the averaged Green's function must have translational symmetry, as does the Green's function for the perfect system, $G^{(o)}$. If we denote the average by $\langle \underline{G} \rangle$ we may define a self energy Σ by $\langle \underline{\Delta}\ \underline{G} \rangle = \Sigma \langle G \rangle$ and from equation (35) write

$$\langle G \rangle = G^{(o)} + G^{(o)}\ \Sigma \langle G \rangle\ . \tag{45}$$

Given the solution for $G^{(o)}$ in equation (18) we find

$$\langle G \rangle = (E - E(\underline{k}) - \Sigma)^{-1}, \tag{46}$$

where, in general, Σ depends on both $\underline{k}$ and E. Depending on our averaging technique we may choose to solve equations in $\langle G \rangle$ or in Σ.

A further insight into the self energy Σ is obtained if we consider the actual perturbation expansion in equation (33) term by term.

$$\langle G_{ij} \rangle = G^{(o)}_{ij} + \sum_k G^{(o)}_{ik} \langle \sum_\ell \Delta_{k\ell} G^{(o)}_{\ell j} \rangle + \sum_{k\ell} G^{(o)}_{ik} \langle \sum_{mn} \Delta_{k\ell} G^{(o)}_{\ell m} \cdot \Delta_{mn} G^{(o)}_{nj} \rangle + \cdots \tag{47}$$

Since the $\underline{G}^{(o)}$ does not depend on the disorder these terms can be removed from the average. Furthermore, terms like $\langle \sum_\ell \Delta_{k\ell}\ G^{(o)}_{\ell j} \rangle$ depend only on k and j and may be written as

$$\equiv \langle \Delta\ (k) \rangle\ G^{(o)}_{kj}\ .$$

Thus we have

$$\langle G_{ij} \rangle = G^{(o)}_{ij} + \sum_k G^{(o)}_{ik} \langle \Delta(k) \rangle\ G^{(o)}_{kj} + \sum_{k\ell} G^{(o)}_{ik} \langle \Delta(k) \Delta(\ell) G^{(o)}_{k\ell} \rangle G^{(o)}_{\ell j} + \cdots \tag{48}$$

There exists a considerable body of literature on a number of schemes for the evaluation of equation (48). Simple truncation of terms in Δ is difficult because of divergencies, term by term, in the expansion. The simplest scheme to avoid this problem is to use the random phase approximation, which ignores terms in the averages in which sites coincide, i.e., we set

$$\langle\Delta(k)\,\Delta(1)\,\Delta(m)\cdots\Delta(f)\rangle = \langle\Delta(k)\rangle\,\langle\Delta(1)\rangle\,\langle\Delta(m)\rangle\cdots\langle\Delta(f)\rangle .$$

Since each term is now site-independent the nth term average is equal to $\langle\Delta\rangle^n$ and we have

$$\langle\underline{G}\rangle = \underline{G}^{(o)}\,(\underline{1} - \langle\Delta\rangle G^{(o)})^{-1} \tag{49}$$

This approximation amounts merely to a simple shift in the energy bands by a constant, Δ, since equation (49) implies that

$$\langle G\rangle = (E - E(k) - \langle\Delta\rangle)^{-1} \tag{50}$$

This so-called virtual crystal approximation (for which $\Sigma = \langle\Delta\rangle$) is successful in cases of very small Δ.

In cases where Δ is not small, we can use a so-called single-site approximation, making use of the T matrix we have already found for isolated defects. We remove all terms in the sums of equation (48) which involve repeated scattering at the same site by defining a single site T matrix by

$$\sum_{k\neq 1} G^{(o)}_{ik}\langle\Delta\,(k)\;\Delta(1)G^{(o)}_{k1}\rangle G^{(o)}_{1j} = \sum_{k\neq 1} G^{(o)}_{ik}\langle T(k)T(1)G^{(o)}_{k1}\rangle G^{(o)}_{1j}, \tag{51}$$

the T (k) are given by $T=\Delta/(1-\Delta G^{(o)}(0))=T_1$, as above at all sites at which there is non-zero deviation, Δ, and are zero elsewhere. Again, as in the virtual crystal approximation, the $G^{(o)}$ terms can be removed from the averages, and we may approximate

$$\langle T(k)T(1)T(m)\ldots T(f)\rangle \sim \langle T(k)\rangle\,\langle T(1)\rangle\,\langle T(m)\rangle - \langle T(f)\rangle \sim \langle T\rangle^n .$$

We must also define an unperturbed function by $\underline{G}^{(o)'} = \underline{G}^{(o)} - G^{(o)}(0)\,\underline{1}$ so that we have

$$\langle\underline{G}\rangle = \langle\underline{G}^{(o)}\rangle + \underline{G}^{(o)}\langle T_1\rangle\underline{G}^{(o)} + \underline{G}^{(o)}\langle T_1 G^{(o)'} T_1\rangle\underline{G}^{(o)} + \underline{G}^{(o)}\langle T_1\underline{G}^{(o)'} T_1\underline{G}^{(o)'} T_1\rangle \cdot\underline{G}^{(o)} + \ldots \tag{52}$$

From this expansion we get the "average T matrix approximation"

$$\langle\underline{G}\rangle = \underline{G}^{(o)} + \underline{G}^{(o)}\,(\langle T_1\rangle\,(\underline{1} - \underline{G}^{(o)'}\,\langle T_1\rangle)^{-1})\,\underline{G}^{(o)} \tag{53}$$

From equation (37) this gives an effective T matrix of the system, in terms of the isolated defect T matrix, T_1, by

$$\underline{T} = \langle T_1 \rangle \, (\underline{1} - \underline{G}^{o'} \langle T_1 \rangle)^{-1}. \tag{54}$$

From the definition of the self-energy Σ in equation (45) we can relate the T matrix to Σ by

$$\underline{\Sigma} = \underline{T} \, (\underline{1} + \underline{G}^{o} \, \underline{T})^{-1}, \tag{55}$$

from which, in the average T matrix approximation, we have

$$\Sigma = \langle T_1 \rangle / (1 + \langle T_1 \rangle \, G^{o} \, (0)). \tag{56}$$

If a fraction f of the sites deviates from average we have

$$\langle T_1 \rangle = f\Delta / (1 - \Delta G^{o}(0)), \tag{57}$$

which gives us

$$\Sigma = f\Delta / (1-f)\Delta G^{o}(0)). \tag{58}$$

The idea of the single-site approximation can be extended, with the concept of an effective medium, to a self-consistent approximation which has come to be known as the coherent potential approximation (CPA). The fundamental physical idea has by now a long history, and has been applied to a large variety of physical problems (for example in the molecular field theory approach to magnetic problems). There are also a number of different mathematical approaches to the fundamental equation of the CPA. We make no attempt at historical accuracy or mathematical rigor here. Extensive bibliographies and reviews are found in references 4 and 5. The physics of the approximation is perhaps simplest stated for the case of a two-component alloy (substitutional randomness). One wishes to replace the average of the A, B alloy by an effective medium determined such that the average fluctuation through the medium is zero. More specifically one replaces each site but one with an unknown coherent potential. One then imbeds at the special site an A or a B component with respective probabilities f and 1-f. One then solves the problem of this single impurity imbedded in an effective medium characterized by the coherent potential. This coherent potential, on the other hand, is determined by the self-consistency requirement that the average scattering (or fluctuation) from the special site is also zero. This almost simple-minded physical picture is also obtainable from rigorous mathematical techniques in perturbation theory, or multiple scattering methods, diagrammatic techniques, so-called cumulant expansion methods, etc.

We may sketch the basic method by postulating an effective medium Green's function G_ℓ such that equation (45) for the averaged Green's function is obeyed, i.e.

$$\underline{G}_{\ell} = \underline{G}^{(o)} + \underline{G}^{(o)} \underline{\Sigma}\, \underline{G}_{\ell}. \tag{59}$$

We may then solve for the unperturbed Green's function in terms of $\underline{G}_{\ell}$,

$$\underline{G}^{(o)} = \underline{G}_{\ell} (\underline{1} + \underline{\Sigma}\underline{G}_{\ell})^{-1} \tag{60}$$

and insert it in equation (35) for the actual Green's functions of the system, G,

$$\underline{G} = \underline{G}^{(o)}(\underline{1} + \underline{\Delta}\underline{G}) = \underline{G}_{\ell}(\underline{1} + \underline{\Sigma}\underline{G}_{\ell})^{-1}(\underline{1} + \underline{\Delta}\underline{G}). \tag{61}$$

Using this one may arrive at an equation analogous to (35), but with the unperturbed propagator being G_{ℓ} of the effective medium, rather than $G^{(o)}$. From equation (61) we have

$$(\underline{1}+\underline{\Sigma}\underline{G}_{\ell})\, \underline{G}_{\ell}^{-1}\underline{G} = \underline{1} + \underline{\Delta}\underline{G} = \underline{G}_{\ell}^{-1}\, \underline{G} + \underline{\Sigma}\underline{G} \tag{62}$$

which gives the required effective medium equation

$$\underline{G} = \underline{G}_{\ell} + \underline{G}_{\ell}\, \underline{\Delta}_{\ell}\, \underline{G}, \tag{63}$$

where the scattering perturbation on the effective medium, $\underline{\Delta}_{\ell} = \underline{\Delta}-\underline{\Sigma}$ is given by $-\Sigma$ at unperturbed sites of the original system, and $\Delta-\Sigma$ at impurity sites of the original system. One can now find the average single site T matrix for this system, analogously to equation (57) for $\langle T_1\rangle$. For the present case $\langle T_1\rangle$ is found by taking the isolated site to have a probability f of being an originally perturbed site (perturbation from the effective medium of $\Delta-\Sigma$) and a probability 1-f of being an originally unperturbed site (perturbation from the effective medium of $-\Sigma$). Thus we have

$$\langle T_1\rangle = (1-f)(-\Sigma)/(1+\Sigma G_{\ell}^{(o)})+f(\Delta-\Sigma)/1-(\Delta-\Sigma)G_{\ell}^{(o)}). \tag{64}$$

Now the fundamental approximation of the CPA can be made in the form of setting $\langle T_1\rangle$ in equation (64) equal to zero. By definition of the effective medium the average scattering by any portion of it should be zero--the CPA approximates this by setting the average scattering from a single site to zero. This leads to the self-consistency equation mentioned above. By multiplying equation (64) by the denominator of the second term and writing the equation over the common denominator $1 + \Sigma G_{\ell}{}^{(o)}$ we obtain

$$\Sigma = f\Delta/(1+(\Sigma-\Delta)G_{\ell}(0)) \tag{65}$$

which is one form of the CPA equation.

We may get further insight into the CPA by considering equation (65) in certain limits in which the approximation is exact. In the limit of extremely dilute impurities, for example, $f \ll 1$, and thus $\Sigma/\Delta \approx f \ll 1$ and $\underline{G}_{\ell} \approx \underline{G}$, we obtain

$$\Sigma \approx f\Delta/(1 - (1 - f)\, \Delta G^{o}(0)) \tag{66}$$

which is merely the equation for the effect of single isolated independent impurities(see equation (58))in an otherwise periodic system. For very small deviations from the perfect, i.e. $\Delta/E(0) \ll 1$ system, we obtain from equation (65) that $\Sigma \approx f\Delta$ which is merely the virtual crystal approximation (see equation (50)).
One may continue to expand in powers of Δ in this weak coupling limit. For example, if the Green's function of the first order expansion is given by

$$G_1 = (E - E(k) - f\Delta)^{-1} \tag{67}$$

then the next order gives

$$\Sigma \approx f\Delta + f(1-f)\Delta^2\ G_1 \tag{68}$$

and so forth. If one goes to the split-band or atomic limit (zero band width) and can also show that equation (65) leads to the exact atomic Green's function.[5, 6] The fact that these various limits are contained exactly in the CPA becomes obvious in the light of the recent demonstration that the CPA is the best possible single site self-consistent approximation for the configuration averaged Green's function.

Attempts to extend or generalize the averaging of the Green's function to include multiple site scattering have been made increasingly often over the past few years. In particular a number of forms of including pair scattering have been produced. Thus far the success of these extensions must be considered as limited, especially since they result in extremely difficult integral formulas which are far harder to use in practical calculation than those of the CPA. Bibliographies for these cluster theories can be found in the previously cited review articles.

IV SUMMARY AND DISCUSSION.

As we mentioned at the outset, we have chosen to stress in our examples the types of disorder more likely to be of use in the treatment of aperiodic polymers . For example we have not specifically mentioned structural disorder in section III, choosing our examples from various types of compositional disorder. This should not be taken to imply that treatment of structural disorder has not been equally vigorous in the literature.

Methods analogous to the CPA of section III, and other simplified practical single-site approximations have been developed for the treatment of amorphous solids and liquids. A detailed description of these techniques is out of place in these talks. Suffice it to say that averages equivalent to those described in section III may be performed for such systems with the aid of knowledge of pair correlation functions for the systems.

The success of the application of any of these techniques to aperiodic polymers must of course lag behind the applications to amorphous solids, metallic alloys, liquid metals, etc. simply because of the difficulty of obtaining an accurate description of the elementary excitations of the perfect (periodic) system. In turn, even the simplest of monomers is a much more difficult object to consider as a repeating unit than those treated by solid state physicists to date. But as calculations of the monomers and hence periodic polymers improve, applications to aperiodic polymers will follow, since the treatment of the aperiodicity <u>per se</u> is of no greater difficulty than in simpler random materials. As an example we may mention the work in progress of Prof. J. Ladik and coworkers on $(SN)_x$ polymers in which they are applying the CPA to $(SN)_x$ with hydrogen impurities.[14]

As we have also mentioned above, we have chosen to ignore the special property of disorder-caused localization, usually called Anderson localization.[15] In 1958 Anderson published his paper on 'the absence of diffusion in certain random lattices'. He showed that if an electron moves in a particular kind of disordered array, at a certain critical value of the disorder the solutions for any energy in the band are no longer extended Bloch-like states but are localized in space. Propagation for such electrons can occur only by thermally activated hopping. It is now recognized that in many different kinds of disordered systems we must also deal with anomalies in the current carrying properties of the states which are wholly or partially of the Anderson type. Such anomalies can also occur for other transport properties as well. Clearly Anderson localization might be of extreme importance for some disordered polymers. We have chosen to ignore it nonetheless because it is a phenomenon that is as yet imperfectly understood even for simple systems and because it remains an object of some considerable controversy. For example a number of workers agree that although the averaged single-particle Green's function directly describes the density of states, it does not describe transport properties,[16] for which two particle Green's functions are required. Other workers, on the other hand, have derived "localization criteria" based on the single particle Green's function. Suffice it to say that the subject is an active and exciting one, and promises to remain so for a while. The existence of a definite mobility edge where the states change from a localized to an extended character has been

well established experimentally by now - perhaps most definitely in the work on field effect transistors.[17]
A thorough and complete understanding of the phenomenon awaits to reward further experimental and theoretical work.

*Supported in part by the City University of New York Faculty Research Award Program.

REFERENCES

1. M. Lax in "Stochastic Differential Equations", SIAM-AMS Proceedings, Vol. VI, pp. 35-95 (1973).
2. R. Bell, Rep. Prog. Phys. 35, 1315 (1972).
3. P. Dean, Rev. Mod. Phys. 44, 127 (1972).
4. N.F. Mott and E.A. Davis "Electronic Processes in Non-Crystalline Materials" (Clarendon, Oxford) (1971).
5. F. Yonezawa and K. Morigaki, Prog. Theor. Phys. Suppl. 53 (1973).
6. A.J. Elliott, J.A. Krumhansl and P.L. Leath, Rev. Mod. Phys. 46, 465 (1974).
7. A.A. Abrikosov, L.P. Gorkov, E. Dzyaloshinsky, "Methods of the Quantum Theory of Fields in Statistical Physics", Prentice-Hall, Englewood Cliffs, New Jersey, 1963.
8. L.P. Kadanoff and G. Baym, "Quantum Statistical Mechanics", Benjamin, New York, 1962.
9. P. Nozieres, "Le Problème a N corps", Dunod, Paris, 1963.
10. D. Pines, "Many Body Problem", Benjamin, New York, 1961.
11. A.A. Maradudin, in "Solid State Physics", ed. by F. Seitz and D. Turnbull (Academic, New York) Vol. 18-19, 1966.
12. R.J. Elliott and D.W. Taylor, Proc. R. Soc. Lond. A 296, 16 (1967).
13. F. Yonezawa, L. Roth and M. Watabe, J. Phys. F: Metal Physics 5, 435 (1975).
14. J. Ladik, Progress report from Lehrstuhl für Theoretische Chemie, Erlangen (1977).
15. P.W. Anderson, Phys. Rev. 109, 1492 (1958).
16. D.J. Thouless, J. Phys. C 3, 1559 (1970).
17. F. Yonezawa, Y. Ishida, F. Martino, S. Asano, Inst. Phys. Conf. Ser. 30 (1977).
18. N.F. Mott, M. Pepper, S. Pollitt, R.H. Wallis and C.J. Adkins. Proc. R. Soc. Lond. A 345, 169 (1975).

ELECTRON STATES OF STRUCTURALLY-DISORDERED CHAINS

W.L. Mc CUBBIN

Department of Physics
Arya-Mehr University
Isfahan, Iran

1. INTRODUCTION

The great majority of calculations of the energy spectra of polymers have been performed on ideal chains. That is to say, they are the band structure calculations on extended periodic chains described by André and Delhalle in this volume.

Now the chains in real polymer specimens are remarkably unwilling to arrange themselves in this manner and it is not obvious that the ideal band structures have any quantitative connection with the properties of actual materials. In the case of polyethylene, for example, a close agreement has been claimed with experiments on photoelectron spectroscopy (1). It is not our present purpose to assess the weight of this claim, but it is worth pointing out that a close quantitative agreement between the theoretical and experimental density of states would present us with the problem of explaining why chain folding and the considerable randomisation of atomic positions that occurs even within spherulites should have such a small effect on the eigenvalue spectrum. Furthermore, it would be important to understand the circumstances in which the effects of randomisation were <u>not</u> small.

The quantum-mechanical properties of structurally-disordered systems (the only kind we shall be concerned with) can be pursued in two directions :

a) the energy spectrum,via a density of states function N(E) and
b) the nature of the eigenfunctions via a range of localisation function R(E).

Success in either direction is strongly circumscribed by the absence of a theorem with the simplifying power of the Bloch-Floquet

J.-M. André et al. (eds.), Quantum Theory of Polymers, 185-198.

theorem. One consequence of this has been that one has often been obliged to use numerical methods in which one hopefully generates a typical segment of the infinite system and then computes its properties exactly. It is to be expected, however, that as experience with Green's function methods grows and our knowledge of the properties of random functions improves powerful analytic methods equivalent to those used for periodic solids will emerge.

The fact that polymer chains possess many topological properties of the line suggests that it might be expedient to seek to map the linear chain problem onto a strictly one-dimensional one. However, it is well known that 1D theory can give not only quantitative errors, but also qualitatively wrong descriptions of certain physical processes. For example, the possibility of an electron to "walk round" a defect does not exist in one dimension. In the past the most common ways of dealing with this dilemna have been either to put all objections aside and study 1D arrays as interesting in their own night or to regard 1D theories as completely irrelevant and study some idealised (usually over-idealised) 3D model. I should like to point out (a) that non-trivial generalisation of the theory of 1D arrays is sometimes possible [(2)] and (b) in the localisation problem it is the longitudinal motion that is of primary interest; the lateral decay of the wave function is physically irrelevant. Intuitively, therefore, we might expect that an appropriate 1D mapping of the linear chain problem would be at least qualitatively correct.

In the following section we shall discuss the basic effects of disorder on the energy bands and then proceed to describe a computation on rotationally-disordered polyethylene. In the final section we shall discuss some of the difficulties that obstruct our understanding of localisation and then proceed to describe a computation of R(E) for a 1D potential derived from polyethylene.

2. EFFECT OF STRUCTURAL DISORDER ON ENERGY BANDS

2.1. *Basic concepts*

It was not universally realised that band broadening and band tailing are distinct phenomena that arise from different kinds of structural disorder. It is therefore worthwhile to give a simple explanation based on the Schrödinger equation for an array of δ-function potential wells,

$$\frac{-d^2\psi}{dx^2} + V(x)\psi = E\psi \qquad (1)$$

where

$$V(x) = -\mu\sum_{x_i} \delta(x-x_i)$$

and the $x_{ij} \equiv x_j - x_i$ are independent random variables.

Quite generally, we can describe the relationship between the amplitudes and phases at two points on the array by a transfer matrix T connecting the state vectors :

$$\Psi(x_2) = T\,\Psi(x_1) \tag{2}$$

where

$$\Psi(x) = \begin{pmatrix} \psi\ (x) \\ \psi'(x) \end{pmatrix}$$

In any region where the potential $V =$ constant , we can write $\psi(x)$ as a superposition of plane waves :

$$\psi(x) = A\,e^{ikx} + B\,e^{-ikx} \tag{3}$$

where

$$k = [2\,m\,(E - V)]^{1/2}$$

We can then define a more useful transfer matrix (for computational purposes) relating <u>coefficients</u> :

$$\begin{pmatrix} A(x_2) \\ B(x_2) \end{pmatrix} = M \begin{pmatrix} A(x_1) \\ B(x_1) \end{pmatrix} \tag{4}$$

It is easily shown that $M = R^{-1}\,TR$ where

$$R = \begin{pmatrix} 1 & 1 \\ ik & -ik \end{pmatrix}$$

Note that this holds for any potential $V(x)$ connecting two points on the array. In the special case of the δ function well the change in amplitude and phase in crossing the well is contained in the transfer matrix

$$M_\delta = \begin{pmatrix} 1-i\lambda & -i\lambda \\ i\lambda & 1+i\lambda \end{pmatrix} \quad ; \quad \lambda = -\frac{\mu}{2k}$$

The phase change caused by the δ-function may clearly result in a change in the number of nodes in ψ leading up to the next well. The relevance of this lies in the fact that the number of nodes per unit length determines the integrated density of states, by means of the following theorem.

<u>*Theorem 1*</u> : If $\alpha(E_1)$ is the number of nodes of a real solution $\psi(x;E_1)$ of the Schrödinger equation in the open interval $0 < x < L$,

then $\alpha(E_1)$ is the number of eigenvalues with energy $-\infty < E < E_1$ for the system of length L.
Counting nodes thus yields an integrated density of states D(E) to which the usual density N(E) is related by N(E) = d D(E)/dE.

Now consider a very long sequence of wells. These will, in circumstances discussed later, exist an energy E_1 such that between E_1 and $E_1 + dE_1$ no new nodes are introduced into the wave function. Clearly E_1 must exist in a band gap. At other energies, the change may be large or small, the value determining the band curvature in that energy interval. (In case this all seems excessively simple-minded, I should mention that some results on the node distributions of random functions have recently been incorporated by Lukes[3] into an exact theory of the density of states in 3D disordered systems).

It is sometimes useful to make a distinction between short-range disorder (SRDO) and long-range disorder (LRDO). Suppose SRDO is incorporated into a 1D array by means of a small constant parameter ε such that successive site separations are given by

$$u_n = a\,(1 + \varepsilon\gamma_n)$$

where $a = \langle x_{n+1} - x_n \rangle$ and γ_n are random numbers. It can be shown that LRO disappears, ie that the uncertainty in the position of the nth site exceeds a when $n > \frac{1}{\varepsilon^2}$. Clearly one may have SRDO while retaining LRO.

The point in making this distinction between SRDO and LRDO is that the effects of each on the eigenvalue spectrum can be separately assessed. An example is the work of Makinson and Roberts[4] on a δ-function array when the spacings u_n were independent random variables with the common distribution

$$f(u) = \frac{3\sqrt{5}}{20\sigma}\left(1 - \frac{(u-1)^2}{5\sigma^2}\right)$$

where $(1-\sigma\sqrt{5})a \leqslant u \leqslant (1+\sigma\sqrt{5})a$

$$\langle u_n \rangle = a \;,\; \mathrm{Var}(u) = \sigma$$

The thick line in fig.(1a) shows the gap between two bands when both SRDO and LRDO were present. As SRDO was progressively smoothed out by averaging over a cluster of cells D(E) retained a tail but retreated to the periodic form. Fig.1b shows the effect of increasing SRDO while retaining long range order (no LRDO). Clearly "disorder" and "band tailing" are not necessarily synonomous.

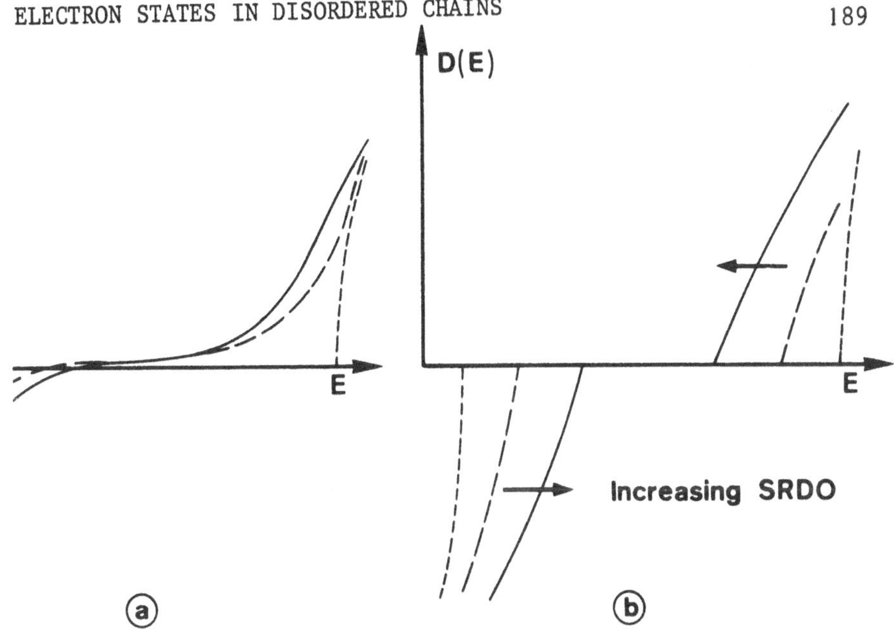

Fig. 1 : Integrated density of states for a disordered array.
In (a) there is LRDO as well as SRDO.
In (b) there is SRDO only.

2.2. *Effect of rotational disorder on the energy bands of polyethylethylene*

Changes of conformation of the polyethylene chain will most readily occur by means of rotations about carbon-carbon bonds. The exercise that suggested itself to us was by how much do the energy bands of the extended periodic chain change when sujbjected to random rotations about main-chain bonds.

This is a difficult problem which, as far as I know, has not been previously attempted. In order to make progress it was necessary to resort to a trick. We decided to exploit the notion that a piece of disordered material of length L is but one (giant) cell in an infinite periodic system. Intuitively we expected that the mistake thus made regarding the boundary conditions would become negligible for sufficiently large L.

In the case of polyethylene we decided that the smallest useful giant cell should contain 6 CH_2 groups. The largest value depended, of course, on the sophistication of the eigenvalue program and the computers available; in our case it turned out to be $12CH_2$ groups. A random number generator produced sets of rotations that were fed into CORGEN, a coordinate generator which also tested whether the final carbon atom in the cell was close enough to the chain axis for the formation of a periodic chain with the correct C-C bond length. The distance between the first and the (M+1)th carbon atom defined the cell length.

This data was then fed into our Extended Hückel program.[5] (A more elaborate program would have incurred the penalty of a severe restriction in the maximum value of M. Furthermore, we were looking for energy shifts and not absolute values).

The rms value of the random rotations, θ_{rms}, was taken as a measure of the degree of disorder. Clearly, this was not as specific a variable as one might have wished since a given θ_{rms} could correspond to chains with quite different distributions of near-neighbour H atom separations (note that bond rotations leave nearest and next-nearest neighbour carbon atom separations unaltered). We were thus obliged to average over small intervals of θ_{rms} in order to obtain the variation of band parameters with degree of disorder. In Table 1 the shift value given for $\overline{\theta}_{rms}$=0.125 is really an average over values for the range 0.10 to 0.15 in θ_{rms}, and similary for the others.

M	$\overline{\theta}_{rms}$			
	0.125	0.175	0.225	0.275
6	0.136	0.265	0.280	0.651
8	0.189	0.064	0.159	0.139
10	0.072	0.140	0.339	0.407
12	0.050	0.706	1.780	-

Table 1 : Upward shifts (eV) of the uppermost valence state for different sizes of giant cell (M) and different degrees of disorder ($\overline{\theta}_{rms}$).

We had hoped that the results for a given $\overline{\theta}_{rms}$ would show an agreable monotonic variation with M, indicating perhaps some asymptotic value. However, the true situation is revealed by an examination of the resulting lengths of the unit cell. In Table 2 we compare the size of the corresponding cell of the fully <u>trans</u> chain (RTF_{cryst}) with the smallest value for the chains investigated.

M	$RTF_{cryst.}$	$RTF_{min.}$	ΔRTF
6	7.620	7.438	-0.182
8	10.160	9.752	-0.408
10	12.700	12.324	-0.376
12	15.240	14.135	-1.005

Table 2 : Comparison of giant cell lengths RTF(Å) for regular and disordered chains.

We see that the M=12 chains have taken advantage of the freedom to achieve a much denser packing. Clearly, in an actual specimen the energy available to drive this process is limited and at present we do not know how closely the degree of disorder assumed here represents that found in real material. However, it seems reasonable to claim that valence edge shifts $\gtrsim$ 1eV will occur in the disordered regions of a polyethylene specimen.

Because of its monatomic main chain and its lack of large side groups, polyethylene must be one of the polymers least sensitive to disorder. Therefore it would appear that structural disorder will have to be taken seriously in future attempts at quantitative comparison of theory and experiments.

3. ELECTRON LOCALISATION IN ID DISORDERED SYSTEMS

3.1. *Basic concepts*

Reference has already been made to the difficulty one has even in defining a range of localisation, let alone calculate it. A discussion will be found in ref.6. Here we simply point that some localisation criteria make no attempt to provide a prescription for calculating the range of localisation. For example, the Anderson-Lehmann criterion requires $\langle\langle |A_s(t)|^2 \rangle\rangle \to 0$ as $t \to \infty$ whenever the amplitude on site s at time 0, $A_s(0) \neq 0$. As is well-known Anderson [7] derived a condition for this to be satisfied for states at the centre of the band and hence, by implication for all states in the band. Fascinating as this problem is, even a rigorous solution would not provide for us an expression for the range of localisation.

Other criteria do not provide a numerically precise concept. For example, the Thouless criterion [8] states that a change to anti-boundary conditions ,

$$u\ \psi\ (x+L_1,\ y+L_2,\ z+L_3) = -\ \psi(x,y,z)$$

should leave the energy eigenvalue unchanged if the state is localised within L. How small a charge counts as zero in actual computations ?

The prescriptive definition that applies most naturally to a 1D system goes as follows. If the envelope of the wave function decays, on average, exponentially from some point x_o according to

$$y = C\ e^{-\lambda|x-x_o|} \qquad (5)$$

then $\frac{2}{\lambda}$ is a measure of the range of localisation, R. There is, of course, a rather-fundamental objection to this definition, viz how does one know that instead of decaying more or less symmetrically from x_o the envelope of the wave function does not spend some time oscillating randomly. To deal with this objection we refer to the calculations of Papantriantafillou and Economou [9]. They calculated an effective range of localisation L_{eff} for the case of substitutional disorder by finding the number of sites contributing to the eigenfunction $|E\rangle$. If the tight binding bandwidth is denoted by 4V and the width of the distribution of site energies ε_i by 2α, they showed that $L_{eff} \approx R = \frac{2}{\lambda}$ when $\alpha \gtrsim 2V$ (large degree of disorder).

If one is working in this limit, the above definition of R becomes respectable and from this point on the theory becomes quite rigorous, thanks to the following theorem.

Furstenberg's limit theorem for non-commuting random variables [10]

Let SL(m,r) be a group of m-dimensional unimodular matrices transforming the real vector space R^m into itself, μ be a measure on SL(m,R), G be the smallest closed subgroup of SL(m,R) containing a support of μ and $x_1, x_2, \ldots x_n$ denote the sequence of G-valued random variables with the common distribution μ . Then, if G be a non-compact subgroup of SL(m,R) such that no subgroup of G of finite index is reducible, $\|x_n \ldots x_1 \omega\|$ grows exponentially as $n \to \infty$ with probability 1 for all ω except for the zero vector.

The relevance of this theorem (the conditions of which are satisfied for disordered chains [11]) is immediately evident when we identify the x_k with the transfer matrices M_k and ω with the initial vector amplitude $\binom{A_1}{A_2} = u$, say. The theorem thus states that for $u \neq \underline{0}$,

$$\lim_{n\to\infty} \frac{1}{n} \ln \| M_n M_{n-1} \; .... \; M_1 u \| = \gamma > 0$$

Now the quantity within bars is just the amplitude vector after n cells, viz. $\binom{A_{n+1}}{B_{n+1}}$. Hence we may write

$$\lim_{n\to\infty} \frac{1}{n} \ln \left[|A_{n+1}|^2 + |B_{n+1}|^2 \right]^{1/2} = \gamma > 0$$

Since $n \simeq n+1$ for large n, a plot of $\ln [|An|^2 + |Bn|^2]$ against n will be a straight line of slope 2γ. Synthesising this and equation (5),

$$R = \frac{2}{\lambda} = \frac{2\langle u_n \rangle}{\gamma} \quad \text{where } \langle u_n \rangle = \text{average cell length} \qquad (6)$$

3.2. *Electron localisation in a 1D potential derived from polyethylene*

The work to be described can be seen either as a rather primitive attempt to say something about the electron states of disordered polymer chains or simply as a treatment of electron localisation in a certain class of 1D potentials.

The method contains a number of steps, which I shall enumerate for clarity.

(1) Definition of the potential $V_{eff}(x)$. The actual one-electron potential of the chain is replaced by an effective potential $V_{eff}(x)$ defined as the total Coulomb potential along the line joining centres of successive C-C bonds. The electronic configuration of each carbon and hydrogen atom was taken to be that of the isolated neutral atom.

$$V_{eff}(x) = \sum_{i=-N}^{N} \sum_{j=1}^{3} \{ V_{nuc}(x-x_{ij}) + V_{elec}(x-x_{ij}) \} \qquad (7)$$

where i labels the contributing CH_2 group and j runs over atoms in the group. It is known from band structure work that the probability amplitude of conduction electrons is low close to the chain axis; consequently the potential (3) will exaggerate the potential energy variation actually experienced by the conduction electron. The assumption of neutral atom charge distributions introduces an error in the same direction. The result (in the case of disordered chains is to over-localise electrons and thereby lead to a lower bound to the true range of localisation).

(2) Definition of the analogue potential $V_a(x)$. Even with these simplifying assumptions it was still not possible to compute the potential for very long chains required in a reasonable time. It was observed, however, that in short chains (say 50 CH_2 groups) the variation of $V_{eff}(x)$ between successive turning points V_{max} and V_{min} was given to close approximation by the simple trigonometric form :

$$V_a(x) = V_{max} - W_1 \left(\sin^2 \frac{\pi x}{\lambda_1} - \frac{1}{10} \sin^2 \frac{2\pi x}{\lambda_1} \right) \quad , \; 0 < x < \frac{\lambda_1}{2}$$

$$V_{min} - W_2 \left(\sin^2 \frac{\pi(x-\lambda 1/2)}{\lambda_2} + \frac{1}{10} \sin^2 \frac{2\pi(x-\lambda 1/2)}{\lambda_2} \right) , \quad \lambda 1/2 < x < \frac{\lambda_1+\lambda_2}{2} \tag{8}$$

where W_1, W_2, λ_1 and λ_2 are parameters to be determined.

An analogue potential may then be generated in a reasonable time for very long chains (typically, 12×10^4 CH_2 groups) from a sequence of potentials (8). The resulting analogue potential possessed all the statistical information of the effective potential $V_{eff}(x)$ leading to a lower bound for R.

(3) Calculation of electron amplitudes. Firstly, divide the array into cells (arbitrarily) at the maxima. If the amplitude in the first cell are given by the state vector $\binom{A_1}{B_1}$ then the amplitudes in the nth cell are given by

$$\begin{pmatrix} A_n \\ B_n \end{pmatrix} = \prod_{k=N-1}^{1} M_k \binom{A_1}{B_1} \tag{9}$$

where M_k is the transfer matrix in the kth cell.

Although Borland[12] showed the formal equivalence of the atom-like potential well V(x) and the δ -function well and also gave the transfer matrix for the δ function, he gave no method for calculating the transfer matrix of the wave general potential. In this work we use the method of Mc Cubbin and Teemul[2] for calculating the M_k.

(4) Calculation of the range of localisation at energy E. Given V(x), the energy E fixes $k = [2m(E-V)]^{1/2}$. This permits the calculation of amplitudes in the nth cell from (9), and hence provides a plot of $\ln\{|An|^2 + |Bn|^2\}$ versus n. The slope γ provides R from (6).

The actual computations involved the following steps.

(a) Introduction of disorder. A regular polyethylene chain was subjected to random changes in successive C-C band lengths and CCC bond angles with no deviations from planarity and no corresponding charges in C-H bonds. The probability densities of the bond length and bon angle deviations were uniform in both cases within limits of 1% in bond length and 20% in bond angle relative to the crystal values.

(b) Testing the accuracy of the analogue potential. V_a and V_{eff} were both calculated at 192 points on a chain of 13 CH_2 groups. The standard deviation of V_a-V_{eff} was 0.00063.

(c) Determination of the parameters W_1, W_2, λ_1 and λ_2. The potential energy is maximum at the centre of each C-C bond. Thus a calculation of $V_{eff}(x)$ at its tuning points for a chain of 501 carbon atoms yielded 1000 values of W and λ. The W and λ distributions were found to be cut-off normal distributions with parameters given in Table 3.

	Max	Min	Mean	St.Dev.
W	1.2700	0.6299	0.9054	0.1052
λ	2.5196	2.2766	2.3992	0.0630

Table 3 : Parameters defining the W and λ distributions (Hartree a.u.)

(d) Generation of disordered chains. Long chains (12,000 carbon atoms) were generated by selecting independently and at random values of W and λ according to the distributions defined in Table 3.

(e) Computation of transfer matrices and plots of $\frac{1}{2}\ln\{|An|^2 + |Bn|^2\}$ vs n. For energies in the band tails, plots showed small-amplitude long-wavelength oscillations about a straight line as found by Bush[13]. That was the reason for producing such long chains. For energies near the centre of the band, the scatter of points was found to be much greater, as shown in Fig.2. However straight lines obtained by least squares fit (higher values being given a greater weighting) still appeared meaningful. From the gradient γ , R was computed for the chosen value of E. Repetition of the last two steps for a number of energies resulted in the function R(E).

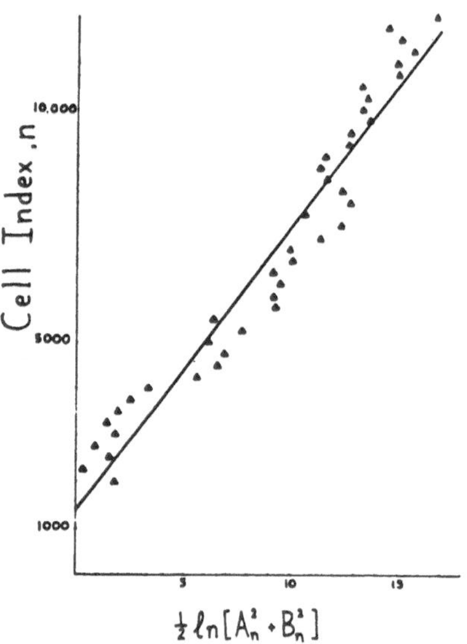

Fig. 2 : Plot of vs.n for E = 10.5 eV.

The function R(E) obtained in the above manner is shown in Fig.3. It shows the expected smoothly-varying tails, since this system as LRDO, but also a quite unexpected oscillating behaviour. Several different chains with the same statistical parameters were investigated and the points were reproduced. The error estimates indicated by the error bars were obtained by varying the starting phase. We note in passing that the errors obtained by Bush lay between 20% and 40%, whereas our largest error was 23% at 10 eV. To the best of our knowledge, therefore, these oscillations are not artefacts of the method.

This finding appears not to have been previously reported and still awaits a satisfactory explanation.

ACKNOWLEDGMENTS

I wish to express my warm thanks to Prof. J.M. André and to Prof. J. Ladik for the opportunity of attending this Summer School. The computational work that produced the new results quoted in the text was carried out mainly by L.L. Mosely of the University of the West Indies, Trinidad, E. Stuart of the University of Wales, Cardiff and A.U. Tang Kai of the Quantum Chemistry Group, Uppsala.

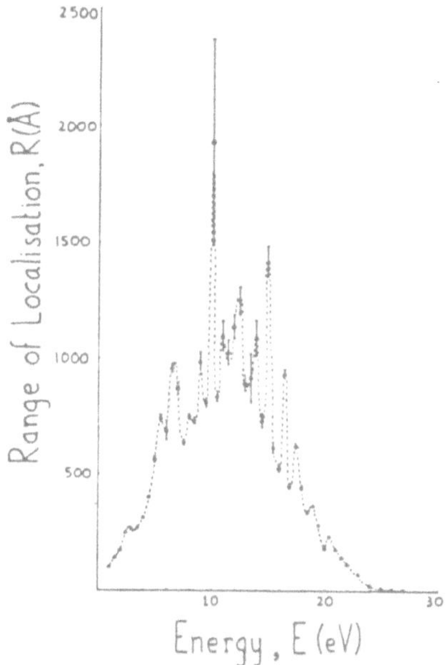

Fig. 3 : Range of localisation as a function of energy for disordered chains with parameters given in Table 1. Points without error bars have a standard deviation of <5%.

REFERENCES

1. J. Delhalle, J.M. André, S. Delhalle, J.J. Pireaux, R. Caudano and J.J. Verbist, J. Chem. Phys., 60, 595 (1974).
2. W.L. Mc Cubbin and F.A. Teemul, Phys. Rev., A6, 2478 (1972).
3. T. Lukes, private communication.
4. R.E.B. Makinson and A.P. Roberts, Australian J. Phys., 13, 437 (1960).
5. W.L. Mc Cubbin and R. Manne, Chem. Phys. Letts., 2, 230 (1968).
6. W.L. Mc Cubbin, Chap.4 in "Transfer and Storage of Energy by Molecules : Vol.4, The Solid State", G.M. Burnett (Ed.), Wiley (1974).
7. P.W. Anderson, Phys. Rev., 109, 1492 (1958).
8. J.T. Edwards and D.J. Thouless, J. Phys., C5, 807 (1972).
9. Papatriantafillou and E.N. Economou, Phys. Rev., B13, 920 (1976).

10. H. Furstenberg, Trans. Amer. Math. Soc., 108, 3 (1963).
11. K. Ishii, Progr. Theor. Phys. Suppl., 53, 77 (1973).
12. R.E. Borland, Proc. Phys. Soc. (London), 78, 926 (1961).
13. R.L. Bush, Phys. Rev., B6, 1182 (1972).

THEORETICAL MODELS OF CHEMISORPTION OF IONS AND SMALL MOLECULES ON POLYMERS

G. DEL RE
Cattedra di Chimica Teorica dell' Università
Via Mezzocannone 4, I-80134 Napoli, Italy -
UER de Physique, Université d'Aix-Marseille I,
Centre St. Charles, F-13003, Marseille, France

ABSTRACT. The application of current theories of chemisorption to polymers is reviewed. Using as a reference a very simple example the concepts of charge transfer, surface molecule, virtual state are introduced. The rôle of sites and of geometrical relaxation, as well as of polarization, is discussed with a view to laying the ground for the interpretation of the time-dependent properties of chemisorbed species: changes in optical spectra, residence times.

1. INTRODUCTION

The adsorption of ions and small molecules on polymer chains has received little theoretical attention, although it is of fundamental importance for many phenomena, in particular ion exchange, the behavior of biological membranes, interface properties of polymeric materials, etc.

In the present notes we shall discuss aspects of the current simple models of adsorption and related phenomena with a special view to the application to polymers. Those models are current in the sense that they already exist or are being developed for solids (surfaces): thus, much of what we shall treat here is either derived or applicable to surface studies; but certain aspects are more typical of polymers in that the latter are essentially one-dimensional structures which, if they are periodic at all, are anyway very rich in the details of individual unit cells. Whereas the solids normally studied by experimentalists and theoreticians in adsorption studies are relatively simple crystals (the most complicated ones being so far alumina and silica), the polymers

J.-M. André et al. (eds.), Quantum Theory of Polymers, 199-227.

to be considered in analyses of experimental facts are more varied if not more complicated structures: think, for instance, of polyphenylalanine. Therefore, even if methods are similar, in the case of polymers emphasis is placed rather on the nature of the monomer than on other features. Nevertheless, the fundamental distinction between highly conjugated ('metallic') and fully saturated ('covalent') systems is here of the same importance as in solid state studies.

In the present notes we shall concentrate on models in two senses: model Hamiltonians intended to simplify the task of understanding the main features of adsorption in quantum mechanical terms; model systems (i.e. specially chosen polymers) intended to have properties resulting from just one or two structural features.

We shall study several quantities, in particular: charge transfer from the admolecule to the polymer; adsorption energy; inductive and mesomeric effects of admolecules on adjoining monomer units; spectral effects of adsorption; residence times and other time-dependent quantities.

2. A PRELIMINARY STUDY: ADSORPTION OF A ONE-ELECTRON ATOM ON A LINEAR FULLY CONJUGATED CHAIN

In order to approach our subject in the simplest way and to introduce the basic questions and concepts we consider a very simple case as a preliminary, and proceed from it to more realistic cases.

The following notation will be used:
A is the adatom, which will represent in general the admolecule
R(A) is the position vector of A
S is a site strongly bound to A, R(S) its position vector
T is a site of the polymer different from S, R(T) its pos. v.
P is the polymer (or, more generally, the substrate)
a refers to A
k, j refer to the polymer
s refers to S, etc.
σ is the spin-state suffix

2.1 The Huckel scheme and the surface molecule

Consider a finite chain, consisting, say, of six atoms connected through their π orbitals.

In the simplest one-electron model, the Huckel MO-LCAO (TB) scheme, the basis will then consist of one $p\pi$ orbital per atom, and will be assumed to be orthogonal. We shall refer to spin orbitals of a given spin σ; it is understood that an entirely symmetric analysis can be carried out for spinorbitals having the opposite spin $(-\sigma)$. The basis consisting of the orbital $|a\rangle$ of A, the orbital $|s\rangle$ of S, the orbitals $|t_1\rangle, \ldots, |t_5\rangle$ of the five sites $T_1 \ldots T_5$.

With the notation given above, the elements of the Huckel Hamiltonian matrix may be written

$$\begin{aligned} &\langle a|\hat{H}|a\rangle = \epsilon_a, \quad \langle s|\hat{H}|s\rangle = \alpha_s, \quad \langle t_j|\hat{H}|t_j\rangle = \alpha, \\ &\langle a|\hat{H}|s\rangle = \beta, \quad \langle s|\hat{H}|t_1\rangle = \langle t_j|\hat{H}|t_{j\pm 1}\rangle = \gamma, \end{aligned} \tag{1}$$

the other elements being zero in the nearest-neighbor approximation, which is part of the Huckel scheme assumptions. The complete Hamiltonian matrix, representing the Hamiltonian operator $\hat{H}$, is thus

$$\mathbf{H} = \begin{vmatrix} \epsilon_a & \beta & 0 & 0 & 0 & 0 & 0 \\ \beta & \alpha_s & \gamma & 0 & 0 & 0 & 0 \\ 0 & \gamma & \alpha & \gamma & 0 & 0 & 0 \\ 0 & 0 & \gamma & \alpha & \gamma & 0 & 0 \\ 0 & 0 & 0 & \gamma & \alpha & \gamma & 0 \\ 0 & 0 & 0 & 0 & \gamma & \alpha & \gamma \\ 0 & 0 & 0 & 0 & 0 & \gamma & \alpha \end{vmatrix} \tag{2}$$

This matrix contains <u>in nuce</u> the whole problem to be treated. Chemisorption is the formation of a bond between A and S associated with β. The general problem of chemisorption at this simple level is the determination of charge transfer and bond energies between A and P.

Before treating the general case of arbitrary values of the parameters defined in (1), consider the question: what are factors determining whether adsorption is a purely 'local' phenomenon or involves 'long range' effects from the polymer? In the former case we shall be dealing with a strictly <u>localized</u> chemisorption bond, and we can speak of a 'surface molecule', a 'surface complex', a 'surface localized bond'. The surface molecule case is one of the limiting cases of the problem with which $\mathbf{H}$ of Eq. (2) is associated. In fact, when $\beta \gg \gamma$, a reasonably accurate solution of the corresponding eigenvalue equation is obtained by neglecting γ, so that the interesting eigenvalues of the system are those of the matrix

$$\mathbf{H}_{AS} = \begin{vmatrix} \epsilon_a & \beta \\ \beta & \alpha_s \end{vmatrix} \tag{3}$$

Thus, when $\gamma \ll \beta$, chemisorption can be described - within the Huckel scheme - by molecular orbitals that are just linear combinations of the atomic orbitals of A and S: if those orbitals are practically zero outside a finite region of space, this corresponds to space localization in addition to describing chemisorption in terms of a (diatomic) molecule AS.

In the limiting case in question, the reduced matrix (3) determines both the chemisorption energy ΔE and the charge transfer q_a - assuming that no charge transfer takes place to the rest of the substrate - :

$$\Delta E = 2\,(e_+ - \tilde{\alpha}) = -2\sqrt{u^2 + \beta^2} \qquad \tilde{\alpha} = (\epsilon_a + \alpha_s)/2, \quad u = (\epsilon_a - \alpha_s)/2 \tag{4a}$$

$$q_a = 2\sin^2\xi - 1 \qquad \left(\xi = \frac{1}{2}\,\mathrm{arctg}\,\frac{\beta}{u}\right) \tag{4b}$$

e_+ being the lower eigenvalue of (3), viz. the energy of the bonding orbital; the antibonding-orbital energy will be denoted e_-.

2.2 Charge transfer and the rôle of the substrate

It might seem that in the limiting case where the bond integral γ between the substrate sites is very small with respect to the bond integral β between the adatom A and the adjacent site S the presence and the nature of the T centers, i.e. of the whole of the substrate, are of no importance whatsoever. This is true only as regards the shift of the energy levels: it is not true as regards electron transfers. In fact, it can be shown that the probability for an electron initially in T_1 to be found at S after a time interval t is given by

$$P_{T_1 \rightarrow S} \approx t^2 |\langle s|\hat{H}|t_1\rangle|^2/h^2 . \tag{5}$$

This expression is obtained from a series expansion limited to first-order terms, and thus is not generally valid; but it defines a time after which, if it were rigorously valid, the probability of finding the electron entirely at S would be unity:

$$t_{max} = \hbar/\langle s|\hat{H}|t_1\rangle . \tag{6}$$

Standard textbooks of quantum mechanics justify the assertion that t_{max} gives a measure of the time taken by an electron to pass from T_1 to S. If $\langle s|\hat{H}|t_1\rangle = \gamma$ is even as small as 0.5 kJ/mole, the time in question is shorter than .15 psec, i.e. practically too small to be observable; yet that value of γ is a thousandfold less than the CC bond energy.

In other words, even if γ is thousands of times less than β, so as to make approximation (3) perfectly acceptable from the energy point of view, the transfer time of an electron from T_1 to S (or between any two adjacent sites of the polymer) is small enough to justify a description of the adsorbate-substrate system as one of two independent ions having reached equilibrium from the point of view of charges. For example, we might have to describe AS as a molecule containing an excess of 0.3 electrons, T_1-...-T_5 as a molecule having a defect of 0.3 electrons.

Equilibrium is reached when electrons find it indifferent to pass from one part of the system to the other, or, *a fortiori*, when they would require energy to perform that transition. The first condition is realized when the highest occupied orbital (HOMO) lies at the same energy for both partners. The second condition is realized when the lowest empty molecular orbital (LUMO) of either partner lies above both HOMO's. Therefore, in the simplest version of the Hückel scheme, two electrons will flow from P to AS if the HOMO of P lies above the LUMO of AS, and the converse; no electron flow will take place if both LUMO's lie above the two HOMO's; one electron will flow from P to AS (or viceversa) if the HOMO of P is degenerate with the LUMO of AS (and viceversa). Fractionary numbers of electrons will be exchanged (in the average) only if the change of the parameters with the elctron populations is taken into account. This brings us out of the simple Hückel scheme proper, and provides a simple illustration of the nature and physical significance of SCF techniques.

2.3 Charge transfer in a simple SCF (Hubbard) scheme

That a study of charge transfer should be carried by a method involving some degree of self-consistency is evident when one reflects that a flow of electron from, say, P to AS will bring about a change in the orbital energies of AS, which will become those of the corresponding anion. Therefore, if two electrons pass from P to AS, the HOMO of AS^{--} may move to energies higher than the LUMO of P^{++}, and electrons will flow back to P; a process which is not physically acceptable, because in fact electrons will flow from P to AS just in the amount required to bring the HOMO of $AS^{-\delta}$ to the same energy as the HOMO of $P^{+\delta}$: a process which, as has been mentioned, will result in a fractionary charge δ transferred from P to AS. The reason why this fractionary charge (in electrons) is not in contrast with physics has been given above: it is actually an average number of electrons δ missing from P and present as an excess in AS, or viceversa.

In short, it will be necessary to take into account the charge dependence of the orbital energies of AS. On the other hand, the charge dependence of the orbital energies of P need be taken into account only if P is small; if it is sufficiently large, one more or one less electron will not affect its energy levels to any important extent. To make our examples simpler, we shall assume here that the number of centers T is very large, and not just five as in Sec. 2.1, so as to take advantage of the above remark.

We introduce the dependence of the orbital energies of AS on the charge via some kind of dependence of ϵ_a and α_s on the electron defects q_a, q_s, of $|a\rangle$ and $|s\rangle$, with respect to some standard population which corresponds to the valence state. That standard population may be taken to be half an electron with spin $\uparrow$ and half an electron with spin $\downarrow$: we assume that the corresponding one-electron energies ϵ_a^o and α_s^o are the arithmetic means of the energies of one electron with spin $\uparrow$, alone and in the presence of an electron with spin $\downarrow$:

$$\epsilon_a^o = \epsilon_a^{oo} + (aa/aa)/2, \qquad \alpha_s^o = \alpha_s^{oo} + (ss/ss)/2, \tag{7}$$

ϵ_a^{oo} being the energy of a single electron in $|a\rangle$, (aa/aa) being the repulsion integral of two electrons in $|a\rangle$, α_s^{oo} and (ss/ss) being the corresponding quantities for $|s\rangle$.

Equations(7) can be considered as special cases of more general equations. For $|a\rangle$, the corresponding general equation is

$$\epsilon_a^{\uparrow} = \epsilon_a^{oo} + (aa/aa) \langle n_{a\downarrow}\rangle \tag{8}$$

where $\langle n_{a\downarrow}\rangle$ is the total number of electrons of spin $\downarrow$ on $|a\rangle$. We get equation (7) when $\langle n_{a\downarrow}\rangle$ is 0.5, the valence-state value. Equations similar to (8) obviously hold for $\epsilon_a^{\downarrow}$, $\alpha_s^{\uparrow}$, $\alpha_s^{\downarrow}$.

If we assume that the states under consideration are closed shell (non-magnetic) ones, we write

$$\langle n_{a\uparrow}\rangle = \langle n_{a\downarrow}\rangle = (1 - q_a)/2, \quad \epsilon_a^{\uparrow} = \epsilon_a^{\downarrow} = \epsilon_a \quad , \tag{9}$$

where the electron defect or net charge in atomic units is

$$q_a = 1 - \sum_j \langle a|j\rangle^2 n_j, \tag{10}$$

$\langle a|j\rangle$ being the coefficient of the atomic orbital $|a\rangle$ in the total molecular orbital $|j\rangle$, whose occupation number is n_j. Using the standard molecular orbital energy defined in Eq. (7), and dropping the spin suffix, Eq. (8) becomes

$$\epsilon_a = \epsilon_a^o - (aa/aa)\, q_a/2, \tag{11a}$$

$$\alpha_s = \alpha_s^o - (ss/ss)\, q_s/2. \tag{11b}$$

Equation (11b) applies in principle also to diagonal matrix elements of **H** that are associated with the T centers. However, if P is very large, the average number of electrons per atom always remains very close to 1, which means $q_t \approx 0$. This is why, as has been mentioned, we can take the diagonal elements of H for the substrate P constant and equal to : the only exception is α_s when S is tightly bound to A.

We have considered explicitly only the case when there is a strong interaction between A and S, so that the system to be treated actually consists of a substrate P including the T sites and a surface molecule AS interacting with a neighboring site T by a weak coupling of the same order of magnitude as the interaction between the T sites: this we shall consider one of two limiting cases. The second limiting case is when no site S exists, for A is very weakly bound with the polymer: we have $\beta \ll \gamma$, and the interaction between A and P does not result from the formation of a bond but simply from charge transfer. For the present problem both cases correspond to the study of the equilibrium conditions of an adsorbed 'phase' (AS or A) with respect to a substrate 'phase' (P without or with S): the rôle of the chemical potential is played here by the HOMO of either phase. In both cases, due to the assumption that P is large, the chemical potential or HOMO or Fermi level of P is just α.

The first case actually divides into two possibilities, because the LUMO of AS may be higher than α, the HOMO being lower: then no electron transfer is to be expected, and the application of equations (11) just amounts to internal self-consistency of AS: this will be called case 1a. Case 1b is found when the opposite situation applies: the LUMO of AS lies below the Fermi level of P, so that electrons will flow from P to AS until the resulting increase in the diagonal elements of $\mathbf{H}_{AS}$ has brought the upper MO of AS (which is now a HOMO) to the Fermi energy.

Case 1a. This is when $\beta \gg \gamma$, $e_- > \alpha$: it is realized when we have a polymer with a non-nearest neighbor π-bond arrangement, as in certain configurations of polyethylene if the CH_2 groups are considered as π-orbital carrying groups as in hyperconjugation. Let us take (in ev)

$$\alpha = \alpha_s^0 = -11.54 \qquad \epsilon_a^0 = -20 \qquad \beta = -4 \quad \gamma \cong -0.2 \qquad (ss/ss) = 10.85 \qquad (aa/aa) = 14 \tag{12}$$

Taking Eqs. (11) to convergency for zero overall charge (computed by diagonalizing (3), of course) the following values are found :

$$\epsilon_a = -17.2005 \quad \alpha_s = -13.7096 \quad q_a = -0.3999 \quad q_s = +0.3999$$
$$e_+ = -19.8193 \qquad e_- = -11.0908$$

In an SCF method such as the present one the chemisorption energy ΔE cannot be given by Eq. (4a) unless complete equivalence of core repulsion energies with exaggerated weight of electron repulsions given by the sum of orbital energies is assumed as in the Hückel method. The complete formula for ΔE is

$$\Delta E = [2e_+ - (aa/aa)(1-q_a)^2/4 - (ss/ss)(1-q_s)^2/4] + - \epsilon_a^o - \alpha_s^o + R \qquad (13)$$

R being the core repulsion energy. If the latter is taken as the repulsion of two unit charges at $3a_o$ (1.59 A), we obtain for the case of the parameters (12)

$$\Delta E = -6.8645 \text{ ev}$$

to be compared with ~-8.1 ev as obtained by the Hückel recipe (4a). Both values are evidently too large to correspond to reality. We shall discuss at some length the reasons and posible remedies to that deficiency, but we warn the reader against dismissing results like the present ones only in view of the superficial consideration that the quantitative aspects are unrealistic. We are dealing with a model which should be used to interpret, explain, and predict trends and qualitative differences: the fact that quantitative agreement is not obtained makes the use of the model more difficult than it would be otherwise; but it is an inevitable price to pay for the simplifications that give the model its interpretative qualities.

Case 1b. This is when $\beta \gg \gamma$, but $e_- < \alpha$ when calculated for neutral AS. Then an electron flow will take place from P to AS. To realize numerically this case, we take the parameters (12) with

$$\epsilon_a^o = - 22 \text{ ev} .$$

Let us call q the total charge of AS, which corresponds to -q electrons in the antibonding orbital of AS. Then we find

$$\text{for } q = 0 \quad e_+ = -20.4885 \quad e_- = -12.0614;$$

$$\begin{aligned} \text{for } q = -0.0844 \quad & e_+ = -19.9848 \quad e_- = -11.5400 = e_{Fermi} \\ & q_a = -0.3489 \quad q_s = -0.2645 \\ & \epsilon_a = -17.1147 \quad \alpha_s = -14.4101 \end{aligned} \qquad (14)$$

The above results for a population of 0.0844 electrons in the antibonding orbital correspond to equilibrium between AS and P under the assumption already mentioned that P is so long that the loss of about a tenth of an electron does not change its level distribution.

The chemisorption energy computed according to Eq. (13) with the same value of R as in case 1a is

$$\Delta E = -6.1694 \text{ ev}$$

which is approximately 66 kJ/mole or 16 kCal/mole less than in case 1a, notwithstanding the lowering of about 46 kCal/mole in the orbital energy of |a>. Strictly speaking, there should be a contribution to energy coming from the electrostatic attraction between AS and P: this is extremely small if the positive charge of P is distributed over its whole length, as has been assumed; if it is assumed to be concentrated on T_1 (an assumption which is not in contradiction with the assumption that the Fermi level is α, for most of the levels come from atoms far removed from S), then ΔE is lowered by ~ 0.0646 ev or 6.24 kJ/mole or 1.49 Kcal/mole. The electrostatic contribution is thus a very small one.

The reason why the charge transfer is so small in the present numerical results is very simple: the standard orbital energy of A is just sufficient to bring e_- below the Fermi level, so that a very small electron excess will restore equilibrium. This situation is nearly always the case with non-ionic adsorbates or surface complexes, because -22 ev is a very large value for neutral atoms.

Case 2. We now consider the opposite limiting case, $\beta \ll \gamma$. We carry out this application on the case of an N-atom polyene, with $\gamma = -2$ ev, $\epsilon_a = -15$ ev, (aa/aa) = 12 ev: A is now approximately a nitrogen atom. If +q is now the charge on A (which now forms the surface molecule all by itself), we have

$$\epsilon_a = \epsilon_a^o - (aa/aa)q/2, \alpha = \alpha^o + (ss/ss)q/2N \quad (\alpha_s = \alpha_t = \alpha) \tag{15}$$

and therefore the highest occupied orbital of P is

$$e_{HOMO}(P) = \alpha^o + (ss/ss)q/2N + 2\gamma\cos[N\pi/2(N+1)]. \tag{16}$$

The condition that

$$\epsilon_a = e_{HOMO}(P)$$

gives an equation for q, from which one obtains

$$q = \frac{2N[(\epsilon_a^o - \alpha^o) - 2\gamma\cos\frac{N\pi}{2(N+1)}]}{N(aa/aa) + (ss/ss)}. \tag{17}$$

This gives the following results:

for N = 6: $q_a = -0.3722$ $q_s = q_t = +0.0620$

$\epsilon_a = -12.7668$ $\alpha = -11.8765$

N = 100: $q_a = -0.5612$ $q_s = q_t = +0.0057$

$\varepsilon_a = -11.6328$ $\alpha = -11.5704$

The limiting value of q_a is - 0.5767.

The above results show that in the present case the charge transfer is much more marked than in the cases discussed before, even though A has a much lower orbital energy. This is related to the fact that β is now much smaller than in case 1: and, as is well known, the polarization of a bond decreases, all other things being equal, when the bond integral increases. A choice of the atomic parameters as in case 1b would give a paradoxical result: a net charge on A of -1.21 electronic charge units: this gives a measure of the validity limits of a model where just one orbital is assigned to the adatom.

In the present case the chemisorption energy arises completely from electrostatic forces. According to a remark made above, and pending further considerations, the most reasonable way to estimate that energy is to assume that the polymer charge, instead of being distributed over the whole system, as is assumed in the values given above, is concentrated at S: then the electrostatic energy is --3.0165 ev, i.e. approximately 291 kJ/mole, or ca. 70 kCal/mole at a distance of $3a_0$: the order of magnitude of a standard bond energy. Thus, it must be concluded that at least a metastable AP complex kept together by an ionic bond is formed in the circumstances described by our parameters and our model Hamiltonian.

2.4 Intermediate case: the virtual level concept.

In order to investigate the intermediate case, when neither β nor γ are small, the best procedure is to place ourselves in a more suitable basis set. So far, we have referred to the atomic orbitals of A and of P, each center being represented by its own atomic orbital. If we want to look at the adsorption phenomenon as the formation of a bond between S and P, it is better to introduce orbitals corresponding to what would be the atomic orbitals of P, if the latter were just a single atom: the basis $|\mathbf{k}\rangle$ of the molecular orbitals of P, treated as a row matrix of N elements $|1\rangle, |2\rangle, \ldots, |N\rangle$. The basis $|\mathbf{k}\rangle$ has the obvious property of diagonalizing the corresponding block of $\mathbf{H}$ of Eq. (2): for the general case of an N-atom chain, if $|\mu\rangle$ is the μ-th atomic orbital, and the atomic orbitals are orthogonal to one another ($\langle\mu|\nu\rangle = \delta_{\mu\nu}$), we have:

$$\langle\mu|k\rangle = \sqrt{2/(N+1)}\ \sin k\varphi \quad (\mu, k=1, \ldots, N;\ \varphi = \frac{\pi}{N+1}) \qquad (18)$$

$$e_k = \alpha + 2\gamma\cos k\varphi. \tag{19}$$

Therefore, in the basis $(a,\mathbf{k})$ consisting of the adatom orbital and of the N substrate orbitals, and taking α as the energy zero, we obtain the new Hamiltonian ($\alpha_s = \alpha$)

$$\mathbf{H}' = \begin{vmatrix} 2u & b\sin\varphi & b\sin 2\varphi & b\sin 3\varphi & \cdots \\ b\sin\varphi & 2\gamma\cos\varphi & 0 & 0 & \cdots \\ b\sin 2\varphi & 0 & 2\gamma\cos 2\varphi & 0 & \cdots \\ b\sin 3\varphi & 0 & 0 & 2\gamma\cos 3\varphi & \cdots \\ \vdots & \vdots & \vdots & \vdots & \vdots \end{vmatrix}$$

$$\text{with} \quad u = (\epsilon_a - \alpha)/2, \quad b = \beta\sqrt{2/(N+1)}. \tag{20}$$

The matrix (20) is easily diagonalized because the eigenvalue equation is obtained from the system

$$\begin{cases} 2uc_{aj} + b\Sigma\, c_{kj}\sin k\varphi = x_j c_{aj}, \\ c_{aj} b\sin k\varphi + 2\gamma c_{kj}\cos k\varphi = x_j c_{kj}, \end{cases} \tag{21}$$

whence

$$c_{kj} = b\sin k\varphi/(x_j - 2\gamma\cos k\varphi) \tag{22}$$

and

$$x_j = 2u + b^2\Sigma\, \sin^2 k\varphi/(x_j - 2\gamma\cos k\varphi); \tag{23}$$

x_j being, of course, $e_j - \alpha$, the general eigenvalue of $\mathbf{H}'$. All the summations are carried over $k = 1,\ldots,N$.

Equation (23) can be shown to coincide with (3) when $\gamma \to 0$, up to a degenerate eigenvalue 0; the proof goes through a series expansion in γ/x_j.

The problem of determining the eigenvalues and eigenvectors of our simple chemisorption model in the general case is thus solved analytically, up to a discussion of Eq. (23), whose structure suggest an iterative method which in fact does not generally converge. We leave to the reader and to later sections of these notes a complete analysis of the above results. Here we consider only the connection between the general case and case 2 of the preceding subsection.

Let us interpret the eigenvalue spectrum of the adsorbate-substrate system as if it resulted from the contributions of A and P separately. In this interpretation ϵ_a is seen as a narrow line transformed by interaction with the substrate into a 'band' whose abscissae are the energies and whose ordinates are the weights $\langle a|j\rangle^2$ of $|a\rangle$ in the MO's whose energies correspond to those abscissae. The 'band' thus constructed has the property that the sum of the ordinates over that part of it that lies below the Fermi level (HOMO energy of the whole system) is half the total population of $|a\rangle$ in Mulliken's sense: thus the picture obtained is equivalent to that of case 2 of the preceding section, with the addition that the single sharp level of A is now replaced by a set of levels contributing to the total energy by fractionary populations $2|\langle a|j\rangle|^2$. Of course, if the virtual level lies entirely below the Fermi energy, $|a\rangle$ will be doubly filled and hence A will behave as a -1 negative ion; if the virtual level lies entirely above the Fermi level A will behave as +1 positive ion.

In other words, in the case when the bond integral between A and the substrate is not negligible, we can still speak of charge transfer to or from A, but the fact that β is not negligibly small involves two effects:

a) broadening of the level of A;
b) shift of the level of A.

To speak of a broadening is another way of stating that the electrons of A occupy energy levels other than ϵ_a, and are in fact distributed around a mean value e_a. To speak of a shift is a way of saying that e_a, the maximum of the 'band', does not lie at ϵ_a^0, nor at ϵ_a. The level shift is essentially the result of the fact that the level of A goes into a molecular orbital energy, as happens for the surface molecule; the broadening is due to the fact that $|a\rangle$ combines with many closely spaced levels of the substrate, and gives rise to many closely spaced levels (for N large enough).

In the limit $N \to \infty$, the virtual level concept is associated with a continuous curve obtained by dividing the ordinates $2|\langle a|j\rangle|^2$ by the mean energy interval $(e_{j+1}-e_{j-1})/2$ and by taking the separation of the energy levels to zero. The curve then represents the so-called 'projected state density' at A, an the electron population of A is represented by the integral of the projected density over the energy between $-\infty$ and the Fermi energy.

3. MODEL ELECTRONIC HAMILTONIANS

The example treated in Sec. 2 raises most of the questions that have to be solved to set up a theory of adsorption on real polymers, at least as far as electronic states are concerned. Before listing those questions in some detail, it is important to consider more explicitly the model on which the quantum scheme used in Sec. 2 rests.

Our mathematical model has consisted in the adoption of a very simple orbital basis; we have carried out on that basis a Hückel and an iterative Hückel treatment using the nearest neighbor approximation. In the iterative case we have applied a self-consistency criterion bearing only on the diagonal elements of the one-electron (Hückel) Hamiltonian matrix. Can all this be stated in a more general form, so as to show what kind of approximations and simplifications it contains with respect to a rigorous theory?

The answer is positive. First of all, the Born-Oppenheimer approximation has been assumed, and the foundations of that approximation are well established and well known. As regards the many-electron Hamiltonian, it is best discussed by using the second-quantization formalism: then it can be written in terms of the matrix elements $<u|\hat{H}^c|v>$ of the core Hamiltonian (the one electron operator representing the kinetic energy and the potential energy of one electron in the field of the atomic cores) between given basis orbitals $|u>$ and $|v>$, and of the two electron integrals (uv/wz) corresponding to electron 1 'hopping' from orbital $|u>$ to orbital $|v>$, and electron 2 'hopping from $|w>$ to $|z>$.

We consider a spinorbital basis $|\mathbf{u}\sigma>$ consisting of the spinorbitals $|u\uparrow>$, $u\downarrow>$,... that can be built from the orbitals $|u>$, $|v>$, $|w>$,... of a given orbital basis $|\mathbf{u}>$; and the many-electron states $|\Phi>$ (Slater determinants in the wavefunction representation) that can be obtained by occupying the given spinorbitals in all possible ways by n electrons. (In our considerations, unless otherwise stated, $|\Phi>$ is always the ground state, and the brackets $< >$ imply an expectation value over that state if no states are indicated.) The spinorbitals have a given order.

Further, we introduce creation and annihilation operators $\hat{a}^+_{u\sigma}$ and $\hat{a}_{u\sigma}$, respectively. They have the following properties:

- $\hat{a}^+_{u\sigma}$ creates an electron in the spinorbital $|u\sigma>$ if it is empty in the (n-1)-electron state to which the creation operator is applied, and multiplies the resulting n-electron state by -1 if the number of electrons occupying spinorbitals preceding $|u\sigma>$ is odd; if $|u\sigma>$ is full the corresponding creation operator gives zero.
- $\hat{a}_{u\sigma}$ has the same properties of $\hat{a}^+_{u\sigma}$ but for the fact that it <u>empties</u> $|u\sigma>$ instead of filling it, and will give zero if $|u\sigma>$ is already empty.

Operators derived from the creation and annihilation ones are: the number operator $\hat{n}_{u\sigma} = \hat{a}^+_{u\sigma}\hat{a}_{u\sigma}$, which multiplies a many electron state by 1 if $|u\sigma>$ is full, by 0 if it is empty; and the hopping operator $\hat{a}^+_{u\sigma}\hat{a}_{v\sigma'}$ $(u\sigma \neq v\sigma')$ which takes an electron from

$|u\sigma\rangle$ to $|v\sigma'\rangle$, if possible, and otherwise gives zero.

With the above notation, and calling R the core repulsion energy, which must always be added to the electronic energy in molecular calculations, we write for the general Hamiltonian $\hat{H}$ referred to the basis $|\mathbf{u}\sigma\rangle$:

$$\hat{H} = R + \sum_{u,v}[\langle u|\hat{H}^c|v\rangle + \frac{1}{2}\sum_{w,z}\sum_{\sigma}(uv/wz)^{\uparrow\sigma}\,\hat{a}^+_{w\sigma}\hat{a}_{z\sigma}]\,\hat{a}^+_{u\uparrow}\hat{a}_{v\uparrow} + \text{same with}\downarrow\text{instead of}\uparrow. \qquad (24)$$

The general Hamiltonian thus written has the advantage that assumptions on the orbital basis and on the one- and two-electron integrals are immediately translated into physical models, insofar as it is legitimate to say that we have a physical model of a system when we can write the corresponding Hamiltonian operator. Of course, it is not normally easy to translate features of a model given in second-quantization formalism into forces and motions: but that is often neither necessary nor desired.

The derivation of simplified models from Eq. (24) is straightforward: an elegant example can be found, for instance, in the work of P.W. Anderson of 1961. In the case of the (a,**k**) basis used to treat an adatom chemisorbed on P, with P so long that no charge dependence (or, more generally, no correlation effect) need be introduced for it, we introduce precisely the Anderson Hamiltonian

$$\hat{H} = R + \sum_{\sigma}\epsilon^{o}_{a}\hat{n}_{a\sigma} + \sum_{k\sigma}\sum e_k\hat{n}_{k\sigma} + \sum_{k\sigma}\sum(v_{ak}\hat{a}^+_{a\sigma}\hat{a}_{k\sigma} + v^*_{ak}\hat{a}^+_{k\sigma}\hat{a}_{a\sigma}) + (aa/aa)\,\hat{n}_{a\uparrow}\hat{n}_{a\downarrow}, \qquad (25)$$

which is obtained by (a) specifying the basis; (b) shifting by some trick part of the two-electron terms to the one-electron part so as to replace $\hat{H}^c$ by an effective Hamiltonian whose matrix representation is given by (20) with the standard valence state orbital energy of A for the first diagonal element (v_{ak} is an off-diagonal element of the first row of (20));(c) droping all two-electron contributions except that of two electrons occupying a> with opposite spins. Note that b and c are compatible insofar as the various parameters are estimated all with reference to standard valence states, which in practice means that they are estimated on the basis of experimental atomic properties, as has been done in Sec. 2.

To extract a one-electron (SCF) Hamiltonian from (25) it is necessary to apply the well known considerations leading to the

Hartree-Fock method. One finds

$$\hat{H}_{SCF} = R' + \hat{H}^{\uparrow} + \hat{H}^{\downarrow}, \tag{26}$$

where R' is an effective core repulsion given by

$$R' = R - (aa/aa)\langle\hat{n}_{a\uparrow}\rangle\langle\hat{n}_{a\downarrow}\rangle, \tag{27}$$

and

$$\hat{H}^{\sigma} = (\epsilon_a^o + (aa/aa)\langle\hat{n}_{a,-\sigma}\rangle)\hat{n}_{a\sigma} + \Sigma_k e_k \hat{n}_{k\sigma} + \\ + \Sigma_k (v_{ak}\hat{a}^+_{a\sigma}\hat{a}_{k\sigma} + v^*_{ak}\hat{a}^+_{k\sigma}\hat{a}_{a\sigma}); \tag{28}$$

$\langle\hat{n}_{a\sigma}\rangle$ is the electron population of the spinorbital $a\sigma\rangle$ in the HF state under consideration: a state defined through the eigenvectors obtained from $\hat{H}^{\sigma}$ when the populations used to evaluate its matrix elements coincide with those computed from the eigenvectors themselves for the set of occupation numbers which corresponds to the particular many-electron state. By diagonalizing the matrix representing $\hat{H}^{\sigma}$ - which is but the matrix (2) with ε_a given by (8) or (11a) - we obtain the orbital energies e_m of the adatom-substrate complex. Note that, except when the so-called DODS scheme suggested by Lowdin is adopted, the Hamiltonians associated with opposite spins coincide.

The chemisorption energy is given by

$$\Delta E = \Sigma^{occ.}_{m,\sigma} e_{m\sigma} + R' - (2\Sigma_k e_k + \epsilon_a^o) \tag{29}$$

(<u>occ</u>. means that the summation is taken only over occupied spin orbitals, and σ is indicated because there might be an odd number of electrons - a situation which demands certain precautions because it corresponds to an 'open shell' problem, but can be treated formally in the same way as the closed shell one in our model calculations).

Another model Hamiltonian useful for chemisorption studies is the Hubbard Hamiltonian. That Hamiltonian is just a generalization of (25), in that the one-center two-electron integrals are retained for all or at least for several atomic orbitals: the 1-electron Hamiltonian matrix will contain several charge dependent diagonal elements since in SCF form the Hubbard Hamiltonian corresponds to Eq.s (11). The chemisorption energy is given by a generalization of (29) similar to Eq. (13).

Examples of the way in which self-consistency is applied in practice have been given in the preceding section. It may be mentioned here that in fact the model Hamiltonians of which the Anderson Hamiltonian is an example are many-electron ones which can be used as such on a many-electron basis, so as to introduce a certain amount of correlation: in that case their use is equivalent to neglect of certain terms in a configuration-interaction calculation, for instance to neglect of interatomic correlation in the Hubbard Hamiltonian case. Of course, the limitations due to the fact that only a few configurations can normally be included in the many-electron basis set add an entirely different set of approximations, just in the same way as the fact that the orbital basis is far from complete represents a limiting feature of one-electron calculations that has nothing to do with modelling of the Hamiltonian. The configuration interaction scheme requires a very special kind of analysis when applied to systems with very large numbers of electrons. We shall not be interested in it here, because we shall not deal extensively with states other than closed-shell ground states. The question remains open and important, however, for the excited states of adsorbed molecules: a very important subject for the study of certain biological mechanisms.

4. NATURE OF ADSORBED SPECIES, ADSORPTION SITES, POLARIZATION

The preliminary example of Sec. 2 has been intended to raise a number of questions by means of a concrete example. Only one of those questions has been somehow answered by the concise presentation of model Hamiltonians given in Sec. 3. Even there, however, it has been more or less apparent that the chemical nature of the system under study, its number of electrons, its geometry are of the greatest importance in determining the choice of the model and the type of computation to be performed, to say nothing of the mechanism and properties of adsorption.

4.1 Nature of the adsorbed species

In Sec. 2 the nature of the adsorbed species A has not been specified, although it was mentioned that A was treated as a neutral atom carrying a single electron. Actually, apart from computational difficulties having to do with the existence of open shells, experimental evidence is concerned mainly with ions and small molecules. Thus, we need be specially interested in either of two possibilities:

A is an ion contributing 0 or 2 electrons;

A forms a molecule AS which then interacts with the rest of the polymer, and must be considered as the real adsorbed species.

Both cases can be treated along the lines of Sec. 2: but, in the case of an ion electrostatic effects and polarization are very important and require special attention; in the case of a molecule, the considerations of Sec. 2 should be extended to cover the case when S is not an atom of the substrate - and, of course, to polyatomic adsorbed molecules.

In the adsorption of ions, electrostatic interaction may be a decisive factor even putting all quantum effects aside: that happens especially when non-vanishing net charges exist in the polymer chain. An idea of the situation that can be found is provided by infinite polyglycine treated by a very simple $\sigma + \pi$ semiempirical method assuming complete localization. The net charges are

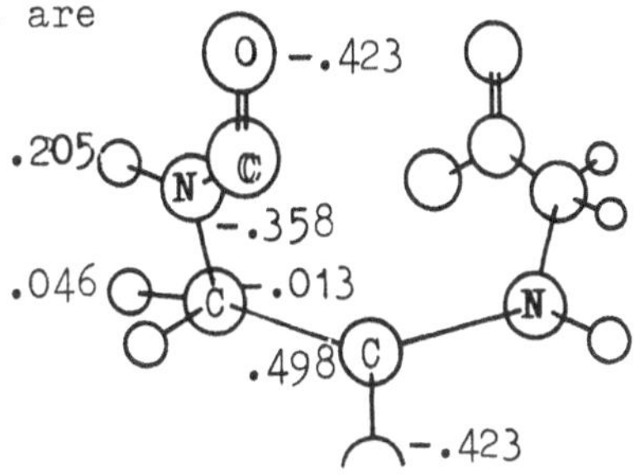

O −.423
.205 N C
−.358
.046 C −.013
.498 C N
−.423

smaller circles: H

Method: Del Re 1958
with Born-v.Karmán cond.
(cf. Biczó et al.)

FIG.1

for C=O q_σ = .209, −.136; q_π = .287, −.287.

This scheme gives a clear picture of the two preferred sites of adsorption of an ion like Na^+: an external site, near the oxygen atom of carbonyl, and an internal one, bridging the two nitrogen atoms. To examine the possible equilibrium positions not only the electrostatic potential but the empirical potentials representing the van der Waals forces should be used to introduce some short range repulsion. At a more refined level it would be necessary to introduce some polarization by taking the conjugation of N with C=O into account and by modifying at least the properties of the nitrogen atoms both with respect to π electrons and to σ electrons. (The polarization effects in question will be briefly discussed in a later subsection.)

The polyglycine example illustrates the problem of ion adsorption and the determination of the most favorable sites in terms of simple methods and models. It is well known that very effective and sophisticated ab initio quantum mechanical methods have been devised for the conformational analysis of polymers with and without adsorbed ions. Those methods are very useful, but an analysis of the present kind present an interest in that it allows separation of various effects: purely electrostatic, van der Waals, π-polarization, σ-polarization, charge transfer, etc.

The adsorption of a molecule is best illustrated by a diatomic molecule like H_2 or CO. The adsorption problem can be treated along the same lines as for an adsorbed ion but for two novel features. Electrostatic interaction takes place here at the dipole level, either because of a permenent dipole, or because an electric dipole moment is induced in the adsorbed molecule by the substrate. Orientation effects appear that could not be present in the case of an ion.

Moreover, interaction with the substrate tends to modify the equilibrium distance and the bond strength of the adsorbed molecule: so that, in addition to classifying possible adsorption sites in terms of their binding power, it is important to associate to them also an equilibrium distance of the adsorbed molecule. A qualitative analysis of this particular aspect of the adsorption problem can be carried out along the following lines. Consider a homonuclear diatomic A_2 (bond length $\underline{r}$) held perpendicular to a substrate by electrostatic forces. The substrate is simulated by a point charge, say a proton H located at a distance $\underline{r}'$ from the nearest atom of A_2. Then the system to be studied is the three-center two-electron linear molecule

$$(H—A—A)^+$$

The corresponding Hamiltonian matrix is

$$\mathbf{H} = \begin{vmatrix} \alpha_H & \beta_{HA} & 0 \\ \beta_{HA} & \alpha_{A_a} & \beta_{AA} \\ 0 & \beta_{AA} & \alpha_{A_b} \end{vmatrix} = \alpha_A \mathbf{I} + \beta^{\circ}_{AA} \begin{vmatrix} u' & g' & 0 \\ g' & u_a & g \\ 0 & g & u_b \end{vmatrix} \tag{30}$$

where $g' = \beta_{HA}/\beta^{\circ}_{AA}$, $u' = (\alpha_H - \alpha^{\circ}_A)/\beta^{\circ}_{AA}$, etc.; α°_A, β°_{AA} being the values of α_A and β_{AA} at the standard equilibrium distance of isolated A_2.

We now introduce: a) the dependency of the diagonal elements of $\mathbf{H}$ on the net charges [cf. Eq.s(11)]:

$$u' = u'^{\,o} - w_H q_H/2, \tag{31}$$

$$u_a = - w_A q_a/2, \qquad u_b = - w_A q_b/2, \tag{32}$$

where w_H and w_A are the one-center two-electron integrals in units β°_{AA}, q_H-1-q_a-q_b is the net charge of H, q_a and q_b are the net charges of the two A atoms, and finally $u'^{\,o}$ is standard valence state energy of the hydrogen atom in units β^{o}_{AA} and with a zero-point energy $\alpha_A{}^{o}$; the dependency of the off-diagonal elements of $\mathbf{H}$ on the bond-orders p_{XY} (sums over the MO's of products of the coefficients of a given pair of AO's by one-another and by the occupation numbers of the corresponding MO's):

$$\begin{aligned} g' &= g'_o \exp(-kp_{HA}), \\ g &= g_o \exp(-kp_{AA}). \end{aligned} \tag{33}$$

These expressions have been introduced and justified by Julg,

Del Re, Barone in a study of border effects in model clusters. They correspond to a linear relationship between bond distances and bond orders. Noting that g_o = exp$\underline{k}$ because the bond order of isolated A_2 is 1, and g must also be 1 at the equilibrium distance of the isolated molecule, we are reduced to the three parameters g_o', k', k, which can be estimated by fitting the above formulas to a few experimental data. The expected geometry is found when the g's calculated from the bond orders coincide with those appearing in the matrix from which the same bond orders have been obtained.

4.2 Inductive (polarization) effects

In Sec. 2 we have seen that, in order to assign some realistic value to the energy of the ionic bond formed by an adatom weakly chemisorbed at one end of a conjugated polymer chain, it is necessary to assume that the chain is not only charged, but polarized. That polarization can be taken into account in a very simple way: the diagonal elements of the effective one-electron matrix associated with the various atoms of the chain may be considered to be affected by the adsorbed atom as in some versions of the Huckel method for heterocycles, viz. by increasing their absolute values proportionally to the α's of the neighboring atoms: the self-consistency procedure is then applied for an initial asymmetric distribution of the diagonal elements of the Hamiltonian matrix of P. An alternative treatment involves addition to the Hamiltonian of a slowly varying electrostatic potential V^{ext}: a very realistic way of seeing thing especially when the adsorption of ions is considered. Then the matrix elements of (2) must be supplemented by elements

$$\delta H_{uv} = \langle u|V^{ext}|v\rangle \qquad (u,v \text{ denote AO's}). \qquad (34)$$

If V^{ext} is really slowly varying and the atomic orbitals form an orthogonal set, (34) gives only contributions to the diagonal part of the Hamiltonian matrix, so that we can represent the effect of V^{ext} as a diagonal matrix whose elements are the mean values of that external potential at the various nuclei.

When it comes to the σ skeleton of a polymer, its polarization can be introduced following a procedure suggested by Del Re in 1958. Call u_X the orbital energy of X (in arbitrary units and with a zero-point such u_X is zero for the isolated hydrogen atom). Then the set of u's of a saturated molecule is given by

$$u_X = u_X^{\circ} + \Sigma_Y\, m_{XY} u_Y, \qquad (35)$$

the m's being polarization constants, equal to zero for non-linked atoms. The charge distribution is found by solving as many two-center two-orbital problems of the type (3) as there are bonds, with suitable values of the bond integrals and with α-values obtained by solving (35) and writing

$$\alpha_X = \alpha_H^{\circ} + u_X\beta \qquad (36)$$

with β the arbitrary unit chosen for the energies.

Owing to the fact that the u-values are independent of the bond integrals, it is possible to introduce in (35) atoms for which the bond integral vanishes, so that their function is just to modify the u-values of the other atoms without actually participating in bond formation. For instance, one can treat the approach of a proton to a polyglycine chain on top of a nitrogen atom (cf. fig. 1) by assigning to it a nonvanishing value of m_{NH^+} varying as a function of the distance. For a value of that m of 0.04, which is about 1/10 of the normal value, and a $u_{H^+} = 0.24$, the other parameters being those used to get the charge distribution of fig. 1, we find for σ charges:

...NH(−.358)–C(.209)(=O −.136)–CH₂(.004)–NH*(−.400)–C(.213)(=O −.136)–CH₂(−.012)–NH(−.358)–C(.209)(=O)–CH₂...

Thus, even for a very weak inductive effect, the computed local polarization is quite significant, although it dies off very rapidly. Note that the change in local charges is such that we can expect - and estimate by current techniques - a very important change in local conformation of the polypeptide chain.

4.3 Modelling of differences in adsorption sites

The above discussion of inductive and mesomeric effects cannot be separated from the question of the adsorption site. In the linear model of Sec. 2 it has been assumed that A is at the end of the chain. What if A is adsorbed at an intermediate site? What if it is adsorbed in a bridge position?

We leave to the reader the detailed analysis of the model of Sec. 2 in these new cases, and give here only an outline of the treatment.

In the case of 'top' adsorption at the center of the chain, we can rewrite the matrix (2), and find immediately that it does not reduce to just interchanging a few rows and columns. To visualize what happens, think of a hydrogen atom adsorbed on top of the plane of an all-trans polyene at a carbon atom: a 1s orbital interacting with the positive lobe of a $2p\pi$ carbon orbital just below it. That carbon atom now plays the rôle of of S, but it is linked to two T sites, say T_{-1} and T_{+1}: we can write, instead of (2), the new matrix

$$\mathbf{H} = \begin{vmatrix} \varepsilon_a & 0 & 0 \ldots & 0 & \beta & 0 \ldots \\ 0 & \alpha & \gamma \ldots & 0 & 0 & 0 \ldots \\ 0 & \gamma & \alpha \ldots & 0 & 0 & 0 \ldots \\ . & 0 & 0 \ldots & \alpha & \gamma & 0 \ldots \\ \beta & 0 & 0 \ldots & \gamma & \alpha_s & \gamma \ldots \\ 0 & 0 & 0 \ldots & 0 & \gamma & \alpha \ldots \\ \vdots & \vdots & \vdots & \vdots & \vdots & \vdots \end{vmatrix} \tag{37}$$

The limiting cases where $\beta \gg \gamma$ or $\beta \ll \gamma$ remain practically unchanged, unless one introduces the inductive effect discussed the preceding subsection. The general case changes because, in the basis $(a,\mathbf{k})$, we have

$$V_{ak} = \langle a|\hat{H}|k\rangle = \sqrt{\frac{2}{N+1}}\ \Sigma_\mu \langle a|\hat{H}|\mu\rangle \sin\frac{k\mu\pi}{N+1} \tag{38}$$

and, if $\langle a|\hat{H}|\mu\rangle = \beta$ for $\mu = m$, and zero otherwise, with the notation of Eq. (20), we obtain

$$\mathbf{H}' = \begin{vmatrix} 2u & b\sin m\varphi & b\sin 2m\varphi & b\sin 3m\varphi & \ldots \\ b\sin m\varphi & 2\gamma\cos\varphi & 0 & 0 & \ldots \\ b\sin 2m\varphi & 0 & 2\gamma\cos 2\varphi & 0 & \ldots \\ b\sin 3m\varphi & 0 & 0 & 2\gamma\cos 3\varphi & \ldots \\ 0 & 0 & \ldots & & \ldots \\ \ldots & \ldots & \ldots & \ldots & \ldots \end{vmatrix} \tag{39}$$

The eigenvalues and eigenvectors are thus given by (22) and (23) with $\sin k\varphi$ (but not $\cos k\varphi$) replaced by $\sin km\varphi$. The changes are not dramatic, because they correspond to second order effects; but they are substantial. For $2u = -3.46$ ev ($\epsilon_a^\circ = -15$ ev), $\beta = -0.2$ ev, $\gamma = -2$ ev, the results obtained for x_j are:

−3.4600	−3.4604	−3.4122	ev
−3.6039	−3.6188	−3.6603	
−2.4940	−2.4868	−2.4917	
−0.8901	−0.8859	−0.8874	
+0.8901	+0.8926	+0.8917	
+2.4940	+2.4951	+2.4943	
+3.6039	+3.6042	+3.6054	

for A and P separated, for A on end site, and for A on site 3 (m=3), respectively.

When the adsorbed atom is in a bridge position over two sites, the situation can be schematized as follows

$$\ldots T_{-2} \text{----} T_{-1} \text{----} S_{-1} \overset{A}{\text{----}} S_{+1} \text{----} T_{+1} \text{----} T_{+2} \ldots$$

The matrix **H** then contains two β's, corresponding to the coupling of A with S_{-1} and S_1: it reduces to a three-by-three block rather than to a two-by-two one when γ tends to zero. We leave to the reader the discussion of this case.

The general problem of comparing different adsorption sites is to be tackled along the lines indicated above. The great difficulty comes from the enormous number of variables that come into play. Think of a polypeptide, or even of polypropylene: an adsorbed atom or ion interacts with many of the substrate atoms, depending on the conformation as well as on their nature. It is practically impossible to take all the parameters into account, and that is why it is useful to carry out a separation of effects - van der Waals, electrostatic, charge transfer, quantum coupling interactions - so as to make a classification of those parameters and to base on it reasonable simplifications.

The relationship between adsorption and change in conformation is a most delicate point. We have mentioned that question when speaking of the deformation of an adsorbed molecule and when speaking of the polarization induced by an ion approaching a polypeptide. The general question is: how does adsorption modify a given conformation of a polymer? Are there just local or also long-range effects? How can we obtain some working assumption on the 'geometrical relaxation' that follows adsorption?

Much can be done to answer that question by present quantum chemical computational methods, possibly supplemented by empirical formulas for certain contributions (like long-range effects due to correlation). However, the best techniques for most purposes are those which try to determine the equilibrium geometries with the help of educated guesses or empirical relations on most parameters, limiting sophisticated analyses to a few selected features. The temptation to use big computers to carry out brute-force highly sophisticated computations should be resisted, because the situations studied would be anyway highly idealized ones: theoretical studies are generally meant to detect the mechanisms of phenomena and to classifying the factors influencing them according to their relative importance; quantitative accuracy is essentially a reliability test subject to the condition that one knows exactly the correspondence between calculated and measured quantities.

4.4 The orbital basis: overlap

Two comments are important as a conclusion of the present cursory examination of the elementary electronic aspects of adsorption.

The distinction between σ and π (localized and delocalized) MO's has not been critically discussed in the preceding considerations. We have referred to the π orbitals of a polyene, but we have also treated a σ system on the example of polyglycine. Actually, the latter provides a fairly complete example of the situation to be expected in general in polymers from the point of view of orbitals. There is a σ-bond framework which provides a backbone whose properties with respect to adsorption are strictly local ones. Then there may be π systems which extend over more than two centers and provide highly polarizable electron groups (the NCO bond); finally, there may be a more or less weak long range interaction, which is not necessarily conjugation of the polyene type, but may arise from hyperconjugation, viz. weak interactions between distant σ and σ or π obitals. The latter, in a polypeptide, might well be interactions between different peptide bonds. We have seen that even very weak γ-values allow at least charge transfer: the long-range nature of that effect (and of polarization) is associated with some 'metallic' quality: in chemical language, we may well speak of a <u>mesomeric</u> effect accompanying the local inductive effect already discussed.

The above considerations require an orbital basis corresponding essentially to the entire $\sigma+\pi$ STO (Slater atomic orbital) basis. A discussion of the characteristics of that basis is out of the scope of the present notes. We just remind the reader of one point which is always somewhat disturbing when the models and examples given above are referred explicitly to a STO basis: the assumed orthogonality of the AO basis, which is certainly not respected by STO's. As a matter of fact, orthogonality is not a necessary assumption: indeed, there are modifications of the definitions and formulas associated with model Hamiltonians that allow explicit inclusion of overlap. At any rate, in most cases it is legitimate to assume that the off-diagonal elements of the one-electron Hamiltonian matrix are proporti nal to overlaps through a constant factor k. Then it is possible to prove that a computation not neglecting overlap reduces to one where orthogonality is assumed by a linear transformation of the eigenvalues

$$e_m \longrightarrow ke_m/(k - e_m) \qquad (40)$$

and similar transformations of the Hamiltonian matrix elements. If k is not too small, the conclusions drawn by neglecting overlap remain valid, except perhaps when it comes to excited states.

5. TIME-DEPENDENT ASPECTS OF ADSORPTION

The model schemes discussed above lay the way not only to the discussion of geometrical effects of chemisorption, but to the theoretical study of time-dependent phenomena associated with it. We shall outline here two aspect of this subject: spectral changes in adsorbed molecules and residence times of adsorbed species.

5.1 Electronic absorption spectra of adsorbed molecules

We shall consider here relatively large adsorbed molecules, because only with such molecules has the consideration of optical spectra a real interest. To fix ideas, we shall think of an azobenzene adsorbed head to tail or by its N=N bridge on a polymer of the helical polypeptide type. Assume that adsorption reduces to a very weak association. The question is: is the UV spectrum of azobenzene affected by adsorption, and how? The answer can be given at various levels of sophistication: we consider here only the qualitative aspects of the simplest level.

The appearance of an UV absorption band results from the manifold of transitions that take place under a flow of white radiation from the Boltzmann distribution of the lower vibrational levels of the ground state (say, S_0) to accessible vibrational states of the first excited-state manifold (say, S_1) having the same spin multiplicity. In the Born-Oppenheimer approximation, the shape of the band and the position of its peak are functions of shapes and positions of the potential energy surfaces of the given molecule in the states S_0 and S_1. The absorption coefficient is proportional to the energy difference and to the square of the electric dipole matrix element $\langle fv'|\underline{\vec{M}}|iv\rangle$ between the final B.O. state $|fv'\rangle$ and the initial B.O. state $|iv\rangle$. A UV band is the envelope of all the transitions corresponding to the given electronic states $|f\rangle$, $|i\rangle$ with variable vibrational parts.

For a given transition $\langle fv'|\underline{\vec{M}}|iv\rangle$ can normally be treated as the product of an electronic factor $\underline{\vec{M}}_e = \langle f|\underline{\vec{M}}|i\rangle$ by a Franck-Condon (FC) factor which is the overlap integral of the vibrational states associated with v' and v, respectively. Now, adsorption may change very little $\underline{\vec{M}}_e$ (in certain circumstances), but will practically always affect the FC factors by affecting certain vibrational modes. For instance, it is not totally unreasonable to imagine that <u>trans</u>-azobenzene is adsorbed on a polypeptide in its normal geometry and with little change in the energy of the electronic states (if it is adsorbed at all): but it is certainly unlikely that the torsional motions of the phenyl groups would then remain exactly as easy as in the free molecule. Therefore, the frequencies and the spacings of the torsional vibrations of the phenyl groups will be different from what they are in the isolated molecule (or, say, in cyclohexane solution) and the shape of the absorption band will be changed because of changes in the FC factors associated with those torsional motions.

The FC factors actually originate from all the vibrational modes of a molecule, and a complete calculation of them would require knowledge of the entire potential energy hypersurfaces associated with the two electronic states: for azobenzene 66 space coordinates each. Therefore, the spectroscopic behavior of an adsorbed molecule can be described only if it is possible to sort out the modes that are affected by adsorption: the modes that are 'coupled' to one another by the field of the substrate. An analysis of this question must be carried out by a careful examination of theoretical models and experimental evidence.

In the example given above it has been assumed that the only role of the substrate is to create a potential modifying the vibrations of an adsorbed molecule. A much more intriguing possibility of change in the absorption spectrum is the coupling between the motions of the molecule and (local or collective) vibrations of the substrate. In this case we deal with a situation where excitation and re-emission are not the only possible processes: energy absorbed by the molecule may be transferred to the substrate by excitation of vibrations of the substrate itself that are degenerate with the excited molecular state created by the exciting light. The bandshape will be affected accordingly, and - at variance the other case - its total intensity will not be conserved.

Also the latter situation cannot be treated theoretically unless a severe selection of the modes involved is carried out. The subsequent quantitative part of the analysis is based on the state-coupling concept. Note first of all that we are no longer considering orbitals nor electronic states, but states of the whole admolecule-substrate system: for the rest, the quantum-mechanical formalism to be adopted follows exactly the same lines as in the electronic case, with introduction of a basis and of a Hamiltonian matrix representing the total Hamiltonian operator. Now consider the following three states:

$|a\rangle$, representing the whole system with energy E above the ground state, the admolecule in an excited 'vibronic' state and the substrate in its ground state;

$|b\rangle$, representing a state of energy E' above the ground state, with the substrate still in its ground state and the admolecule in a different excited vibronic state;

$|c\rangle$, a state of energy E", with the molecule in its ground state, and the substrate vibrationally excited.

The three states just defined are stationary states of the whole system when there is no interaction between the admolecule and the substrate, but may have to be linearly combined to give stationary states if $E \approx E' \approx E''$ and if there is such an interaction. Now suppose that an excitation by light can only take place to one of the three basis states, say $|b\rangle$: then excitation with a frequency corresponding to a stationary state $|\bar{E}\rangle$ of energy $\bar{E}$ will take place with a probability that depends on the weight of $|b\rangle$ in $|\bar{E}\rangle$. That weight is the modulus square of the coefficient of $|b\rangle$ in the linear combination representing $|\bar{E}\rangle$: that linear combination can be found by diagonalizing the matrix representing the total (electronic plus nuclear) Hamiltonian operator

$$\hat{H}_{total} = \hat{H}_{admol} + \hat{H}_{substr} + \hat{V}, \tag{41}$$

i.e.

$$\mathbf{H} = \begin{vmatrix} E & \langle a|\hat{V}|b\rangle & \langle a|\hat{V}|c\rangle \\ \langle b|\hat{V}|a\rangle & E' & \langle b|\hat{V}|c\rangle \\ \langle c|\hat{V}|a\rangle & \langle c|\hat{V}|b\rangle & E'' \end{vmatrix}, \tag{42}$$

exactly as in the Huckel method.

5.2 Residence times

Residence times of adsorbed species are important for the theoretical interpretation of the properties of materials like ion exchange resins: the equilibrium between adsorbed ions and ions in solution is obviously related to the mean lifetime of an adsorbed ion. From the theoretical point of view this problem is one where time dependence comes into play explicitly, and not indirectly as in the case of optical spectra.

The process to be studied can be represented as a combination--dissociation process:

$$A + P \rightarrow AP \rightarrow A + P \tag{43}$$

where A is the adion, P the substrate. The process (43) contains, first of all, the statement that at least one of the external degrees of freedom of isolated A becomes an internal degree of freedom of the system AP (a consideration left in the shadow in Sec. 5.1). The situation can be visualized by thinking of H^+ or OH_3^+ approaching an oxygen atom of the polypeptide of Fig. 1. In the case of H^+ it is easily acceptable that the process (43) can be described by a single parameter R, the $H^+ \ldots O$ distance, even though it is not clear how changes in bond lengths and force constants of C=O and other bonds should be treated. In the case of OH_3^+ the situation is even more complicated, because the orientation of approach and the rôle of rotation of the hydronium ion may be quite important.

In short, before even asking how residence times could be evaluated, the possibility of isolating a single reaction coordinate to describe the process (43) must be studied, as well as the way in which the other degrees of freedom should be taken into account. We give here only the answer to that question, without any attempt to justification. If motion along a special coordinate, say R, is much slower than all the other motions, it is actually possible to write a time-dependent problem associated only with that coordinate: the other degrees of freedom are assigned stationary states, and build up an effective time-dependent operator $\Delta\hat{H}(t)$ which adds to $\hat{H}_0$, the kinetic and potential energies already associated with R in the total Hamiltonian. Thus, calling $|\Phi(t)\rangle$ the state of (A,P) at the time t, we can write the equation

$$i\hbar \frac{\partial}{\partial t}|\Phi(t)\rangle = [\hat{H}_0 + \Delta\hat{H}(t)]\,|\Phi(t)\rangle. \qquad (44)$$

This equation is defined for any state of (A,P). We can choose a state which at t = 0 corresponds to AP:

$$|\Phi(0)\rangle = |AP\rangle. \qquad (45)$$

Using (45) and (44) we can now try to answer the question: why and after how long will $|AP\rangle$ become $|A+P\rangle$?

The answer is quite simple. The states $|AP\rangle$ and $|A+P\rangle$ are coupled states of the (A,P) system: there is an off-diagonal element β of the Hamiltonian operator appearing in (44) that connects them. Suppose that we are actually dealing with a two state system: then the problem to be solved reduces to the diagonalization of the 2×2 matrix representing $\hat{H}_0 + \Delta\hat{H}$ (if the latter can be treated as a constant, or to a slightly more complicated procedure in general) over the two states $|AP\rangle$ and $|A+P\rangle$. If the latter are degenerate and $\Delta\hat{H}$ is virtually constant over the time intervals of physical interest, the probability of finding $|A+P\rangle$ in $|\Phi(t)\rangle$ at any given time t is a periodic function of time going from 0 to 1 in a time

$$T_{diss} = \pi\hbar/2\beta. \qquad (46)$$

This time can be taken as the residence time.

The above reasoning is extremely simplified, not only because it assumes that the two basis states are degenerate and that the Hamiltonian of (44) is constant in time, but because, if $|AP\rangle$ can indeed be treated as a single vibrational state of A linked to P, A+P is not a single state, but a set of very closely spaced states - in fact a continuum if we do not assume that we are working in a finite vessel - corresponding to the translations of A. Nevertheless, (46) can be accepted as a good estimate of the residence time if we select for A+P only that state which has exactly the same energy as AP and corresponds to an appropriate direction of flight. It can be proved that the most important consequence of taking into account the fact that A+P is a continuum or a quasi continuum is that the decay of AP , instead of being periodic, becomes exponential.

In either case, numerical results can only be obtained if some potential energy surface is available for describing the situation of A in the vicinity of P: so that we are thrown back to the application of simple models of the type discussed in the preceding sections.

REFERENCES

We do not give here a detailed list of references, but some suggestions for further reading.

To Sec. 2.

1) D.M. Newns, Phys. Rev. 178, 1123 (1969).
2) T.B. Grimley, CRC Solid St. Sci. 6, 239 (1976).
3) T.B. Grimley, J. Vac. Sci. Techn. 8, 31 (1971).
4) J. Koutecký, Adv. Chem. Phys. 9, 85 (1965); Trans. Fer. Soc. 54, 1038 (1958).
5) J. Koutecký, Progress in Surface and Membrane Sci. 11, 1 (1976).

To Sec. 3.

1) P.W. Anderson, Phys. Rev. 124, 41 (1961).
2) B.J. Thorpe, Surf. Sci. 33, 306 (1972).
3) J. Koutecký, Adv. Chem. Phys. 9, 85 (1965); Trans. Fer. Soc. 54, 1038 (1958).
4) J. Koutecký, Progress in Surface and Membrane Sci. 11, 1 (1976).

To Sec. 4.

1) G. Del Re, J. Chem. Soc. 1958, 4031.
2) G. Biczó, M. Kertész, S. Suhai, Z. Chem. 15, 203 &1975).
3) E. Clementi, H. Popkie, J. Chem. Phys. 57, 1077 (1972).
4) A. Julg, G. Del Re, V. Barone, Phil. Mag. 35, 517 (1977).
5) A. Warshel, S. Lifson, J. Chem. Phys. 53, 582 (1970).
6) H. Müller, Ch. Opitz, Z. Physik. Chem. 257, 482 (1976).
7) T.B. Grimley, J. Phys. C: Solid St. Phys. 3, 1934 (1970).
8) G. Del Re, in "Quantum Science". J.L. Calais et al., eds., Plenum Press New York, 1976 p. 53.

9) J. Koutecký, Adv. Chem. Phys. 9, 85 (1965); Trans. Fer. Soc. 54, 1038 (1958).
10) J. Koutecký, Progress in Surface and Membrane Sci. 11, 1 (1976).

To Sec. 5.
1) A. Lami, G. Del Re, Techn. Note 1977; cf. A. Wittkowski, W. Moffitt, J. Chem. Phys. 33, 872 (1970); J.S. Briges, A. Herzenberg, Mol. Phys. 23, 203 (1972).
2) G. Del Re, A. Lami, Bull. Soc. Chim. Belg. 85, 995 (1976).
3) R.A. Marcus, J. Chem. Phys. 45, 4493 (1966).
4) R.D. Levine, Quantum Mechanics of Molecular Rate Processes, Clarendon Press, Oxford, 1969.
5) R.A. Van Santen, J. Chem. Phys. 57, 5418 (1973).
6) J.L. Houben, O. Pieroni, A. Fissi, F. Ciardelli, Biopolymers (in press) (1977).

HYDROGEN BONDING BETWEEN MOLECULES OR IONS AND IN MOLECULAR CRYSTALS [+)]

P. Schuster

Lehrstuhl für Theoretische Chemie und Strahlenchemie der Universität Wien

I. INTRODUCTION

In molecular physics scientists are inclined to interpret interactions between atoms, ions or molecules by a set of "forces" which usually are grouped into different classes: weak and strong, specific and unspecific, etc. Quantum mechanics, on the other hand, tells us that there is no unique way to split a given energy of interaction into "physically" meaningful contributions as long as the charge distributions of individual atoms, ions or molecules are spread over unlimited regions due to the exponential decay of their wave functions (see eg. Ahlrichs [1976]). The source of these ambiguities is the mutual penetration of electron density distributions of the interacting subunits which can be expressed quantitatively in terms of overlap integrals.

Several attempts have been made to justify and define energy partitioning in intermolecular associations at least under certain conditions. The perturbational approach, one of the oldest attempts of this kind (Eisenschitz and London [1930]) has been reactivated recently. An alternative approach starts out from SCF calculations on the complexes: the energy of interaction is split within the frame of model assumptions.

In this contribution we are concerned with one of the most complicated kind of intermolecular interactions: the hydrogen bond. Using concrete examples of small complexes and molecular crystals built from small units we try to review the present stage of the quantum chemical approach to this phenomenon. Additionally, we shall present the results of SCF energy partitioning in order

---

+) This contribution consists of parts of a review by Schuster 1977 .

J.-M. André et al. (eds.), Quantum Theory of Polymers, 229-255.

to be able to see the hydrogen bond within a more general concept in intermolecular interactions.

2. ENERGIES OF INTERACTION AND VIBRATIONAL FREQUENCIES OF HYDROGEN BONDED COMPLEXES

Within the frame of MO theory intermolecular interactions can be studied in a straight forward way by the "supermolecule" approach. A Hamiltonian is constructed for the complete system which consists of two or more atoms, molecules or ions, and the Schrödinger equation is solved by the standard approximation techniques. For a binary complex, AB we obtain:

$$H\Psi_{AB} = E_{AB}\Psi_{AB}; \qquad H = H_A^o + H_B^o + V_{AB}^o \tag{1}$$

H consists of the Hamiltonians of the isolated subsystems H_A^o and H_B^o and the interaction term V_{AB}^o. The energies of interaction are obtained as differences:

$$\Delta E = E_{AB} - (E_A^o + E_B^o) \tag{2}$$

E_A^o and E_B^o are total energies obtained by independent calculations of the isolated subsystems. In order to be consistent the same approximation method as used in the calculation of the complex has to be applied again. The major problem of the supermolecule approach to intermolecular forces can be visualized immediately from eq. (2). E_{AB} and $E_A^o+E_B^o$ are huge numbers compared to their difference. Consequently, no MO calculation on intermolecular interactions could be successful before the numerical techniques had reached sufficiently high accuracy.

2.1 SCF-results

The results of SCF calculations on energy surfaces of hydrogen bonded complexes have been reviewed recently (Kollman and Allen [1972], Schuster [1976]). Therefore we will discuss here only a few representative examples showing all important and characteristic features of ordinary hydrogen bonds: $(HF)_2$, $(H_2O)_2$ and $F^-.H_2O$. In order to be able to recognize the differences, in case there are any, between hydrogen bonded complexes and other strong intermolecular interactions we will compare some results also with those obtained on a typical ion-molecule complex, Li^+OH_2 (see also Schuster et al. [1975], Schuster, Lischka and Beyer [1976]). Two very recent and extensive studies on hydrogen bonded complexes should be mentioned here too: calculations on a large variety of complexes including hydrogen bonded structures involving π-electron systems and compounds of third row elements (HCl, H_2S, PH_3, HCP and H_2CS) too (Kollman et al. [1975]) and a systematic study on basis set effects using various medium size basis sets (Dill et al. [1975]). Finally, we will refer the interested reader also to a recent study on symmetric intramolecular hydrogen bonds (Isaacson and Morokuma [1975]).

$(HF)_2$ is the simplest example for a hydrogen bonded complex between two neutral molecules. Since it consists of two linear molecules the whole energy surface is four dimensional (fig.1) if we assume frozen geometries of both HF molecules. A representative calculation covering the energetically important parts is within the range of present computational possibilities. Furthermore the most stable linear arrangement of all four atoms has almost the same energy as the tilted true equilibrium geometry. Linear geometry can be assumed therefore for model calculations without losing all relevance for the real system. Due to higher symmetry the computations on linear arrangements are less time consuming, and more accurate treatments are accessible. Additionally, $(HF)_2$ and (H_2O) are the only two small hydrogen bonded complexes for which direct experimental data are available.

The most extensive calculations on linear $(HF)_2$ have been performed by Lischka [1974]. The most complete investigations of the energy surface have been presented by Yarkony et al. [1974], and by Dierksen and Kraemer [1970]. The equilibrium structure obtained in the latter two investigations is roughly the same (table 1). Variation of the HF bond length (R_{HF}) in the proton donor molecule leads to very small elongation only and therefore the rigid molecule approach of Yarkony et al. [1974] seems to be well justified. The calculated FF distance (R_{FF}) comes very close to the experimental value obtained by Dyke, Howard and Klemperer [1972] using a molecular beam - electric resonance technique: R_{FF} = 2.79 Å. Variation of the angle Θ_B leads to a very flat minimum around $\Theta_B = 50^o$ which is only 0.1 kcal/mole more stable than the completely linear structure. Variation of Θ_A finally shows that the equilibrium geometry of the hydrogen bond is slightly tilted: $\Theta_A \approx 355^o$. As far as structural data for $(HF)_2$ are available the agreement with the SCF results is excellent indeed.

Coming now to the absolute value of the hydrogen bond energy we realize immediately that much more care is needed than in case of the equilibrium structure. Although Diercksen and Kraemer [1970] as well as Yarkony et al. [1974] used rather extended basis sets, we find that the energy of interaction is changed substantially - decrease of about 20% - when a second polarization function with a small exponent is added to the basis set. Similar basis set extension has an important influence on some calculated properties of the isolated molecules as well. In case of HF and H_2O the addition of flat polarization functions to the basis set reduces dipole moments and increases polarizabilities. In a very qualitative way we can estimate the influence of these corrections on the energy of interaction in a

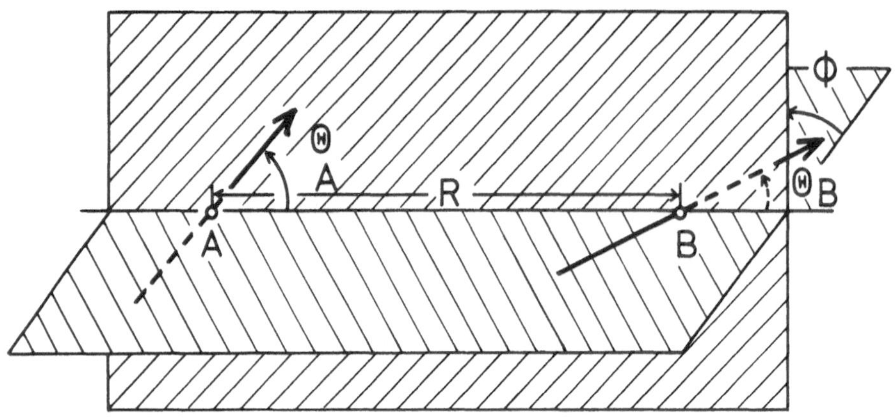

Fig. 1. Four intermolecular coordinates (R, ϑ_A, ϑ_B and φ) characterizing the geometry of a complex formed from two linear molecules A and B.

Table 1: Results of *ab initio* calculations on $(HF)_2$

Basis set (GTO's)	Method	Equilibrium geometry$^{+)}$ R_{FF}(Å)	ΔR_{HF}(Å)$^{+)}$	ΘA	ΘB	ΔE (kcal/mole)	Ref.
(9,5,1/4,1)—[4,2,1/1,2]	SCF	2.80		3.55	52°	-4.63	a
(11,7,1/6,1)—[5,4,1/3,1]	SCF	2.85		0°++)	40°	-4.50	b
(11,7,1/6,1)—[5,4,1/3,1]	SCF	2.90		0°++)	0°++)	-4.38	b
(11,7,2/6,1)—[7,4,2/4,1]	SCF	2.90		0°++)	0°++)	-3.46	c
	IEPA	2.89	0.003	0°++)	0°++)	-3.25	c
	CEPA	2.89	0.003	0°++)	0°++)	-3.36	c
	EXP	2.79	0.002	0°	60-70°	-6±1.5	d,e

+) According to fig.1: "A" represents the proton donor molecule. ΔR_{HF} is the elongation of the central FH bond on hydrogen bond formation. The planar arrangement ($\phi=0^\circ$) was found to be most stable at the equilibrium value of R_{FF} (see Ref. a).

++) Constrained values

References: a YARKONY et al. [1974], b DIERCKSEN and KRAEMER. 1970, c LISCHKA [1974], d DYKE, HOWARD and KLEMPERER [1972], e see SCHUSTER [1976a].

complex between two polarizable dipole molecules A and B. According to the classical expression the electrostatic contribution is proportional to the square of the dipole moment (3) and will decrease in absolute value with decreasing $|\vec{\mu}|$.

$$\Delta E^{CL}_{COU} = (1/R^3)\{\vec{\mu}_A \cdot \vec{\mu}_B - (3/R^2)\ (\vec{\mu}_A \cdot \vec{R})\ (\vec{\mu}_B \cdot \vec{R})\} + \ldots \quad (3)$$

The polarization energy is proportional to the product of the polarizability and the square of $|\vec{\mu}|$:

$$\Delta E^{CL}_{POL} = (1/R^6)\{\alpha_A \cdot |\vec{\mu}_B|^2 + \alpha_B |\vec{\mu}_A|^2\} + \ldots^{+)} \quad (4)$$

On basis set extension one of these two factors will increase, whereas the other will decrease. Since both relative changes fall into the same order of magnitude we can assume the product remains roughly constant. Additionally, around the energy minimum of the complex the polarization energy is substantially smaller in absolute value than the electrostatic contribution. At the equilibrium geometry of the complex we expect therefore a net destabilizing effect of basis set extension as it is observed indeed. In other parts of the energy surface the electrostatic contribution may have opposite sign and in contrary we expect a lowering of the energy of interaction.

A second source of errors, most important for small, especially minimum basis sets, concerns mutual basis set extension by the orbitals of the other molecules in the complex. An upper limit of error caused thereby can be estimated by the function counterpoise method of Boys and Bernardi [1970], which consists in calculations of the isolated molecules using the basis set of the whole complex. These basis set superposition errors have been investigated for many hydrogen bonded complexes (Meunier, Levy and Berthier [1973], Johansson, Kollman and Rothenberg [1973], Diercksen, Kraemer and Roos [1975]). In case of the most extended basis set used in a calculation on $(HF)_2$ this error amounts roughly one tenth of a kcal/mole, but may be substantially larger for smaller basis sets (table 2).

The results obtained in some extended calculations on $(H_2O)_2$ are summarized in table 3. On the whole the situation is essentially the same as in $(HF)_2$ besides the fact that due to additional degrees of freedom a complete investigation of the energy surface requires much more computer time. Popkie, Kistenmacher and Clementi [1973], Diercksen [1969, 1971] and Hankins, Moskowitz and Stillinger [1970a, b; 1973] have made the most extensive SCF calculations on this surface. Only in Diercksen's [1971] paper the deformation of the proton donor molecule in the field of the proton acceptor

+) For this simplified estimate spherical symmetry of the polarizability tensor has been assumed.

Table 2: Basis set superposition errors in calculations on $(HF)_2$ and $(H_2O)_2$

Complex	Basis set (GTO's)	Method	Energies in kcal/mole $E_A^{++)}-E_o$	$E_B^{++)}-E_o$	$E_A+E_B-2E_o$	Ref.
$(HF)_2$	STO-3G$^{+)}$			-1.97	-1.97	a
linear, R_{FF}=2.91 Å,	(8,4,1/4,1)—[5,3,1/3,1]	SCF	-0.19	-0.77	-0.96	b
R_{HF}=0.917 Å	(11,7,2/6,1)—[7,4,2/4,1]	SCF	-0.025	-0.044	-0.07	b
$(H_2O)_2$, Θ_B=54°	STO-3G$^{+)}$	SCF		-2.82	-2.85	a
R_{OO}=3.00 Å	(11,7,1/6,1)—[5,4,2/3,1]	SCF-CI	-0.39	-0.42	-0.81	c
		SCF-CI	-0.77	-0.77	-1.54	c

+) See HEHRE, STEWART and POPLE [1969] and DITCHFIELD, HEHRE and POPLE [1971]

++) E_A and E_B represent the energies of the proton acceptor and proton donor molecule calculated with the basis set of the whole complex. E_o is the energy of the isolated monomer.

References: a JOHANSSON, KOLLMAN and ROTHENBERG [1973], b LISCHKA [1974]
c DIERCKSEN, KRAEMER and ROOS [1975].

is reported. Similar to the result on $(HF)_2$ a small elongation of the central OH bond was obtained. Again no experimental information on this change in geometry is available. The data obtained from various structures of ice are not comparable to vapor phase results, as we will see in section 2.3. The other predictions of SCF calculations on the structure of $(H_2O)_2$ agree well with the measurement of Dyke and Muenter [1974]. The calculated energies of interaction again are highly sensitive to changes in the basis sets applied. For both dimers $(HF)_2$ and $(H_2O)_2$ the experimental data on the energy of complex formation are at least as uncertain as the theoretical results.

IR spectra of complexes with hydrogen bonds between neutral molecules have been calculated by ab initio methods as well (see e.g. Janoschek [1976]). The most extensive theoretical study on the vibrational spectrum of $(H_2O)_2$ has been performed recently by Curtiss and Pople [1975]. For the major part of their investigations they used a small basis set without polarization functions, which is well known to overemphasize the hydrogen bond energy in $(H_2O)_2$ (see Schuster [1976]). Nevertheless the general features of the experimental spectrum measured by matrix isolation technique in solid N_2 (Tursi and Nixon [1970], Van Thiel, Becker and Pimentel [1975]) are described correctly (table 4). The most serious errors, of course are due to the weakness of SCF calculations in reproducing stretching force constants correctly.

Table 3: Results of ab initio calculations on $(H_2O)_2$

Basis set (GTO's)	Method	Equilibrium geometry+) R_{OO}(Å)	ΔR_{OH}(Å)	Θ_B	ΔE (kcal/mole)	Ref.
(11,5,1/4,1)→[5,3,1/2,1]	SCF	3.00		40°	-5.0	a
(11,7,1/6,1)→[5,4,1/3,1]	SCF	3.00	0.002		-4.84	b
(11,7,1/6.1)→[5,4,1/3,1]	SCF	2.99		41.5^o	-5.14	c
	SCF-CI	2.92		42.6^o	-6.05	c
(13,8,2,1/5,2,1)→[8,5,2,1/2,1]	SCF	3.00		0°++)	-3.67	d
	SCF	3.00		30°++)	-3.90	d
(11,7,1/6,1)→[4,3,1/2,1]	SCF	3.01		40°	-4.55	e
	SCF-CI	2.98		40°	-5.63	e
	EXP	2.98±0.04		$60^o \pm 10^o$	-5.0	f,g

+) Linear or almost linear hydrogen bonds O-H··O. C_S symmetry is assumed. Θ_B denotes the angle between the C_2 axis of the proton acceptor molecule "B" and the hydrogen bond axis. ΔR_{OH} is the elongation of the central OH bond on hydrogen bond formation.

++) Constrained values

References: a HANKINS, MOSKOWITZ and STILLINGER [1970a,b, 1973] b DIERCKSEN [1971], c DIERCKSEN, KRAEMER and ROOS [1975], d POPKIE, KISTENMACHER and CLEMENTI [1973], e MATSUOKA, CLEMENTI and YOSHIMINE [1976], f DYKE and MUENTER [1974], g see SCHUSTER [1976a].

All wavenumbers for stretching vibrations obtained by Curtiss and Pople [1975] are too high by 350-400 cm^{-1}. Similar results have been reported on $(HF)_2$ (table 4). In this case we should note that small basis sets - we will use one of them later in crystal orbital calculations - sometimes are useful for predictions of roughly correct geometries and force constants due to error compensation. In general, the shifts to lower frequencies of stretching vibrations of HX bonds involved in hydrogen bonding are much smaller in isolated dimers than those observed in associated liquids and hydrogen bonded crystals.

Table 4: Calculated force constants and IR frequencies for $(HF)_2$ and $(H_2O)_2$

Molecule or complex	Basis set (GTO's)	Method	XH Stretching modes[+)] f_{XH} (mdyn/Å)	Wave numbers ν'_i (cm^{-1})	XX Stretching mode[+)] f_{XX} (mdyn/Å)	HXH Bending modes[+)] $f_{HXH} \cdot R_e^2$ (mdyn/A)	Wave numbers (cm^{-1})	Ref.
HF	(8,4/4)→[4,3/3]	SCF	9,42					a
	(11,7,2/6,1)→[7,4,2/4,1]	SCF	11.20					b
		IEPA	9.27					b
		CEPA	9.92					b
		EXP	9.66					b
$(HF)_2$	(8,4/4)→[4,3/3]	SCF	9.24,9.41		0.23			a
	(11,7,2/6,1)→[7,4,2/4,1]	SCF	11.0		0.095			b
		IEPA	9.18		0.101			b
		CEPA	9.83		0.104			b
H_2O	4-31G[++)]	SCF	9.05	3960,4098		0.856	1767	c
		EXP,VAPOR	8.43	3657,3756		0.768	1595	c
		EXP,N_2MATRIX[+++)]	7.52	3632,3725		0.758	1597	c
$(H_2O)_2$	4-31G[++)]	SCF	8.66,9.15,9.15,9.15	9307,3979,4085,4121		0.898,0.858	1813,1771	c
		EXP,N_2MATRIX[+++)]	7.04,7.44,7.50,7.54	3548,3626,3698,3714		0.787,0.756	1618,1600	c
	(11,7,1/6,1)→[5,4,1/3,1]	SCF			0.14			e
		SCF-CI			0.16			e

+) For complexes the harmonic force constants for pure XH or XX stretching or HXH bending motions are given. In case one value is presented only it refers to the force constant of the central XH bond involved in hydrogen bonding.

++) See DITCHFIELD, HEHRE and POPLE [1971], and HEHRE, DITCHFIELD and POPLE [1972]

+++) The values of the force constants are not corrected for the anharmonic effects.

References: a KARPFEN and SCHUSTER [1976], b LISCHKA [1974], c CURTISS and POPLE [1975] d TURSI and NIXON [1970]
e DIERKCSEN, KRAEMER and ROOS [1975]

One of the most remarkable effects of hydrogen bonding in IR spectra is the strong increase in band intensities of HX stretching modes. The intensity of a vibrational mode (Q_v) is proportional to the square of the corresponding dipole moment derivative. For the increase of intensity on complex formation we obtain:

$$I_R = \{|\partial\vec{\mu}/\partial Q_v|_{AH.B}\}^2/\{|\partial\vec{\mu}/\partial Q_v|_{AH}\}^2 \qquad (5)$$

An experimental value of I_R=12 has been reported by Van Thiel, Becker and Pimentel [1957], whereas the more recent results of Tursi and Nixon [1970] indicate a much smaller value. In his extended SCF calculation Diercksen [1971] obtained a value of I_R=5.3 in contrast to some investigations with small and medium size basis sets which suggested a larger increase in intensities (Kollman and Allen [1972]).

2.2 Correlation effects

Electron correlation was found to be important for accurate calculations on energy surfaces of hydrogen bonds. In general Hartree-Fock calculations cannot account for dispersion energies and therefore a stabilizing contribution to the energy of interaction is always missing in SCF calculations. According to the asymptotic formulas derived from intermolecular perturbation theory (see e.g. Margenau and Kestner [1969]) dispersion energies depend on molecular polarizabilities and ionization potentials but not on charges or multipole moments. Neglect of dispersion forces, therefore, is most serious in case of weakly interacting systems where the total energy of interaction is small. Among the examples discussed here we expect the contribution of dispersion energy to be much more important in associations between neutral molecules than in ion-dipole complexes.

Besides its principal importance for dispersion forces electron correlation contributes to intermolecular energy surfaces also in an indirect way by changing the properties of the isolated subsystems. Appropriately, we can distinguish both effects by splitting the change in correlation energy into an intermolecular part - corresponding to the dispersion forces - and an intramolecular part. For the interpretation of the "intramolecular" contributions of electron correlation we make use again of the classical formulas in eq. (3) and (4). In H_2O and HF electron correlation reduces the dipole moments and increases the polarizabilities. Again we can estimate a small net effect of the intramolecular contribution. In fact this term is destabilizing in $(HF)_2$ (Lischka [1974]), whereas it was found to be almost zero in $(H_2O)_2$ (Matsuoka, Clementi and Yoshimine [1976]) at the equilibrium structures. Superimposing now the intermolecular contributions we obtain an almost negligible contribution of the electron correlation to the total energy of interaction in $(HF)_2$, whereas a net stabilizing effect is observed in $(H_2O)_2$ (table 5). Electron correlation effects in $F^-.H_2O$ and $Li^+.OH_2$ are of the same order of magnitude as in $(HF)_2$ or $(H_2O)_2$ and can be analysed in exactly the same way. Due to the larger energy of interaction, correlation contributions are relatively less important here.

Correlation effects on vibrational spectra of ordinary hydrogen bonded complexes mainly consist in a correction of the force constants for stretching vibrations without changing the relative shifts of the individual normal modes.

Table 5: Intra- and intermolecular correlation effects in hydrogen bonded or ion-molecule complexes near the equilibrium geometries.

Complex	Basis set (GTO's)	Method	Energies in kcal/mole			Ref.
			ΔE^{INTRA}_{COR}	ΔE^{INTER}_{COR}	ΔE_{COR}	
$(HF)_2$	(11,7,2/6,1)—[7,4,2/4,1]	CEPA	0.67	-0.71	0.09	a
$(H_2O)_2$	(11,7,1/6,1)—[5,4,1/3,1]	SCF-CI			-0.87	b
	(11,7,1/6,1)—[4,3,1/2,1]	SCF-CI	-0.41	-1.20	-1.61	c
$(FHF)^-$	(8,5,1/4,1) — [4,3,1/2,1]	CEPA	8.22	-10.04	-1.82	d
$F^-.H_2O$	(11,7,1/6,1)—[5,4,1/3,1]	SCF-CI			-1.94	b
$Li^+.OH_2$	(11,7,1/6,1)—[5,4,1/3,1] Li:(11,2)—[5,2	SCF-CI			1.16	b

References: a LISCHKA [1974], b DIERCKSEN, KRAEMER and ROOS [1975], c MATSUOKA, CLEMENTI and YOSHIMINE [1976], d KEIL and AHLRICHS [1976].

2.3 Large clusters and hydrogen bonded crystals

Clusters and crystals of molecules forming hydrogen bonds have been calculated by semiempirical and ab initio SCF methods in the past. A review of the work done in this field has been given recently (Schuster [1976]). We recommend it here as a source for further references. There are mainly two reasons for studying clusters of many molecules:

1. Previous investigations indicated that energies of interaction in hydrogen bonded systems lack pairwise additivity. The knowledge of presence or absence of strong three body potentials or even higher order interactions is of primary importance for theoretical studies on associated liquids by statistical mechanics or numerical methods like Monte Carlo or molecular dynamics calculations. Semiempirical as well as ab initio calculations with

small basis sets tend to overemphasize these cooperative effects and turned out to be not reliable enough for relevant predictions on energies of interactions. The most extensive ab initio studies on aggregates of H_2O molecules up to $(H_2O)_4$ have been performed by Hankins, Moskowitz and Stillinger [1970a, b; 1973] and by Lentz and Scheraga [1973]. They performed also a partitioning of the energies of interaction into two body and three body contributions, thereby giving an idea of the relative importance of three body forces in different regions of the energy surface. All these calculations were performed at frozen monomer geometries.

2. A second aspect of investigations on clusters or crystals of hydrogen bonded monomers is even more interesting for our aims here: how do molecular geometries and force constants or vibrational frequencies change, when we proceed from an isolated dimer to higher aggregates? From previously mentioned, very accurate studies on dimers and from the available experimental data we are inclined to conclude that the typical features of hydrogen bonding can be observed in the dimer to a very small extent only. The more drastic differences by far are found by a comparison of the dimer and condensed phase. Although an explanation of this phenomenon is of primary importance for any general theory of hydrogen bonding, almost no really conclusive studies have been reported in this field. The main reason for this lack of extensive investigations is evident: any attempt to calculate accurate equilibrium geometries and vibrational spectra of clusters containing three or more molecules by ab initio methods has to face enormous computational problems which lie outside our present possibilities. Furthermore three, ten or even one hundred molecules do not represent an appropriate model for an associated liquid. Since we have no efficient formalism for the treatment of liquids in statistical mechanics, it seems hopeless at present to try an ab initio calculation of average structures and vibrational spectra of liquid water or hydrogen fluoride.

The changes in molecular geometries and vibrational spectra mentioned above do also occur in hydrogen bonded crystals. In most cases the differences between dimer and crystal are even larger than those between dimer and liquid. An ideal crystal is characterized by perfect translational symmetry and therefore only few degrees of freedom contribute significantly to the vibrational or phonon spectrum. Many theoretical methods for investigations on molecular crystals have been proposed and applied to hydrogen bonded solids. For the studies we have in mind here the crystal orbital (CO) method developed by Del Re, Ladik and Biczó [1967] as well as André [1969] seemed to be most appropriate. This method can be interpreted as a straight forward extension of MO SCF theory to infinite systems. The whole crystal is treated as a supermolecule. Translational symmetry is introduced by the Born-von Karman periodic conditions. Usually one considers only nearest neighbor interactions, i.e. only the interactions within and between two neighboring elementary cells are taken into account. Karpfen et al. [1974] have shown, however, that for crystals formed from polar molecules interactions with further elementary cells have to be taken into account as well, if a reasonably correct description is desired.

The $(HF)_\infty$ crystals is especially suited for model studies, since it consists of rather weakly bound infinite chains and can be represented therefore by models consisting of a single infinite chain of HF molecules only.

Calculations on $(HF)_\infty$ in the one-dimensional chain approximation and as three dimensional crystals have been performed with the tight binding method by Bassani, Pietronero and Resta [1973] and Pietronero and Lipari [1975]. An ab initio crystal orbital calculation on one-dimensional $(HF)_\infty$ was presented recently by Kertesz, Koller and Azman [1975]. In all these calculations the geometry cell was kept fixed. Crystal orbital calculations with extensive geometry variation have been performed in our laboratory (Karpfen and Schuster [1976]). Additionally, these calculations were extended to second and third neighbors. For technical and methodological reasons a rather small basis set has to be used. The results obtained from calculations on HF and $(HF)_2$ have been mentioned already. For the purpose of comparison they are shown together with the results from CO calculations in table 6.

Interestingly, we obtained fairly good results for both molecular geometries and force constants. As far as these quantities are concerned the results obtained for the dimer were satisfactory as well. In comparison to more extended calculations the energy of interaction is predicted too large as we must expect for basis sets of this size.

Coming to the results for the infinite, linear chain $(HF)_\infty$ finally, we notice convergence of all calculated results with increasing number of neighbors explicitely included in the calculation. Furthermore, the previously mentioned changes in force constants and molecular geometries caused by formation of a condensed phase are predicted correctly here. We can attribute therefore the changes observed experimentally to the formation of long chains or perhaps also networks of hydrogen bonds. Asking for an explanation of the difference between hydrogen bonds in a dimer and in an infinite chain we recall the potential curves for proton transfer along the hydrogen bond. Keeping the distance R_{FF} fixed at its equilibrium value we expect and find a highly asymmetric curve with a single minimum in the dimer. In the infinite chain there are, of course, an infinite number of independent and coupled proton motions. We regard the simultaneous in-phase-transfer of all protons which gives rise to the smallest possible barrier for proton transfer. For this mode we obtain a symmetric double minimum potential with an energy barrier of $\Delta E^{+}=12.8$ kcal/mole. For the static properties of the crystal the presence of a second minimum has a number of consequences: compared to the dimer the potential curve for proton motion becomes flatter around the minimum, the hydrogen bond

Table 6: Crystal orbital (CO) SCF calculations on a linear chain of (HF)[a)]

Structure	R_{HF}(Å)	R_{FF}(Å)	f_{HF}(mdyn/Å)	$\Delta E^{b)}$(kcal/mole)	$q_H^{c)}$	$q_F^{c)}$
HF calc.	.917	-	9.42	-	.530	9.470
exp.	.917	-	9.66	-		
$(HF)_2$ calc.	.919	2.704	9.23	-7.49	.512	9.511
exp.		2.79±0.5		-6±1		
n=1	.934	2.485	7.10	-10.18	.468	9.532
(HF) n=2	.936	2.491	6.96	-11.26	.444	9.556
n=3	.937	2.490	6.85	-11.71	.437	9.536
crystal, exp.	.950	2.49	6.52[e)] 5.24[e)]			

a) Basis set: (8,4/4) contracted to 4,3/3, for the exponents and contraction coefficients see HUZINAGA [1970]. Ref. KARPFEN and SCHUSTER [1976].

b) The energy values refer to the monomer: $\Delta E\left[(HF)_n\right] = 1/n\left\{E\left[(HF)_n\right] - E(HF)\right\}$

c) Electron densities derived from Mulliken overlap populations

d) n represents the number of neighboring elementary cells up to which the interactions are calculated explicitly

e) The values correspond to the antisymmetric and symmetric FH stretching mode.

becomes stronger, the FH bond distance increases and the force constant decreases. Although more extensive calculations and investigations on other structures are necessary before final conclusions can be drawn, we think that the effects on structures, stabilities and vibrational spectra observed in hydrogen bonded crystals can be reproduced and explained within the frame of Hartree-Fock MO theory for crystals.

3. ENERGY PARTITIONING AND ITS ASYMPTOTIC BEHAVIOR AT LARGE INTERMOLECULAR DISTANCES

In our discussion we have shown so far that ab initio MO theory can account for all the phenomena which are associated with hydrogen bonds by the experimentalists. In general we have been interested in quantitative predictions on structures and spectra. By continuing this approach we might come finally in a position to be able to calculate all kinds of observable quantities for molecular complexes with high precision if necessary, provided we have sufficiently large supply with computer time. On the other hand we have not learnt very much yet concerning the nature of intermolecular forces, or in particular the nature of hydrogen bonds. An approach, alternative to the supermolecule method tries to obtain the energy of interaction as a sum of individual contributions which can be identified with the expressions of the classical theory of intermolecular forces by their asymptotic behavior (see e.g. Hirschfelder, Curtiss and Bird [1954] or Margenau and Kestner [1969]). In general, there are the three well known long range contributions, the electrostatic or Coulomb energy (ΔE_{COU}), the polarization energy (ΔE_{POL}) and the dispersion energy (ΔE_{DIS}), which give rise to the characteristic individual terms in an R^{-n} expansion series of the energy of interaction at molecular distances (R), where the overlap between the interacting systems is negligibly small. The procedures can be subdivided into three groups depending on the way how they approach the ultimately identical goal:

1. The straight forward application of perturbation theory to intermolecular interactions. Starting from given solutions of the Schrödinger equations of the isolated subsystems - for almost all concrete examples only approximative solutions are available - individual contributions to the energy of interaction are calculated by different modifications of Raleigh-Schrödinger perturbation theory.

2. The treatment of intermolecular forces as a variational problem by configuration interaction methods. Starting again from MO's of the isolated subsystems the calculation of the energy of interaction is formulated as a configuration interaction problem. The lowest eigenvalue of the CI matrix is calculated then by a perturbation method which allows to separate the intermolecular energy into different contributions.

3. SCF energy partitioning. The energy of interaction obtained

by a full SCF calculation is split into various contributions by different model assumptions.

The procedures mentioned in 2. and 3. become identical at the SCF level, i.e. in case only singly excited configurations are applied in the CI calculation. As mentioned already we shall concentrate on SCF energy partitioning here.

Partitioning of SCF energies of interaction into electrostatic (ΔE_{COU}), exchange (ΔE_{EX}) and delocalization (ΔE_{DEL}) energies has been proposed originally by Kollman and Allen [1970] and Dreyfus and Pullman [1970a, b]. The main idea was to learn more about the nature of intermolecular interactions and to obtain a more rigorous basis for the construction of model potentials. Dreyfus and Pullman [1970a, b] were interested particularly in a test for the reliability of predictions made by electrostatic models (see e.g. Scrocco and Tomasi [1973]). Morokuma [1971] extended this concept to a separation of the delocalization energy into polarization (ΔE_{POL}) and charge transfer (ΔE_{CHT}) contributions. The procedure can be explained most easily by an assumption of four normalized trial wave functions to describe the complex:

1. a simple Hartree product built from the SCF wave functions of the isolated molecules

$$\Psi_1 = \Psi_A^o \cdot \Psi_B^o \tag{6}$$

2. an antisymmetric product of these two wave functions

$$\Psi_2 = A\ (\Psi_A^o \Psi_B^o) \tag{7}$$

3. a product of the two wavefunctions optimized individually in the electrostatic field of the other subsystem

$$\Psi_3 = \Psi_A \cdot \Psi_B \tag{8}$$

4. the ordinary antisymmetric and optimized SCF wavefunction of the whole complex

$$\Psi_4 = \Psi_{AB} \ . \tag{9}$$

For all four wavefunctions energy expectation values are calculated now and the individual contributions to the total energy of interaction are obtained as differences:

$$\Delta E_i = E_i - (E_A^o + E_B^o); \qquad E_i = \int \Psi_i^+ H \Psi_i \, d\tau \tag{1o}$$

According to physical considerations electrostatic, exchange, polarization and charge transfer energies are built from this set of ΔE_i values (Morokuma [1971], Schuster [1976]):

$$\Delta E_{COU} = \Delta E_1 \tag{11}$$

$$\Delta E_{EX} = \Delta E_2 - \Delta E_1 \tag{12}$$

$$\Delta E_{POL} = \Delta E_3 - \Delta E_1 \tag{13}$$

$$\Delta E_{CHT} = \Delta E_4 + \Delta E_1 - \Delta E_2 - \Delta E_3 \tag{14}$$

The total SCF energy, of course, is the sum of all four individual contributions:

$$\Delta E_{SCF} = \Delta E_4 = \Delta E_{COU} + \Delta E_{EX} + \Delta E_{POL} + \Delta E_{CHT} \tag{15}$$

SCF energy partitioning has been performed for a number of intermolecular complexes. Recently, results from calculations using extended basis sets have been reported (Schuster, Lischka and Beyer [1976]; Beyer, Lischka and Schuster [1977]). With the basis set chosen, the electrostatic properties and polarizabilities of the isolated systems are predicted within error limits of less than 20%. Furthermore, the superposition error of basis sets in the complex can be expected to be rather small when basis sets of this quality are used. The molecular geometries applied in these calculations are very close to the most stable structures. The values obtained thereby are shown in table 7. Some of the most important features are essentially the same as derived before from perturbation theory. The electrostatic contribution (ΔE_{COU}) dominates in all complexes investigated here. In $Li^+.OH_2$, $F^-.H_2O$ and $(H_2O)_2$ the first order exchange contribution was found to be next important and much larger in absolute value than the remaining second order terms. In $Na^+.OH_2$ and $Cl^-.H_2O$, ΔE_{EX} and the second order energy are of the same order of magnitude. According to the relatively small intermolecular distances at equilibrium, overlap is rather large in systems forming hydrogen bonds. Apparently, exchange repulsion is stronger in these complexes too. Both ΔE_{COU} and ΔE_{POL} depend strongly on the intermolecular distance at the equilibrium geometry and decrease substantially in absolute value if we substitute a third row element for one of the second row ($Cl^- < F^-$ or $Na^+ < Li^+$). The charge transfer term, however, does not show a physically reasonable behavior. In contrast to the general destabilizing effect we would expect from perturbation theory, ΔE_{CHT} is repulsive in cation-molecule complexes. Nevertheless, there is a significant difference in comparison to hydrogen bonded systems where the charge transfer term remains always negative. A possible reason for the unusual behavior of ΔE_{CHT} will be discussed later.

Electrostatic potentials very often are represented by multipole expansion which leads to an expansion of electrostatic energies in powers of R^{-n}. Recently, Ahlrichs [1976] gave a proof that this

Table 7: Energy partitioning in some molecular complexes+)

Complex	Basis set (GTO's)	Method++)	Energies in kcal/mole					Ref.
			ΔE_{COU}	ΔE_{EX}	ΔE_{POL}	ΔE_{CHT}	ΔE_{DIS}	
$(H_2O)_2$	(11,7,1/6,1)→[7,5,1/4,1]	PT	-7.1	4.9	-1.6	-	-1.5	a
		SCF-EP	-7.3	4.5	-0.7	-1.2	-	b,c
$Li^-.OH_2$	(11,7,1/6,1)→[4,3,1/2,1] Li:(7,1)→[3,1]	SCF-EP	-41.5	15.3	-12.8	5.2	-	b,c
$Na^+.OH_2$	(9,5,1/4)→[5,3,1/3] Na:(11,7)→[6,4]	SCF-EP	-30.3	7.0	-5.1	1.1	-	b,c
$F^-.HOH$	(9,5,1/4,1)→[5,3,1/3,1] F:(9,5,1)→[5,3,1]	SCF-EP	-34.7	26.0	-7.2	-7.2	-	b,c
$Cl^-.HOH$	(9,5,1/4)→[5,3,1/3] Cl^-:(17,10,1)→[9,6,1]	SCF-EP	-14.6	7.0	-1.6	-5.7	-	b,c

+) Equilibrium geometries of the complexes have been used.

++) PT = perturbation theory, SCF-EP = SCF energy partitioning

References: a JEZIORSKI and VAN HEMERT [1976] , b SCHUSTER, LISCHKA and BEYER [1976]
c BEYER, LISCHKA and SCHUSTER [1976] .

<u>Table 8:</u> Multipole expansion of electrostatic energies in $(H_2O)_2$ and $Li^+.OH_2$

Complex	R_{OX}(Å)	Σ_1 +)	Σ_2 +)	Σ_3 +)	Σ_4 +)	ϕ +)	ΔE_{COU}	Ref.
		Energies in kcal/mole						
$(H_2O)_2$	3.00 ++)	-3.28	-3.63	-3.68	-	-	-7.12	a
	4.76	-0.82	-0.87	-0.88	-	-	-1.12	a
	7.94	-0.18	-0.18	-0.18	-	-	-0.21	a
$Li^+.OH_2$	1.80 ++)	-46.48	-47.42	-40.35	-44.47	-42.69	-44.23	b,c
	4.23	-8.74	-8.82	-8.57	-8.63	-8.63	-8.63	b,c

+) Σ_n represents the sum over all contributions up to the multipole corresponding to L=n; ϕ is the electrostatic potential of the water molecule, which becomes identical with ΔE_{COU} if the Li^+ cation is substituted by a point charge.

++) Intermolecular distance close to equilibrium geometry.

References: a JEZIORSKI and VAN HEMERT [1976], b SCHUSTER et al. [1975]
c BEYER, LISCHKA and SCHUSTER [1977]

expansion gives rise to an asymptotically convergent series. It seems interesting therefore to test the properties of the multipole expansion series of ΔE_{COU} for two concrete examples - a complex between two neutral molecules $(H_2O)_2$, and between a cation and a molecule, $Li^+.OH_2$. Around the energy minimum an entirely different situation is encountered in the two cases: in $(H_2O)_2$ convergence is very poor, whereas the four term series gives already satisfactory agreement with the exact value in $Li^+.OH_2$ (table 8). At large intermolecular distances, as expected, the series converge very fast to the corresponding exact electrostatic energies in both examples. A similar investigation on ΔE_{POL} again showed the correct asymptotic behavior (Beyer, Lischka and Schuster [1977]).

Comparing the results from intramolecular perturbation theory (Jeziorski and Van Hemert [1976]) and SCF energy partitioning on the complex $(H_2O)_2$, we find almost the same numerical values for the first order energies ΔE_{COU} and ΔE_{EX}. This finding is not unexpected since the ultimate expressions from which the energies are obtained are identical in both methods and very similar basis sets have been used by Jeziorski and Van Hemert [1976] and Beyer, Lischka and Schuster [1977]. A comparison of second order contributions, on the other hand, is much more difficult. The polarization energy derived from the perturbational approach apparently is larger in absolute value than the corresponding contribution obtained from SCF energy partitioning. Jaziorski and Van Hemert [1976] pointed out, that in their treatment ΔE_{POL} most probably contains major portions of the charge transfer terms in SCF energy partitioning. In fact, the sum of both contributions in the SCF calculations, ΔE_{DEL},

$$\Delta E_{DEL} = \Delta E_{POL} + \Delta E_{CHT} \qquad (16)$$

is not too different from the corresponding quantity in the perturbational approach. The dispersion energy, ΔE_{DIS}, is derived from double excitations exclusively and, therefore, cannot be obtained by an SCF approach.

The SCF energy partitioning as described by eqs. (6-16) was found to suffer from four disadvantages mainly:

1. The results obtained are largely basis set dependent.
2. A direct comparison with intermolecular perturbation theory is rather difficult as far as second and higher order effects are concerned.
3. The charge transfer contribution, ΔE_{CHT}, which is simply the difference between the SCF energy of interaction and all the other contributions is a complex and by no means a clearly defined quantity.
4. A deficiency which, of course, is also present in the perturbational approach but has not been mentioned so far: This kind of SCF energy partitioning is not applicable directly to complexes which differ appreciably in geometry from the isolated

subsystems.

In the following paragraphs we will refer briefly to these four points and mention some suggestions which have been made recently. In order to investigate the basis set dependence of Morokuma's SCF energy partitioning method, Beyer, Lischka and Schuster [1977] have calculated the individual contributions for $Li^+.OH_2$ and $(H_2O)_2$ at the equilibrium geometries with a number of different basis sets. The results obtained are summarized in table 9. ΔE_{COU} and ΔE_{POL} roughly reflect the changes in the calculated properties of the isolated subsystems. According to the calculated values of dipole moments and polarizabilities of the H_2O molecule we expect ΔE_{COU} to be too large and ΔE_{POL} to be too small in absolute values in most calculations using small or medium size basis sets. This in fact is the result obtained from SCF energy partitioning. The numerical values again stress the importance of flat polarization functions for a correct reproduction of intermolecular energies. The first order exchange contribution (ΔE_{EX}), on the other hand, seems to be somewhat less sensitive. Substantial errors are found only in calculations with very small basis sets. The charge transfer energy (ΔE_{CHT}) finally was found to be the most critical term. Despite the already mentioned fact that this contribution is repulsive in large basis set calculations on $Li^+.OH_2$ and $Na^+.OH_2$, the numerical values are extremely basis set dependent. In general, small basis sets give more negative values. As an extensive investigation on $Li^+.OH_2$ shows, ΔE_{CHT} may also change sign on basis set extension. Recalling our previous discussion on basis set effects in MO calculations on intermolecular energies, we have to attribute these changes to the basis set superposition error discussed in section 2.1. Errors of this kind will enter ΔE_{CHT} in case of SCF energies partitioned according to Morokuma's scheme. Furthermore, we can realize another unusual behavior of ΔE_{CHT} when we regard the dependence of this quantity on intermolecular distance: ΔE_{CHT} is repulsive at short distances, changes sign and finally approaches the zero energy line from negative values (Beyer, Lischka and Schuster [1977]).

In order to overcome the difficulties concerning the definition of ΔE_{CHT}, Kitaura and Morokuma [1976] proposed a new energy partitioning scheme for MO-SCF calculations on intermolecular interactions, which relates the former formalism to the CI approach discussed in the previous section. In fact the energy decomposition method proposed becomes identical with the more general procedure of Daudey, Claverie and Malrieu [1974] in case the CI is restricted to single excitation only. Another more technical difference concerns the evaluation of the energy eigenvalues: Kitaura and Morokuma [1976] diagonalize the CI matrices, whereas Daudey, Claverie and Malrieu [1974] use perturbation theory for this purpose. Energy partitioning in the new approach is performed by systematic neglect of several off-diagonal elements of

Table 9: Basis set dependence of SCF energy partitioning in $(H_2O)_2$ and $Li^+.OH_2$

Complex	Basis set (GTO's)	R_{OX}(Å)	Energies in kcal/mole							Ref.
			ΔE	ΔE_{COU}	ΔE_{EX}	$\Delta E^{(1)}$	ΔE_{POL}	ΔE_{CHT}	ΔE_{DEL}	
$(H_2O)_2$	STO[+)]	2.78	-6.6	-8.0	9.9	1.9	-0.3	-8.2	-8.5	a
	4-31G[++)]	2.78	-7.7	-12.8	9.3	-3.6	-0.7	-3.4	-4.1	b
	4-31G[++)]	2.98	-7.7	-9.0	4.2	-4.8	-0.5	-2.5	-3.0	b
	(11,7,1/6.1)—[7,5,1/4,1]	3.00	-4.7	-7.3	4.5	-2.8	-0.7	-1.2	-1.9	c
$Li^+.OH_2$	(7,3/3)—[4,2/2]	1.80	-45.9	-46.5	8.4	-38.1	-6.7	-1.0	-7.8	c,d
	Li:(6)—[3]									
	(9,5/4)—[5,3/3]	1.80	-43.0	-52.0	15.1	-36.9	-9.8	3.8	-6.1	c,d
	Li:(7)—[3]									
	(9,5,1/4)—[5,3,1/3]	1.80	-37.3	-44.2	14.9	-29.4	-11.3	3.4	-7.9	c,d
	Li:(7,1)—[3,1]									
	(11,7,1/6,1)—[4,3,1/2,1]	1.80	-33.9	-41.7	15.3	-26.3	-12.8	5.2	-7.6	c,d
	Li:(7,1)—[3,1]									
	(11,7,3/6,2)—[7,5,3/4,2]	1.80	-33.7	-	-	-22.0	-	-	-11.7	c,d
	Li:(7,1)—[3,1]									

+) Minimal Slater basis set

++) See DITCHFIELD, HEHRE and POPLE [1971] and HEHRE, DITCHFIELD and POPLE [1972]

References: a MOROKUMA [1971], b KITAURA and MOROKUMA [1976], c SCHUSTER, LISCHKA and BEYER [1976]
d BEYER, LISCHKA and SCHUSTER [1977].

the CI matrix and electron exchange operators. These offdiagonal elements are classified according to the MO's from which they are formed. The MO's in turn are grouped according to their provenance, i.e. whether they belong to subsystem A or B and whether they belong to the subspace of occupied or virtual orbitals (fig. 2)

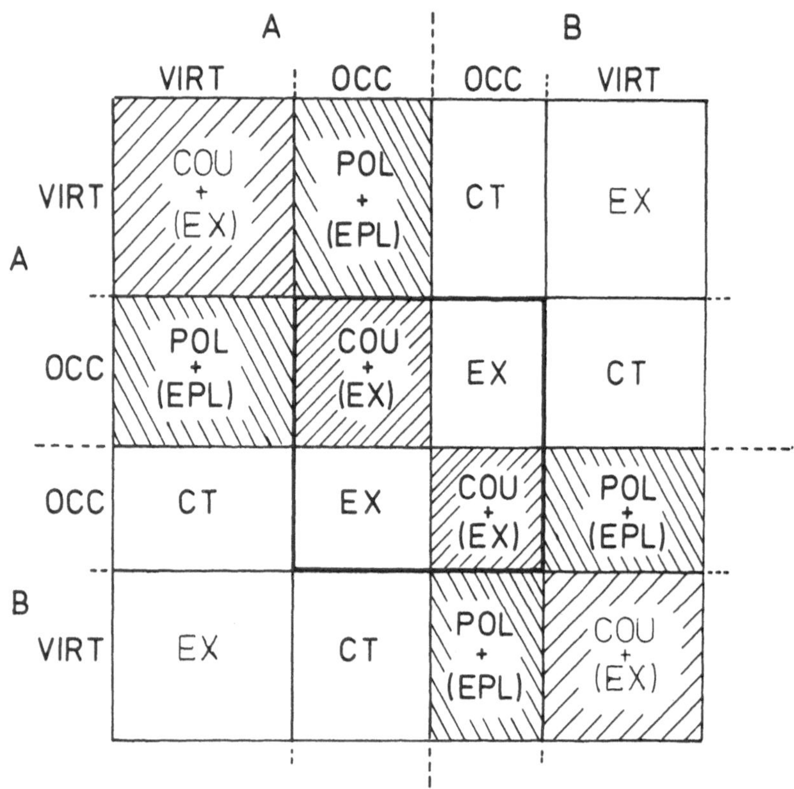

Fig. 2. SCF energy partitioning as a CI problem (for all details see Kitaura and Morokuma, 1976).

The major results of the new energy partitioning scheme is further decomposition of the former charge transfer energy into a "true" large transfer term (ΔE_{CT}) and exchange polarization energy (ΔE_{EPL}) and the remaining coupling term between the different classes of matrix elements (ΔE_{MIX}). The whole partitioning scheme reads now as follows:

$$\Delta E_{SCF} = \Delta E_{COU} + \Delta E_{EX} + \Delta E_{POL} + \Delta E_{EPL} + \Delta E_{MIX} \tag{17a}$$

or

$$\Delta E_{CHT} = \Delta E_{CT} + \Delta E_{EPL} + \Delta E_{MIX} \qquad (17b)$$

The results obtained in case of $(H_2O)_2$ are shown in table 10.

Table 10: Decomposition of the charge transfer energy ΔE_{CHT} in $(H_2O)_2$ (KITAURA and MOROKUMA [1976])

	Energies in kcal/mole +)			
R_{OO} (Å)	ΔE_{CT}	ΔE_{EPL}	ΔE_{MIX}	ΔE_{CHT}
2.78	-2.78	-2.95	2.36	-3.37
2.98	-2.11	-0.40	0.06	-2.45
3.18	-1.74	0.02	-0.18	-1.90

+) A 4-31G basis set (DITCHFIELD, HEHRE and POPLE [1971], HEHRE, DITCHFIELD and POPLE [1972]) has been used.

Interestingly, the "true" charge transfer energy (ΔE_{CT}) is not very different from the fromer ΔE_{CHT} values. Accordingly, ΔE_{EPL} and ΔE_{MIX} have similar absolute values but different sign. Again we realize that these two terms are changing sign with increasing molecular distance. Nevertheless, extensions of this partitioning scheme for ΔE_{CHT} to other complexes and larger basis sets are necessary before definite conclusions on usefulness and applicability of this method can be drawn.

4. CONCLUSIONS AND FURTHER ASPECTS

Ab initio MO calculations on hydrogen bonded dimers between small molecules have reached a stage of high accuracy which allows to predict energies and structures of complexes within the error limits of experimental measurements. In many cases, where the experiments are very difficult to perform, the theoretical results represent the only reliable source of information. The investigations of heterodimers in the vapor phase may serve as an example. In some examples IR and NMR spectra have been calculated by ab initio techniques as well and the agreement with experimental data is satisfactory. On the whole, we can say that the hydrogen bond in dimers of small molecules in their electronic ground states is well understood. At present, the major limitation of the ab initio approach concerns the immense need of computer time which prevents an extension of accurate numerical calculations to larger entities like complexes between larger molecules or higher aggregates simulating local sections of liquids or crystals. Besides the fact that the calculation of the individual point of the energy surface requires much more computer time in higher aggregates, the configurational freedom is much larger too and the point by point approach seems to be hopeless. In fact, the problem to describe the properties of larger clusters of molecules or of hydrogen bonded liquids and crystals has not been solved yet in a satisfactory manner and still remains a striking challenge to theoreticians. At present the attempts to calculate properties of molecular crystals by ab initio SCF methods seem to be promising.

Numerical results on intermolecular complexes obtained with minimum and most of the other small basis sets are rather discouraging. As we have seen, the properties of isolated molecules are often reproduced very poorly. Even more seriously, basis set superposition errors are large. Qualitative results on unknown systems have to be treated with great care therefore. Most probably, appropriate semiclassical models like consistent force fields which can provide analytic expressions for energy surfaces, will turn out to represent a cheaper and more extendable approach to study interactions between larger molecules than the straight forward application of expensive ab initio calculations using minimal or small basis sets.

Most of the theoretical treatments subsequent to the calculation of energy surfaces need analytical expressions. Energy partitioning may help here to derive useful functions for this purpose. More emphasis will have to be put on statistical mechanics in the future. To give just one example, recently Braun and Leidecker [1974] calculated rotational vibrational spectra of $(H_2O)_2$ based on ab initio potential surfaces and made predictions on the temperature dependence of the equilibrium constant for dimerization of H_2O as well as the second virial coefficient of water vapor.

Monte Carlo and molecular dynamics calculations, although very time consuming, present a possibility to proceed towards an understanding of the liquid state and solute-solvent interactions. Important work has been done in this field already. The ultimate test for the usefulness of ab initio energy surfaces in this approach, nevertheless is still missing. The importance of three or higher body interactions in associated liquids and solutions represents a question open to further investigations.

ACKNOWLEDGMENTS

The author is indebted to many colleagues and especially to all his coworkers for numerous stimulating discussions. Support for our investigations in this field has been provided by the Austrian "Fonds zur Förderung der wissenschaftlichen Forschung" (Project No. 1217 and 2756). Last but not least the assistance of Mrs. J. Dura with the preparation of the manuscript is gratefully acknowledged.

REFERENCES

1. R. Ahlrichs, Theoret. Chim. Acta 41, 7 (1976).
2. J.M. André, J. Chem. Phys. 50, 1536 (1969).
3. E. Bassani, L. Pietronero and R. Resta, J. Phys. C: Solid State Phys. 6, 2133 (1973).
4. A. Beyer, H. Lischka and P. Schuster, to be published (1977).
5. S.F. Boys and F. Bernardi, Mol. Phys. 19, 558 (1970).
6. C. Braun and H. Leidecker, J. Chem. Phys. 61, 3104 (1974).
7. L.A. Curtiss and J.A. Pople, J. Mol. Spectroscopy, 55, 1 (1975).
8. J.P. Daudey, P. Claverie and J.P. Malrieu, Int. J. Quant. Chem. 8, 1 (1974).
9. G. Del Re, J. Ladik and G. Biczó, Phys. Rev. 155, 997 (1967).
10. G.H.F. Diercksen, Chem. Phys. Lett. 4, 373 (1969).
11. G.H.F. Diercksen, Theoret. Chim. Acta 21, 335 (1971).
12. G.H.F. Diercksen and W.P. Kraemer, Chem. Phys. Lett. 6, 419 (1970).
13. G.H.F. Diercksen, W.P. Kraemer and B.O. Roos, Theoret. Chim. Acta 36, 249 (1975).
14. J.D. Dill, L.C. Allen, W.C. Topp and J.A. Pople, J. Am. Chem. Soc. 97, 7220 (1975).
15. R. Ditchfield, W.J. Hehre and J.A. Pople, J. Chem. Phys. 54, 724 (1971).
16. M. Dreyfus and A. Pullman, Theoret. Chim. Acta 19, 20 (1970a).
17. M. Dreyfus and A. Pullman, Compt. Rend. Acad. Sci. (Paris) 271, 457 (1970b).
18. T.R. Dyke, B.J. Howard and W. Klemperer, J. Chem. Phys. 56, 2442 (1972).
19. T.R. Dyke and J.S. Muenter, J. Chem. Phys. 60, 2929 (1974).
20. H. Eisenschitz and F. Loudon, Z. Physik 60, 491 (1930).
21. D. Hankins, J.W. Moskowitz and F.H. Stillinger, Chem. Phys. Lett. 4, 527 (1970a).
22. D. Hankins, J.W. Moskowitz and F.H. Stillinger, J. Chem. Phys. 53, 4544 (1970b).
23. D. Hankins, J.W. Moskowitz and F.H. Stillinger, J. Chem. Phys. 59, 995 (1973).
24. W.J. Hehre, R.F. Stewart and J.A. Pople, J. Chem. Phys. 51, 2657 (1969).

25. W.J. Hehre, R. Ditchfield and J.A. Pople, J. Chem. Phys. 56, 2257 (1972).
26. J.O. Hirschfelder, C.F. Curtiss and R.B. Bird, Molecular Theory of Gases and Liquids, J. Wiley, New York, pp. 835-1103 (1954).
27. S. Huzinaga, Tables of Atomic Functions, Alberta Univ. (1970).
28. A.D. Isaacson and K. Morokuma, J. Am. Chem. Soc. 97, 4454 (1975).
29. R. Janoschek, Calculated Vibrational Spectra of Hydrogen Bonded Systems, in Schuster, Zundel and Sandorfy, North Holland, Amsterdam, pp. 165-216 (1976).
30. B. Jeziroski and M. Van Hemert, Mol. Phys. 31, 713 (1976).
31. A. Johansson, P. Kollman and S. Rothenberg, Theoret. Chim. Acta 29, 167 (1973).
32. A. Karpfen, J. Ladik, P. Russegger, P. Schuster and S. Suhai, Theoret. Chim. Acta 34, 115 (1974).
33. A. Karpfen and P. Schuster, Chem. Phys. Lett. 44, 459 (1976).
34. F. Keil and R. Ahlrichs, J. Am. Chem. Soc. 98, 4787 (1976).
35. M. Kertesz, J. Koller and A. Azman, Chem. Phys. Lett. 36, 576 (1975).
36. K. Kitaura and K. Morokuma, Intern. J. Quant. Chem. 10, 325 (1976).
37. P.A. Kollman and L.C. Allen, Theoret. Chim. Acta. 18, 399 (1970).
38. P.A. Kollman and L.C. Allen, Chem. Rev. 72, 283 (1972).
39. P.A. Kollman, J. McKelvey, A. Johansson and S. Rothenberg, J. Am. Chem. Soc. 97, 955 (1975).
40. B.R. Lentz and H.A. Scheraga, J. Chem. Phys. 58, 5296 (1973).
41. H. Lischka, J. Am. Chem. Soc. 96,4716 (1974).
42. H. Margenau and R. Kestner, Theory of Intermolecular Forces, 2nd. ed., Pergamon Press, Oxford, pp. 15-298 (1969).
43. O. Matsuoka, E. Clementi and M. Yoshimine, J. Chem. Phys. 64, 1351 (1976).
44. A. Meunier, B. Levy and G. Bertier, Theoret. Chim. Acta, 29, 49 (1973).
45. K. Morokuma, J. Chem. Phys. 55, 1236 (1971).
46. L. Pietronero and N.O. Lipari, J. Chem. Phys. 62, 1796 (1975).
47. H. Popkie, H, Kistenmacher and E. Clementi, J. Chem. Phys. 59, 1325 (1973).
48. P. Schuster, W. Jakubetz and W. Marius, Topics in Current Chemistry 60, 1-107 (1975).
49. P. Schuster, W. Marius, A. Pullman and H. Berthold, Theoret. Chim. Acta 40, 323 (1975).
50. P. Schuster, Energy Surfaces for Hydrogen Bonded Systems, in Schuster, Zundel and Sandorfy eds. The Hydrogen Bond, North Holland, Amsterdam, pp. 25-163 (1976).
51. P. Schuster, H. Lischka and A. Beyer, Ab initio Studies on Hydrogen Bonding and Ion Molecule Complexes, in I.G. Csizmadia ed., Progress in Theoretical Organic Chemistry, Vol. 2, Elsevier, Amsterdam, pp. 89-105 (1976).

52. P. Schuster, G. Zundel and C. Sandorfy, eds. The Hydrogen Bond - Recent Developments in Theory and Experiment, North Holland, Amsterdam (1976).
53. P. Schuster, The Fine Structure of the Hydrogen Bond, in B. Pullman, ed. Intermolecular Interactions: From Diatomics to Biopolymers (Perspectives in Quantum Chemistry and Biochemistry, Vol. II, J. Wiley, London, pp. 363-432 (1977).
54. E. Scrocco and J. Tomasi, Topics in Current Chemistry, 42, 95-170 (1973).
55. A.J. Tursi and E.R. Nixon, J. Chem. Phys. 52, 4858 (1970).
56. M. Van Thiel, E.D. Becker and G. Pimentel, J. Chem. Phys. 27, 486 (1957).
57. D.R. Yarkony, S.V. O'Neill, H.F. Schaefer III, C.P. Baskin and C.F. Bender, J. Chem. Phys. 60, 855 (1974).

ELECTRONIC STRUCTURE OF BIOPOLYMERS

János J. Ladik

Lehrstuhl für Theoretische Chemie der Friedrich Alexander Universtität Erlangen-Nürnberg, 852 Erlangen, FRG and Laboratory of the National Foundation for Cancer Research at the Chair of Theoretical Chemistry, University Erlangen-Nürnberg

ABSTRACT. The results of an ab initio SCF LCAO crystal orbital calculation of polycytosine are described. The band structure obtained has been corrected also for the exciton correction between an excited electron and the remaining hole applying the so-called ÔÂÔ formalism and for long-range correlation effects (using the electron polaron model). On the basis of these results an estimation has been performed for the band structure of DNA.

The ab initio band structure of polyglycine and the interaction of this periodic protein model with different electron acceptors is presented. Finally, on the basis of the electronic structure of DNA and of proteins different possible mechanisms of chemical carcinogenesis are briefly described.

1. INTRODUCTION

In previous publications semiempirical SCF LCAO CO (crystal orbital) calculations of different periodic DNA models (either in the π-electron or in the all-valence-electrons approximations) have been reported [1]. A review of the used PPP, CNDO/2 and MINDO/2, respectively, CO methods has been given in the book containing the material of the previous NATO ASI on polymers [2]. According to the results obtained in the case of homopolynucleotides (like polycytosine) or polybasepairs (like poly(G-C)) the valence and conduction bands have a width of 0.2-0.3 eV, while in the cases of the more complicated periodic DNA models (like a single stranded poly(GT)=GTGTGT... chain or the double stranded

J.-M. André et al. (eds.), Quantum Theory of Polymers, 257-278.

poly $\binom{GA}{GT}$ = $\begin{matrix} G & A & G & A & G & A \\ | & | & | & | & | & | \\ C & T & C & T & C & T \end{matrix}$ ... system) the widths of the corresponding bands are by one order of magnitude smaller. An all valence electron calculation of the sugar-phosphate chain of DNA (poly(SP)) has given again band widths of 0.2-0.3 eV [3]. In subsequent publications the possibility of charge transfer from the nucleotide base region of DNA to the low-lying conduction band of its sugar-phosphate backbone [4] or from the polypeptide chain of a nucleoprotein again to the poly(SP) chain of DNA [5] was pointed out. Further also estimations were made for the broadening effect of the inhomogeneous electric field due to the K^{+}-PO_4^{-} double layer of DNA on its energy bands based on simple model calculations [5].

In the first part of the present paper the results of the first *ab initio* SCF LCAO band structure calculation of a simple periodic DNA model (polyC) [6] will be reviewed. Further it will be shown how it is possible after applying also a correction for the electron-hole interaction (exciton effect) for the empty bands and taking into account long-range correlation effects in an approximate way (electron polaron model) to make an estimation of the band structure of DNA [7].

The results of the recently performed (CNDO/2- and MINDO/2 CO) all-valence electron band structure calculations of simple one- and two-dimensional periodic polypeptide models (polyformamide [8] and polyglycine [9]) have been discussed also in ref. [2]. As next step *ab initio* SCF LCAO CO calculations have been performed for one-dimensional polyglycine chains [10]. The discussion of the results obtained and of the possibility of charge transfer between these chains and different electron acceptors [11] will form the second part of this paper.

In the subsequent third part possible mechanisms of chemical carcinogenesis based on the electronic structure of DNA, and of proteins will be discussed. In connection with this problem an outlook will be given about the possibilities of further improved calculations for the electronic structure of biopolymers and it will be shown how these results can be applied for the interpretation of their different physical and chemical properties which determine their biological functions.

2. DNA

Applying the *ab initio* SCF LCAO CO method (for the formalism see [12][13] to the single stranded periodic DNA model polycytosine (polyC) its band structure was calculated [6][7]. Since in the case if a helix one comes from one unit to the next one not by a simple translation but by a combined symmetry operation (translation along the helix axis and rotation around it), one has to rotate together with the molecules also the atomic orbitals centered on the atoms of the molecules. This means that if we choose the z axis along the main axis of the helix, AO-s which are directed or have component in the xy plane have to be rotated. Since in this first <u>ab initio</u> calculation of a periodic DNA model only a minimal basis set constructed from Gaussian lobe functions [14] was used, only the contracted $2p_x$ and $2p_y$ Gaussians had to be rotated by 36° together with the molecules on which they were centered.* For the geometry used we refer to the recent structural data of DNA B [16].

In building the Fock-matrix the "nearest neighbor's interactions" approximation in the strict sence was applied, i.e. only those one- and two-electron integrals were calculated in which no such two orbitals occurred which were centered on atoms belonging to non-neighboring cells. It should be emphasized that in this calculation the non-local correct Hartree-Fock exchange term was used without any approximation to it. In this way all the integrals in absolute value larger than the threshold value of 10^{-8} at. u. were calculated and nine different points of k in the first Brillouin zone were taken into account in the numerical integration of the matrix elements $p_{r,s}$ (q) ($q=q_1-q_2=0,1$) (see equ. (54) of [13]).

The calculated energy band structure consists of forty-five bands, of which twenty-nine are completely filled. The correspondence between the invidial MO levels (ε^{MO}) and between the bands is always unambigous as it can be seen from Table I. where the lower and upper limits (ε^{CO}_{min}, and ε^{CO}_{max}.) as well as the widths of the three highest filled and two lowest unfilled bands are shown. By inspecting the wave functions one can still define quasi-π-type bands (as was the case previously for all-valence electron bands [2]) although the symmetry of the original MO's is broken in polyC in consequence of the stacked arrangement of the units. Though the eight such quasi-π-type bands are located mainly around the Fermi level, some σ bands occur between the π bands (see Table I) and thus the condition of σ-π separation is not fulfilled.

---

* On the basis of group theoretical considerations it is not difficult to show [15] that if we have a combined symmetry operation in a chain or in a crystal we have to perform on the local atomic coordinate systems in the different elementary cells the same symmetry operations as on the molecules to preserve the cyclic property of the Hamiltonian hypermatrix of the chain and crystal (this is necessary for its block-diagonalization [12]). Thus in the case of a helix we have to rotate the basis functions together with the molecules.

Table I.

The physically most important energy bands around the Fermi level of polycytocine. The original MO levels (ε^{MO}), the band minima and maxima ($\varepsilon^{CO}_{min.}$ and $\varepsilon^{CO}_{max.}$) with their location (ka) and the corresponding bandwidths are given in eV.

ε^{MO}(type)	$\varepsilon^{CO}_{min.}$ ($k_{min.}$ a)	$\varepsilon^{CO}_{max.}$ ($k_{max.}$ a)	$\Delta\varepsilon^{CO}$
4.585 (π)	4.813 (0)	5.129 (π)	0.316
1.929 (π)	1.535 (0)	2.775 (π)	1.240
-9.766 (π)$^{+}$	-9.665 (π)	-9.113 (0)	0.552
-11.488 (σ)	-11.115 (0)	-10.898 (π)	0.217
-11.562 (π)	-11.774 (0)	-11.251 (π)	0.523

$^{+}$ Highest filled level.

One should point out that the physically most important valence- and conduction bands resulting from the present ab initio calculation are much broader (~0.5 and ~1.2 eV, respectively) than those obtained with different semiempirical crystal orbital methods (0.1-0.3 eV [1],[2]). If this trend will be valid also for other periodic DNA models and also for calculations with larger basis sets, it will be inevitable to revise the previous considerations of the transport properties of DNA [17] on the basis of the new band structures. In connection with the positions of the bands it should be mentioned that though the description of the filled bands seems to be satisfactory (the theoretical ionization potential of 9.11 eV obtained with the aid of Koopmans' theorem agrees quite well with the experimental value of 8.90 eV [18]), the gap between the valence band and the conduction band is obviously too large ($\Delta E_g \approx 10.65$ eV).

Since this too large gap is first of all due to the failure of the Hartree-Fock method in describing virtual levels with the aid of a V^N potential instead of the appropriate V^{N-1} potential [19], as next step we recalculated the virtual levels of cytocine (and those of adenine, thymine and guanine) by using a corrected Fock operator for the excited levels. This corrected Fock operator is [20]

$$\hat{F}' = \hat{F} + \hat{O}\hat{A}\hat{O}, \tag{1}$$

where the projection operator $\hat{O}$ projects into the subspace of the virtual orbitals

$$\hat{O}=\hat{1}-\hat{\varrho}=\hat{1}-\sum_{i=1}^{n^*}|\varphi_i^{HF}\rangle\langle\varphi_i^{HF}| = \sum_{a=n^*+1}^{m}|\varphi_a^{HF}\rangle\langle\varphi^{HF}| = \sum_{a=n^*}^{m}|\psi_a\rangle\langle\psi_a| \tag{2}$$

The Hartree-Fock orbitals $|\varphi_i^{HF}\rangle$ are defined by the equation

$$\hat{F}|\varphi_i^{HF}\rangle = \varepsilon_i^{HF}|\varphi_i^{HF}\rangle, \tag{3}$$

while the orbitals $|\psi_a\rangle$ are eigenfunctions of the modified operator $\hat{F}'$:

$$(\hat{F}+\hat{O}\hat{A}\hat{O})|\psi_a\rangle = \varepsilon_a'|\psi_a\rangle. \tag{4}$$

In the case of a singlet-singlet $i \rightarrow a$ excitation we choose our operator $\hat{A}$ as

$$\hat{A}_i(1) = -\langle\varphi_i^{HF}(2)|\frac{1}{r_{12}}(1-2\hat{P}_{12})|\varphi_1^{HF}(2)\rangle, \tag{5}$$

where the operator $\hat{P}_{12}$ exchanges electrons 1 and 2. In this way the excited electron on level 'a' sees the correct N-1 body potential V^{N-1} with the hole on the i-th level. Expanding the orbitals $|\psi_a\rangle$ in terms of the Hartree-Fock orbitals $|\varphi_i^{HF}\rangle$,

$$|\psi_a\rangle = \sum_{i=1}^{m} c_{ai}|\varphi_i^{HF}\rangle \tag{6}$$

we obtain the matrix equation

$$\underline{\underline{F}}'\,\underline{c}_a = \varepsilon_a'\,\underline{c}_a \tag{7}$$

for the determination of the corrected virtual levels. The corrected excitation energy (difference of the total energies in the excited and in the ground state) can be obtained in this way directly as the difference of the corrected one-electron energies [20]

$$^1\Delta E_{i\rightarrow a} = E_{i\rightarrow a} - E_G = \varepsilon_a' - \varepsilon_i. \tag{8}$$

For the four nucleotide bases a minimal basis set <u>ab intitio</u> SCF LCAO MO calculation was performed [21] using the same basis set as the one applied for the polyC band structure calculation (see above). The virtual levels have been corrected applying the above described $\hat{O}\hat{A}\hat{O}$ formalism and assuming the singlet-singlet excitation occurred from the highest filled MO [21]. In Table II the HOMO and the uncorrected and corrected LEMO levels, respectively, of the four nucleotide bases are presented [21].

Table II.

The highest filled and lowest empty MO levels of the four nucleotide bases (in eV-s). The LEMO levels are the virtual Hartree-Fock levels and the ones obtained with the ÔÂÔ formalism assuming a singlet-singlet transition starting from the highest filled level.

Base	C	G	T	A
HOMO level	-9.768	-8.209	-9.602	-9.288
LEMO level (uncorrected)	1.931	3.612	3.203	2.332
LEMO level (corrected)	-2.816	-3.495	-4.448	-3.582
corrected gap	6.952	4.714	5.154	5.706
Exp. $n^{*} \to n^{*}+1$ excitation energy [22]	4.3	4.2	4.7	4.9

As we can see from this Table already this simple procedure (which does not take into account the change of the correlation energy in consequence of the excitation and takes care only on the major part of the rearrangement in the excited levels but it does not describe this effect for the levels occupied in the ground state) gives with the exception of C a tolerably good approximation to the corresponding experimental vapor-phase first singlet-singlet transition energies.

One can assume in a good approximation that an excitation occurs locally on a single molecule also in a polyC chain. Therefore, after performing an ÔÂÔ calculation for the sinlge molecule one can shift the center of the empty bands to the corresponding corrected positions of the virtual levels. Assuming that the first exciton band of polyC has the same width as its conduction band, one obtains so an approximate exciton band. Applying the ÔÂÔ procedure for the excitation from the HOMO level one gets in this way a corrected gap between the valence and conduction bands. One should point out, however, that by the help of this simple procedure one does not obtain the correct width of the exciton band and one does not learn anything about its dispersion. To answer these questions one has to use a more sophisticated description of the excited states of the polymer applying for instance the intermediate exciton model [23]. Without performing these calculations it seems to be prematurate to compare the corrected band structure of a polynucleotide with its experimental spectra (for a more detailed discussion of this problem see [21]).

As further step the effect of long-range correlation was approximately taken into account applying the electron polaron model in the form worked out by Kunz and Collins [24]. According to this model which corrects only the valence and conduction bands one obtains the following k-dependent energy shifts for them:

$$\Delta E_{cond.}(\vec{k}) = \sum_{\vec{K}}^{1.B.Z.} \frac{|V_{\vec{K}}^{n^*\to n^*+1}|^2}{\varepsilon_{cond.}^{HF}(\vec{k}) - {}^1\Delta E_{n^*\to n^*+1} - \varepsilon_{cond.}^{HF}(\vec{k}-\vec{K})} \quad (9)$$

$$\Delta E_{val.}(\vec{k}) = \sum_{\vec{K}}^{1.B.Z.} \frac{|V_{\vec{K}}^{n^*\to n^*+1}|^2}{\varepsilon_{val.}^{HF}(\vec{k}) - {}^1\Delta E_{n^*\to n^*} - \varepsilon_{val.}^{HF}(\vec{k}-\vec{K})} \quad (10)$$

where ${}^1\Delta E_{n^*\to n^*+1}$ is the energy of excitation from the valence band to the first exciton band (which can be approximated with the aid of the ÔÃÔ procedure; see equ. (8)), and $\varepsilon_{cond.}^{HF}(\vec{k})$ and $\varepsilon_{val.}^{HF}(\vec{k})$ are the corresponding Hartree-Fock eigenvalues which one obtains from an <u>ab initio</u> SCF LCAO CO calculation for a linear chain. Furthermore

$$V_{\vec{K}}^{n^*\to n^*+1}$$

is given by

$$V_{\vec{K}}^{n^*\to n^*+1} = \frac{ie}{|\vec{K}|}\left[\frac{2\pi\, {}^1\Delta E_{n^*\to n^*+1}(1-1/\varepsilon_\infty)}{V_c}\right]^{1/2}$$

$$\cdot \int |\phi^{wannier}(\vec{r})|^2 e^{i\vec{K}\vec{r}}\, d\vec{r} \approx \frac{ie}{|\vec{K}|}\left[\frac{2\pi\, {}^1\Delta E_{n^*\to n^*+1}(1-1/\varepsilon_\infty)}{V_c}\right]^{1/2} \quad (11)$$

(Since the integral occurring in equ. (11) is in a good approximation equal to 1 [24], the matrix element $V_{\vec{K}}^{n^*\to n^*+1}$ is simplified to the expression given on the right-hand side of equ. (11).) Finally ε_∞ is the high-frequency dielectric constant of the system.

In the actual calculations [21] we have substituted the summations over the vector $\vec{K}$ by an integration over $\vec{K}$. In that case we obtain a factor of $V_c/8\pi^3$ before the integral and in this way V_c (the crystal volume) falls out from equ.-s (9) and (10).

The value of ε_∞ for polyC has been estimated [21] on the basis of the expression

$$\frac{\varepsilon_\infty - 1}{\varepsilon_\infty + 2} = \frac{4}{3}\alpha_\perp \frac{1}{v_m} , \tag{12}$$

where $\alpha_\perp$ is the polarization of the cytosine molecule in the direction perpendicular to the molecular plane and v_m is the volume of a cytosine molecule. For $\alpha_\perp$ we have used the previously calculated value of 1.33 Å^3 [25] and for v_m the value of $v_m = 3.36.\pi.1.75^2 = 32.33$ Å^3 (3.36 Å is the stacking distance in polyC, 1.75 is the estimated molecular radius in the plane of the molecule) has been taken. With these values $\varepsilon_\infty \approx 1.5$ has been obtained from equ. (12).

In Table III the conduction and valence bands of polyC corrected for long-range correlation with the aid of the electron polaron model are given.

Table III.

The conduction and valence bands of polyC corrected with the help of the electron polaron model with $\varepsilon_\infty = 1.5$ (in eV-s).

	$\varepsilon^{CO}_{min.}$ $(k_{min.}a)$	$\varepsilon^{CO}_{max.}$ $(k_{max.}a)$	$\Delta\varepsilon^{CO}$
conduction band	1.116 (0)	2.274 (π)	1.158
valence band	-9.190 (π)	-8.674 (0)	0.516

We can see from this Table that by taking into account the long-range correlation in an approximate way the original Hartree-Fock energy gap of 10.647 eV (see Table I) decreased to 9.790 eV and the widths of the conduction- and valence bands, respectively, decreased also by about 10 per cent.

As mentioned previously one can obtain a rough approximation for the first exciton bands of the four homopolynucleotides, if one takes as the center of them the first virtual levels of the single bases corrected by the ÔÂÔ procedure+. Assuming that these

+Actually these band centers have been shifted downwards in all the four cases by 0.236 eV (this value is the difference between the Hartree-Fock LEMO level and the center of the correlated conduction band of polyC; compare Tables II and III) with respect to their LEMO levels obtained with the help of the ÔÂÔ method.

first exciton bands have the same widths in all the four cases as the conduction band of polyC corrected for long-range correlation, we obtain an estimate for these bands. In a similar way we can get an estimate for the valence bands also for the other three homopolynucleotides, if we assume that these bands have the same widths as the correlated valence band of polyC and put their centers at their HOMO levels + 0.836 eV (this upward shift corresponds to the difference between the HOMO level of cytosine and the center of the correlated valence band of polyC). The first exciton and valence bands of the four homopolynucleotides estimated with the aid of the described procedure are given in Table IV.

Table IV.

The estimated first exciton and valence bands of the four homopolynucleotides with inclusion of long-range correlation effects (in eV-s).

		polyC	polyG	polyT	polyA
Exciton band	Center of band	-3.052	-3.731	-4.684	-3.816
	Bandwidth	1.158	1.158	1.158	1.158
Valence band	Center of band	-8.932	-7.373	-8.766	-8.452
	Bandwidth	0.516	0.516	0.516	0.516
Gap		5.043	2.805	3.245	3.799

The comparison of the one-electron level schemes (obtained with the aid of different semiempirical methods [26] and by *ab initio* calculations [27])of the G-C and A-T base pairs with those of the constituent single bases shows that in all cases the one-electron levels of a nucleotide base pair can be obtained in a good approximation as superposition of the corresponding level schemes of the constituent molecules. Therefore, one would expect that the energy band structures of the poly(base pairs) could be also approximated if we superpose the band structures of the corresponding homopolynucleotides. Applying this procedure to the calculated and estimated correlated exciton and valence bands of polyC, and of the other three homopolynucleotides, respectively, we arrive at the estimated band structures of poly(G-C) and poly(A-T) given in Table V. In this Table we have given at each band, in

Table V.

The physically interesting bands of poly(G-C) and poly(A-T) estimated by superposition of the corresponding bands of the homopolynucleotides given in Table IV (in eV-s).

	poly(G-C) $\varepsilon_{min.}$	poly(G-C) $\varepsilon_{max.}$	Type	poly(A-T) $\varepsilon_{min.}$	poly(A-T) $\varepsilon_{max.}$	Type
2.(n^*+2) exciton band	-3.630	-2.473	polyC	-4.395	-3.237	polyA
1.(n^*+1) exciton band	-4.310	-3.152	polyG	-5.263	-4.105	polyT
Valence(n^*) band	-7.631	-7.115	polyG	-8.710	-8.194	polyA
n^*-1 band	-9.190	-8.674	polyC	-9.024	-8.508	polyT
gap	2.805			2.931		

the column "Type" that homopolynucleotide from which it originates. We can see that the bands coming from polyC and polyG are well separated in the ground state of the poly(G-C) system. On the other hand both the exciton and the valence bands of poly(A-T) and the exciton bands of poly(G-C) overlap (as indicated by braces), and therefore in these cases the superposition of the band structures becomes questionable.[+] In the last row of the Table ("Gap") the smallest possible excitation energy of the poly(base pairs) is given. As we can see this is an intrabase type excitation from the upper limit of the valence band of polyG to the lowest exciton level belonging again to polyG in the case of poly(G-C). In the case of poly(A-T), however, the highest filled and lowest exciton levels come from polyA and polyT, respectively, and therefore the first transition of this poly (base-pair) is of interbase type.

[+]It should be mentioned, however, that if one would perform a direct calculation of the exciton bands, most probably their widths would be essentially smaller than the width of the conduction band of polyC (~1.16eV) used in the estimation of the exciton bands and thus most probably the overlap of the exciton bands(0.3-05.eV) would disappear.

3. PROTEINS

Recently ab initio SCF LCAO CO band structure computations [12,13] have been performed [10] on the simplest periodic protein model, on polyglycine. For these calculations the same contracted Gaussian-lobe minimal basis has been applied as for polycytosine [14]. For the geometry of the polypeptide chain the parallel-chain β pleated sheet conformation was taken using the atomic coordinates of Pauling and Corey [28]. The obtained Hartree-Fock band structures were again corrected for long-range correlation effects using the formalism of the electron polaron model [24] in the form described in the previous point of this paper. In the latter calculation the value of 3.5 has been applied for ε_∞ as has been recommended by Breen and Flory [29]. For the excitation energy ${}^1\Delta E_{n^* \to n^*+1}$ occurring in equ.(11) the value of 8.95 eV has been applied. (This is with the help of the $\hat{O}\hat{A}\hat{O}$ procedure corrected HF value (see equ. 8) of a double unit of polyglycine H_2N-CO-CH_2-NH-CO-CH_3 [11]).

In Table VI we give the two highest filled and lowest unfilled energy bands of polyglycine obtained in the Hartree-Fock approximation. The numbers written in parenthesis in the case of the valence and conduction bands of the main chain are the values corrected for long-range correlation. The Table contains the results for the main polypeptide chain of polyglycine and also for the hydrogen-bonded (see Fig. 1) chain perpendicular to the main polypeptide chains.

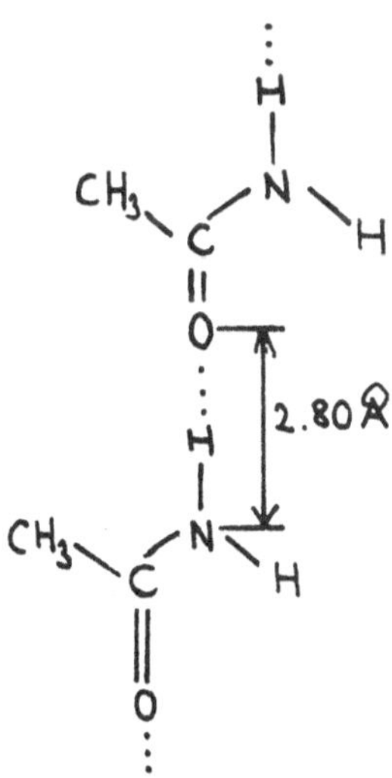

Fig. 1

The hydrogen-bonded chain perpendicular to the main polypeptide chain of polyglycine (schematic).

Table VI.

The physically most interesting bands of polyglycine calculated with the ab initio SCF LCAO CO method (in eV-s). The conduction and valence bands of the main chain are corrected for long-range correlation effects (numbers in parenthesis).

	polyglycine main chain			polyglycine H-bonded perpendicular chain		
	$\varepsilon^{CO}_{min.}(k_{min.}a)$	$\varepsilon^{CO}_{max.}(k_{max.}a)$	width	$\varepsilon^{CO}_{min.}(k_{min.}a)$	$\varepsilon^{CO}_{max.}(k_{max.}a)$	width.
n^*+2 band	9.167(π)	10.860(o)	1.693	6.748(o)	7.836(π)	0.088
conduction band	3.817($3\pi/8$) (3.102)	5.195(π) (4.357)	1.378 (1.255)	3.755(o)	3.896	0.141
valence band	-11.252(o) (-10.373)	-9.154($\pi/2$) (-8.465)	2.098 (1.908)	-11.382(π)	-11.091(0)	0.291
n^*-1 band	-12.802(o)	-12.065($\pi/2$)	0.737	-12.895(o)	-12.517(0)	0.378

As we can see from the Table in the case of the main chain the bands are rather broad (the width of the valence band is ~2.1 eV and that of the conduction band is ~1.4 eV), while the bandwidths of the hydrogen-bonded chain are by one order of magnitude smaller. These results agree qualitatively with those of previous semiempirical all valence electron calculations [8,9]. The long-range correlation correction decreases the band widths - as in the case of polyC in the previous point - again by about 10 per cent. It is interesting to point out that in the case of the main chain the dispersion of the bands does not show a simple behaviour.

The Hartree-Fock gap is for the main chain ~12.4 eV (which decreases to ~11.6 eV due to the long-range correlation), while for the hydrogen-bonded chain its value is (due to the smaller bandwidths) still larger (~14.8 eV). If we shift the center of the conduction band of the main chain to -1.543 eV (the value obtained [11] for the double unit of polyglycine with the help of the ÔÂÔ procedure) and take into account simultaneously the long-range correlation correction (as we did in the case of polyC) we obtain for the gap between the valence band and the first exciton band the approximate value of -1.543-1.255/2-(-8.465)=6.295 eV From the point of view of the conduction mechanism of proteins, however, not the latter value, buth rather the gap of ~ 11.6 eV is important. We can see that this gap is far too large to enable intrinsic conductivity in proteins.

In this respect the finding of Szent-Györgyi [30] that unsaturated ketones or dicarbonyls act as electron acceptors against proteins is of great importance. According to Szent-Györgyi for instance methylglyoxal molecules can act as electron acceptors against a peptide bond and in this way they can produce positive holes in the valence band of a protein transforming it in this way from an insulator to a conductor. Recent _ab initio_ SCF MO supermolecule calculations of a formamide- and of a glyoxal molecule (the former one served as model for the peptide groups in a protein) have shown that if the molecular planes are parallel at 1.80 Å intermolecular distance the charge transfer is from formamide to glyoxal 0.2 e in the ground state of the system [31]. This[+] and the measurements of Pethig and Szent-Györgyi [32] according to which both the d.c. and the low frequency a.c. conductivity of casein increases, if methylglyoxal is added to it, strongly supports the conduction mechanism proposed by Szent-Györgyi.

[+] Independently from this supermolecule calculation one should mention that the with the ÔÂÔ procedure corrected HF LEMO levels of glyoxal and methylglyoxal lie at $\sim$-8.2 and at $\sim$-7.9 eV, respectively [11]. Comparing these values with the upper limit of the valence band of polyglycine (main chain, corrected for long-range correlation) which is $\sim$-8.5 eV (see Table VI), we can see that the highest filled level of the protein chain and the corrected LEMO levels of the acceptors nearly coincide which shows the possibility of a strong charge transfer. In this respect one should mention that the HF LEMO levels of glyoxal and methylglyoxal were in such a way corrected that an excitation from the HOMO levels to the LEMO levels of these molecules was assumed. In reality during charge transfer the transition occurs from the highest filled level of the donor (in this case the protein) to the LEMO of the acceptor and so the operator $\hat{A}$ in the ÔÂÔ procedure had to be defined with the help of $\varphi_i^{HF}(2)$ wavefunctions (see equ.(5)) which belong to the _donor_ and _not_ to the _acceptor_. Therefore the downward shift of the LEMO-s of glyoxal and methylglyoxal, respectively, will be certainly smaller in this case (charge transfer), than by the intramolecular excitations. In the case of a relative geometry as mentioned above for the supermolecule calculation, one would not expect a larger decrease of this shift, than 2-3 eV-s still making a partial charge transfer from polyglycine to these acceptors very probable.

4. POSSIBLE IMPROVEMENTS OF THE QUANTUM MECHANICAL INVESTIGATION OF BIOPOLYMERS; APPLICATION TO THE INVESTIGATION OF DIFFERENT MICROPHYSICAL MECHANISMS OF CHEMICAL CARCINOGENESIS

In this section we shall outline the further steps which can be done to improve the investigation of the electronic structure of biopolymers. In this way one will be able to interpret of course also much better their different physical and chemical properties which underlie their biological functions. To demonstrate this we choose from the numerous possibilities the problem of chemical carcinogenesis. So we try to show in this section how a fair knowledge of the electronic structure of biopolymers like DNA and proteins could give rather important contributions to the understanding of the different mechanisms of chemical carcinogenesis on the molecular level.

Hypotheses for Tumor Development on the Molecular Level

There are several proposed theories of tumor development on the molecular level. According to the mutagenic model through the accumulation of randomly distributed mutations in the DNA molecule of a cell the cancerous information can develope by chance. This cancerous information will be then transcribed to messenger RNA and translated to proteins. The difficulty with this hypothesis is that unless the cancerous information in DNA is very unspecific it has a rather low probability that it will be developed by chance through random mutations.

A large amount of experimental evidence points to the so-called tumor virus theory. It is well established namely, that if this so-called oncogenic (tumor) viruses are taken out from a cancerous cell and are injected into the same kind of normal cells, they very quickly cause cancer [33]. On the other hand it seems rather improbable that most of these tumor viruses are of external origin. (A recent statistical analysis of the different causes of tumor occurrance has shown that less than 5 % of the tumor occurrences is due to external viruses [34])

According to our present knowledge the most realistic hypothesis for tumor development on the molecular level is the so-called "reading error theory" proposed first by Bush [35]. According to this theory the cancerous information is already contained in the DNA molecules of the normal cells but it cannot be expressed because this dangerous part (or parts) of DNA is suppressed by suitable suppressor proteins. If a carcinogen binds to the suppressor protein, according to Bush this protein will be released and so the cancerous part of DNA will become free to transcribe its information to RNA which will be then translated to proteins. Recent experimental findings have shown, however, that practically all chemical carcinogens bind also directly to DNA [36]. It was established for instance that the ultimate metabolite of 3, 4-benzopyrene is bound to the amino group of guanine. The same compound and other carcinogenic hydrocarbons besides binding to guanine, bind to adenine, and to cytosine but not to thymine, showing that the ultimates of all of them most probably bind to the amino groups of these nucleotide bases. It was further established that the metabolites of carcinogenic hydrocarbons can bind also to the phosphate groups of DNA. It was found also that a completely different class of chemical carcinogens, the so-called alkylating agents, alkylate the N_7 atoms of the purine-type nucleotide bases. So we have to assume modifying Bush's original hypothesis that carcinogens bound directly to DNA can cause also the release of suppressor proteins.

Even if we accept the reading error theory there remain several open questions. First of all it is unclear how specific is this so-called cancerous information in DNA. We know that in differentiated cells of higher organisms the major part of the genetic information is suppressed by proteins (nucleohistones) while in an undifferentiated embrionic cell at the early stage of the development very probably the overwhelming majority of the genes are free. So the question arises whether there exists a specific cancerous information, or many of the genes, suppressed in a differentiated cell can act as cancer genes. (The latter possibility is in close relation to Szent-Györgyi's point of view according to which cancerous cells and undifferentiated cells are very similar to each other [30].) If we accept for a moment the existence of specific cancerous genes located at certain parts of the DNA macromolecule, the question arises whether it is really necessary to assume that chemical carcinogens bind to these parts of DNA or the effect of their binding to any part of DNA can be transmitted to those regions where the cancerous genes are situated.

The already mentioned analysis of the relative probabilities of different causes of cancer [34] has also shown that the cause of tumor development is in over 90 per cent of the cases of chemical origin. Therefore, one has to concentrate first of all on the understanding of the mechanism of action of chemical carcinogens. As it was mentioned above, carcinogens can bind directly to DNA. Therefore, if one accepts the most realistically looking "reading error" theory of tumor development, one should try to answer the question, how the binding of different chemical carcinogens to different types of DNA constituents can cause the release of the suppressor proteins from those parts of DNA which contain the cancerous information. In approaching the solution of this problem one has to consider basically two possibilities: (1) The carcinogens bound to certain parts of DNA can cause only the release of such suppressor proteins which are bound to the same DNA sections (local effects);(2) carcinogens bound to DNA can also cause the release of suppressor proteins bound to other parts of DNA (non-local effects). The understanding of both possible effects (especially the non-local effect) requires a fair knowledge of the electronic and vibrational structure of

DNA and proteins including their transport properties. Finally, the interaction between DNA and proteins in a nucleoprotein has to be studied in detail.

Possible Local Effects of Carcinogens

On suitably chosen model systems, like the ultimate of 3, 4 benzopyrene or N_7- alkylated purine rings, first of all the change of the charge distribution of the nucleotide bases caused by the carcinogen binding can be investigated. Special emphasis should be put on the possibility of charge transfer between the carcinogen and DNA constituents (not necessarily that DNA constituent to which the carcinogen is chemically bound but more probably the neighboring ones; there is also a possibility for charge transfer between the carcinogen and a, according to the primary structure far lying constituent of DNA if the tertiary structure allows a bending back of this DNA part to the position of the carcinogen.) For these investigations *ab initio* SCF LCAO MO supermolecule calculations could be performed. Finally, the major part of the correlation in the ground state of these molecules could be calculated (for instance with the aid of suitable pair correlation or cluster expansion methods) to assess its effect on the above mentioned phenomena.

One should carefully study how the above mentioned changes in the charge distribution can effect the Watson-Crick base pairing during DNA duplication. In other words, one should look into the possibility whether the charge distribution changes due to the binding of the carcinogens could cause mispairing and, in this way, point mutations.

Finally, one should study, especially in the case of bulky carcinogens, the probable conformational changes which their binding can cause in DNA. For these investigations both the *ab initio* Hartree-Fock and the PCILO method (both in its semi-empirical cand *ab initio* version) could be used. In these studies besides carcinogens bound to nucleotide bases also the possibility of their binding to the phosphate groups of DNA should be included. These conformational studies can again throw some light on the problem of the possible change of the genetic information in DNA due to conformational changes induced by the binding of carcinogens. Further performing a normal mode analysis one could look also into the change of vibrations of those DNA constituents to which the carcinogens are bound. Finally one can think also on local changes in the tertiary structure of DNA (and/or proteins) at the spot where the carcinogen ist bound.

Possible Non-local Effects of Carcinogens Bound to DNA

One has to keep in mind also the already mentioned possibility that a chemical carcinogen bound to a certain point of DNA can exert a long-range effect and interfere in this way with the DNA-protein interaction in another part of DNA. To investigate this possibility one has to perform first of all *ab initio* Hartree-Fock energy band structure calculations [12] for the other simple periodic DNA models (polyA, polyT, polyG, poly(SP)). To correct the resulting too large excitation energies the electron-hole interaction should be taken into account using again the $\hat{O}\hat{A}\hat{O}$ method [20].

To obtain a fair description of the electronic structure of the different periodic DNA models one has to take into account both the long range and short range correlation effects. For the former case the already described electron polaron model [24] could be used while for the latter case (short range correlation in the ground state of an infinite system) either a cluster expansion method [37] or suitably chosen pair correlation schemes [38] could be adapted.

To approximate better the band structure of real DNA one has to take into account also the aperiodicity of this macromolecule. As a first step starting from the density of states of different periodic models one could calculate the density of states of the aperiodic chain with the help of the coherent potential approximation (CPA) [39]. To take into account the localized levels due to the local perturbation (including possible changes of conformation) in the band structure caused by the binding of a carcinogen to DNA the *ab initio* SCF resolvent method could be applied [40].

To obtain a more realistic description of the electronic structure of DNA one has to take into account also the effects of bound water and ions. Further one should not forget that due to the double layer formed by the PO_4^- and K^+ ions the electrons of DNA (with the exception in its symmetry axis) feel a strong inhomogeneous electric field which very probably considerably influences their distribution and with it the band structure as previous semiempirical band structure calculations have shown it [41].

One has to perform the above described investigations also on the protein macromolecules. Here as compared to DNA we have two additional complications. First of all since the number of different amino acids in proteins instead of 4 (the number of different nucleotide bases) is 20, the aperiodicity problem becomes still more important than in the case of DNA. To solve this rather intricate problem one can try to use besides the already mentioned CPA approximation different forms of cluster methods [42]. A further complication arises by the fact that a protein (either in the α-helix or in the β-pleated sheet) forms as already mentioned in the introduction a two-dimensional periodic system due to the interactions along its chemically bound polypeptide chain and due to the hydrogen bonded interactions between different polypeptide chains (or between different segments of the same chain). As previous semiempirical band structure calculations have shown it to obtain a realistic band structure one has to take into account both interactions simultaneously [8,9]. One should point out in this respect that the several times mentioned *ab initio* crystal orbital method [12,13] is valid also for two-and three-dimensional periodic systems, so the treatment of the two-dimensional protein problem would not cause larger difficulties.

In the investigations of the electronic structure of biopolymers one should also look at the possibility of the existence of collective electronic states (Mott insulator states, Peierls instabilities, charge and spin density waves, plasmon-type states, excitonic insulator states, and possible new types of collective states) using many-body techniques [43]. The effects of the non-linear terms (both electronic and vibrational) in the many-body equations need to be included also in these investigations on biopolymers. It is very well possible that collective states and the mentioned non-linear terms play an important role in energy and charge transfer in biopolymers as it was pointed out in the case of delocalized non-linear vibrations (solitons) by Davydov [44]. Therefore, they could be very important in long-range effects caused by the binding of a carcinogen.

In order to calculate the vibrations from first principles, one has to know the forces between the ions mediated by the electrons. These forces can be expressed as functions of the density response of the electronic system to the displacement of the ions. The matrix describing this linear response can be expressed in terms of the electron energies and wave functions calculated at the equilibrium position of the ions. The method of Sham [45] could be applied for these calculations [46]. This method is particularly applicable for systems with localized electronic wave functions (like biopolymers) [47]. It should be noted that the calculation of the density response already includes that part of the electron-phonon interactions which is due to the potential energy of the ions. The calculation yields the phonon dispersion as well as the vibrational wave functions. Using the calculated electronic and vibrational wave functions of biopolymers one can calculate also the still missing part of the electron-phonon interaction which is due to the kinetic energy of the ions.

Having in this way the complete electron-phonon interaction one can calculate using time dependent perturbation theory and an appropriate solution of the Boltzmann transport equation [48] the different transport properties (mean free paths, mobilities and specific conductivities) of the electrons and holes in different biopolymers. These calculated quantities could be directly compared with experiments if enough sophisticated mobility and conductivity measurements are performed for DNA and proteins (using high frequency a.c., well defined and oriented samples, etc.).

DNA-protein Interaction and its Possible Change due to the Binding of a Carcinogen

The investigation of the DNA-protein interaction in nucleoproteins plays a key role in the understanding of the blocking and deblocking of the genetic information. For these investigations one has to collect primarily the present knowledge about the geometrical structure of nucleoproteins (nucleohistone + DNA complexes) based on X-ray diffraction experiments.

For the theoretical treatment of the interaction between DNA and a protein one has to investigate first of all the possibility of charge transfer between these two polymers, which determines whether the valence bands of these systems become partially filled (in the case of a charge transfer) or remain completely filled. Namely, one can show easily on the basis of perturbation theory that the polarization and dispersion interaction terms are larger between such polymers which have partially filled bands than between those with completely filled ones [49]$^{+}$. For these investigations as first step one could use DNA base-amino acid or phosphate-amino acid supermolecule calculations. As next step one could perform DNA-protein superchain calculations to establish better the amount of transferred charge.

Knowing the relative geometry and the amount of transferred charge between the DNA double helix and a nucleohistone chain, one can perform direct calculations for the interaction energy between them. For this either perturbation theoretical schemes [50] or the newly developed mutually consistent field (MCF) method [51] could be used in a suitably adapted form. One should mention that in the DNA-protein interaction the surrounding water molecules, ions and the charged units both of DNA and proteins (PO_4^- groups in DNA and positively charged arginine molecules in the nucleohistone) very probably play a crucial role and therefore have to be included in the investigations.

After performing all the described interaction calculations between DNA and protein chains one can look into the changes caused in these interactions by the binding of a chemical carcinogen to DNA. Namely, if the carcinogen can cause a charge transfer, this would change the position of the Fermi level of the macromolecular chains and in this way it would change also the interaction energy between them. Further possible local conformational changes caused by the carcinogen would certainly cause first of all local changes in the interaction between DNA and that protein which is directly

---

$^{+}$ One should mention that this result is in complete agreement with Szent-Györgyi's cancer theory [30] according to which if proteins due to the lack of charge transfer become insulators, higher structures cannot be formed and the cell reverts to the primitive "cancerous state". Namely if the biopolymers are insulators, they have only completey filled or completely empty bands, and therefore the interactions between them become really weaker. So one can imagine that charge transfer plays a very important role also in the regulation of the DNA-protein interaction. In this way blocking or deblocking of the cancerous information of DNA may very well depend also on the conduction properties of the proteins attached to these parts of DNA.

bound to the region where the carcinogen attack has occurred. Further one can think also on non-local effects of a local conformational change (first of all by a delocalized perturbation of the vibrations of the chain [44] or by causing changes in the tertiary structure). A third possibility for non-local interference with the DNA-protein interaction is offered by the probable formation of collective states which arise from the solutions of the many-body equations (see above).

Concluding Remarks

It is clear that all the proposed investigations require a rather large effort. On the other hand in our opinion this is the only way to get a deeper insight on the submolecular level into the different possible effects and mechanisms caused by chemical carcinogens bound to DNA. One hopes to clear up whether a carcinogen can interfere with the DNA-protein interaction only locally or one has to take into account also non-local effects. Since chemically rather different carcinogens bound to chemically different parts of DNA may have the same effect by deblocking the previously blocked cancerous information in DNA (if the reading error theory of tumor development is right), one suspects that the second case is the more probable one. If one really could find out on the basis of the proposed quantum mechanical investigations how different carcinogens bound to DNA could cause the deblocking of the previously masked cancerous genetic information, one would make a very important step towards the understanding of the detailed mechanism of tumor development.

ACKNOWLEDGMENT

The author should like to express his sincere gratitude to the NATO Scientific Affairs Division and to the University of Namur, whose sponsorship made it possible to organize the Advanced Study Institute on Electronic Structure and Properties of Polymers.

He is further very much indebted to Professor A. Szent-Györgyi and to Drs. T.C. Collins, K. Laki, J. Čižek, V.E. Van Doren and S. Suhai for the many inspiring and interesting discussions.

REFERENCES

1. For reviews of these band structure calculations see: J. Ladik, Int. J. Quant. Chem. 4, 307 (1971); J. Ladik in Advances in Quantum Chemistry, vol. 7, P.O. Löwdin, Ed. Academic Press, New York, p. 377 (1973); J. Ladik, Int. J. Quant. Chem. QBS1, 651 (1974).
2. J. Ladik in "Electronic Structure of Polymers and Molecular Crystals" J.-M. André and J. Ladik, Eds., Plenum Press, New York-London, p. 663 (1975).
3. S. Suhai, Biopolymers 13, 1739 (1974).

4. J. Ladik, Int. J. Quant. Chem. QBS 1, 65 (1974).
5. J. Ladik, Int. J. Quant. Chem. QBS 2, 133 (1975).
6. S. Suhai, Ch. Merkel and J. Ladik, Phys. Lett. 61A, 487 (1977).
7. J. Ladik, S. Suhai, P. Otto and T.C. Collins, Int. J. Quant. Chem. QBS 4, (accepted).
8. S. Suhai and J. Ladik, Theoret. Chim. Acta (Berlin) 28, 27 (1973).
9. S. Suhai and J. Ladik, Acta Chim. Hung. Ac. Sci. 82, 67 (1974); S. Suhai, Theoret. Chim. Acta (Berlin) 34, 157 (1974).
10. S. Suhai and J. Ladik (to be published).
11. S. Suhai, T.C. Collins and J. Ladik, Biopolymers (submitted).
12. G. Del Re, J. Ladik and G. Biczó, Phys. Rev. 155, 997 (1967; J.M. André, L. Gouverneur and G. Leroy, Int. J. Quant. Chem. 1, 420 and 451 (1967); R.N. Euwema, D.L. Wilhite and G.T. Surrat, Phys. Rev. B7, 818 (1973).
13. J. Ladik in "Electronic Structure of Polymers and Molecular Crystals", J.-M. André and J. Ladik, Eds. Plenum Press, New York-London, p. 23 (1975).
14. B. Mely and A. Pullman, Theoret. Chim. Acta (Berlin) 13, 278 (2969); S. Huzinaga, J. Chem. Phys. 42, 1293 (1965).
15. Ch. Merkel "Elektronische Eigenschaften von Molekülkristallen" (Electronic Properties of Molecular Crystals) Thesis, Technical University Munich, 1977.
16. S.Arnott, S.D. Dover and A.J. Wonacott, Acta Cryst. B25, 2192 (1969).
17. S. Suhai in "Electronic Structure and Properties of Polymers" J.-M. André, J. Delhalle and J. Ladik, Eds., Reidel Publ. Co., Dordrecht-Boston, (1978).
18. C. Lifschitz, E.D. Bergmann and B. Pullman, Tetrahedran Letters 46, 4583 (1967).
19. W. Hunt and W.A. Goddard, III. Chem. Phys. Lett. 3, 414 (1969); H.P. Kelly, Adv. Chem. Phys. 14, 129 (1969).
20. T.C. Collins and A.B. Kunz, Int. J. Quant. Chem. S8, 437 (1974); T.C. Collins, A.B. Kunz and P.W. Deutsch, Phys. Lett. A10, 1034 (1974).
21. J. Ladik, S. Suhai, P. Otto and T.C. Collins, Int. J. Quant. Chem. QBS 4, (1977).
22. L.B. Clark, G.G. Peschel and I. Tinoco, Int. J. Phys. Chem. 69, 3615 (1965).
23. See for instance: R.S. Knox, Theory of Excitons, Solid State Physics, H. Ehrenreich, F. Seitz and D. Turnbull, Eds., Academic Press, New York-London S5, p. 59 (1963).
24. A.B. Kunz, Phys. Rev. B6, 606 (1972); J.T. Devreese, A.B. Kunz and T.C. Collins, Solid State Comm., 11, 673 (1972); T.C. Collins in "Electronic Structure of Polymers and Molecular Crystals", J.-M. André and J. Ladik, Eds., Plenum Press, New York-London, p. 405 (1975.
25. L. Seprödi, G. Biczó and J. Ladik, Int. J. Quant. Chem. 3, 62 (1969).

26. T.A. Hoffmann and J. Ladik, Adv. Chem. Phys. 7, 184 (1964); R. Rein and J. Ladik, J. Chem. Phys. 40, 2466 (1964).
27. E. Clementi, J.-M. André, M.-Cl. André, D. Klim and D. Hahn, Acta Phys. Acad. Sci. Hung. 27, 493 (1969); E. Clementi, J. Mehl and W. von Niessen, J. Chem. Phys. 54, 508 (1971).
28. L. Pauling and R.B. Corey, Proc. Natl. Acad. Sci. USA 39, 253 (1953).
29. D.A. Breen, P.J. Flory, J. Am. Chem. Soc. 87, 279 (1965).
30. A. Szent-Györgyi, Int. J. Quant. Chem. QBS 3, 45 (1976); A. Szent-Györgyi, Bioenergetics 4, 535 (1973); A. Szent-Györgyi, "Electronic Biology and Cancer", Marcel Dekker Inc., New York-Basel, (1976).
31. P. Otto, S. Suhai and J. Ladik, Int. J. Quant. Chem. QBS 4, (accepted).
32. R. Pethig and A. Szent-Györgyi, Proc. Natl. Ac. Sci. USA 74, 226 (1977); R. Pethig, personal communication.
33. I. Berenblum, "Carcinogenesis as a Biological Problem", North Holland Publishing Co., Amsterdam-Oxford, Chapt. 4, p. 211 (1973).
34. Quoted by P.O. Löwdin, Lecture on the 4th Int. Symp. on Quant. Biol., Sanibel Insland (1977).
35. H. Bush, "Biochemistry of the Cancer Cell", Academic Press, New York, p. 292 (1962).
36. R. Daudel in "Mutagenesis and Chemical Carcinogenesis", P. Daudel, R. Daudel, Y. Moulé and F. Zajadela, Eds., C.N.R.S., Paris (1977).
37. J. Čižek in Adv. Chem. Phys., R. Lefebvre and C. Moser, Eds., Interscience, London-New York, vol. 14, p. 35 (1969).
38. See for instance: R.K. Nesbet, ibid, p.1; O.K. Sinanoglu, ibid, p.237.
39. P. Soven, Phys. Rev. 156, (1967); B. Velicky, S. Kirkpatrick and H. Ehrenreich, ibid, 175, 747 (1968); S. Kirkpartrick, B. Velicky and H. Ehrenreich, ibid, B1, 3250 (1970).
40. J. Ladik and M. Seel, Phys. Rev. B13, 5338 (1976).
41. A. Karpfen and J. Ladik (unpublished results).
42. M. Seel, T.C. Collins and J. Ladik (to be published).
43. T.C. Collins, (private communication).
44. A.S. Davydov, Studia Biophys. (Berlin) 62 (1977).
45. L. Sham, Phys. Rev. B6, 3584 (1972).
46. V.E. Van Doren, (private communication).
47. P.E. Van Camp, V.E. Van Doren and J.T. Devreese, Phys. Rev. B (in the press).
48. S. Suhai, J. Chem. Phys. 57, 5599 (1972).
49. K. Laki and J. Ladik, Int. J. Quant. Chem. QBS 3, 51 (1976).
50. J.N. Murrell, M. Randić and D.R. Williams, Proc. Roy. Soc. A284, 566 (1965).
51. P. Otto and J. Ladik, Chem. Phys. 8, 192 (1975); ibid, 19, 209 (1977).

ENERGY BAND STRUCTURE OF HIGHLY CONDUCTING POLYMERS

János J. Ladik

Lehrstuhl für Theoretische Chemie
der Friedrich Alexander Universität Erlangen-Nürnberg
852 Erlangen, FRG

ABSTRACT. Ab initio SCF LCAO crystal orbital calculation (with no approximation for the exchange) in the fourth neighbor's interactions approximation of the one-dimensional $(SN)_x$ chain resulted in a width of $\sim$4.0 eV for the half-filled band. A coherent potential approximation (CPA) calculation of the $(SN)_x$ and $(S^N_H)_x$ mixed chain is in progress.

MINDO/3 crystal orbital calculations of the neutral stacked TCNQ and TTF chains have given a band width of 0.56 eV for the conduction band of TCNQ (in good agreement with previous MINDO/2 calculation) and 0.52 eV for the valence band of TTF.

Finally the necessary next steps in the theoretical investigation of the electronic structure of these highly conducting polymers are discussed.

1. INTRODUCTION

In the last 5 years considerable interest has been raised by highly conducting polymers. The $(SN)_x$ (polysulphurnitride) system is a highly anisotropic metal at higher temperatures and becomes superconductive below 0.26°K [1]. The quasi one-dimensional TCNQ (7,7', 8,8'-tetracyanoquinodimethane)-TTF (2,2'-bis-1,3 dithiole) charge transfer system[+] has a peak in its conductivity at

---

[+] There is a whole class of charge transfer systems in which TCNQ is cocrystallized with another partner. Such systems are for instance the $TCNQ^-$-NMP^+(N-methylphenazinium) the $TCNQ^-$-K^+ system etc. In this paper, however, only the TCNQ-TTF system will be discussed.

J.-M. André et al. (eds.), Quantum Theory of Polymers, 279-288.

58^oK [2] which has been interpreted as a possible Fröhlich-type superconductive fluctuation [3]. Further it is rather probable that under this temperature a Peierls metal-insulator transition [4] occurs. Due to these interesting physical properties the TCNQ-TTF system has been the subject of a large number of experimental investigations [5]. A third group of highly conducting polymers is formed by the Krogmann's salts [like KCP($K_2Pt(CN)_4$ $Br_{0.3}$ $3H_2O$)] [6] which, however, will not be discussed here.

To interpret the different physical properties of these polymers one needs a fair knowledge of their electronic structure. For that purpose it is not enough to investigate only the constituents (unit cells) of these systems, but one has to perform also band structure calculations for the corresponding infinite chains and 3-dimensional crystals, respectively.

For the one-dimensional $(SN)_x$ chain a couple of semiempirical and approximate (using for instance simplified atomic potentials) _ab initio_ calculations have been performed [7], followed by two minimal basis set _ab initio_ calculations [8]. There is in the literature also a non-self consistent OPW calculation performed for the 3-dimensional system [9]. In this paper the results of a double ζ _ab initio_ SCF LCAO CO calculation performed for the 1-dimensional $(SN)_x$ chain will be reviewed. This will be followed by a density of states calculation of the $(SN)_x$ and $(\overset{SN}{\underset{H}{|}})_x$ chain+.

In the case of the TCNQ-TTF system there are different semiempirical [11] and _ab initio_ [12] calculations for the constituents of these stacked chains including a recent _ab initio_ calculation which takes a TCNQ-TTF molecule pair as a supermolecule to find out the amount of transferred charge in the case of different relative geometries [13].

There are also CNDO/2 crystal orbital calculations for neutral and charged TCNQ and TTF stacked chains [14] and for the TCNQ chain the calculations have been repeated with the aid of the MINDO/2 CO [14] method which seemed to provide a more realistic band structure than the CNDO/2 CO method. Due to the lack of parametrization for the S atoms the MINDO calculations could not be performed in 1974 for the TTF stack. Since that time MINDO/3 parameters habe become available [15] also for S, a MINDO/3 crystal orbital programm was developed and has been applied to the neutral TCNQ and TTF chains at different stacking distances. The results of these calculations will be summarized in the second point of this paper.

Finally in the concluding section of this work an outlook will be given about the next steps which will be performed to obtain a better description of the electronic structure (and with the help of it of their interesting physical properties) of the

---

+ At IBM San Jose between 5 and 10 mol. per cent hydrogen has been found recently [10].

discussed highly conducting polymers.

2. $(SN)_x$

The *ab initio* SCF LCAO crystal orbital (CO) formalism has been described previously [16]. For the calculations [17] a double ζ (split-shell) type basis set i.e. 10 and 18 contracted AO-s was applied on atoms N and S, respectively. The AO-s themselves were taken as contractions of Gaussian lobes using 25 and 46 primitive functions for N and S, respectively. The orbital exponents and contraction coefficients weren taken from Roos and Siegbahn [18]. In constructing the elements of the Fock matrix until fourth neighbor's (SN units) interactions have been taken into account. In this way the resulting band structures (in eV-s) were stable for three decimals. All multi-center integrals including the correct HF exchange have been taken into account if their value exceeded the threshold of 10^{-8} a.u.

Since the geometrical structure of the $(SN)_x$ chains is still a matter of controversy the calculations for the linear zig-zag chains have been performed using both published sets of structural data. According to an electron diffraction analysis [19] there is one SN unit per elementary cell along the chain with alternating S-N and N-S bond distances of 1.73 and 1.55 Å, respectively, with a valence angle of 113.5°. X-ray measurements [20] have given on the other hand a structure with two SN molecules in the unit cell and with alternating bound lengths and valence angles of 1.63 Å, 1.59 Å, 120° and 106°, respectively. By performing the calculations on these zig-zag structures (that is on chains not with simple translations but with combined symmetry operations) the contracted basis functions have been rotated together with the units (for more details see [21]).

The calculations [17] have shown that the second more complex structure has a total energy per SN unit which is by 0.32 eV lower. This makes the second structure more probable, but one should not forget that the interchain interactions may be quite different in the two structures. (The differences in the shortest interchain distances are 0.3-0.4 Å.) Therefore only proper two- and three-dimensional calculations can clarify this problem. In this respect it should be mentioned that the already performed 3D OPW calculation [9] cannot be considered - in the author's opinion - as very much relevant, because due to the nature of the OPW method the chemical details of the $(SN)_x$ system most probably are washed away.

The band structures obtained for the two geometries do not differ from each other in essential details. In the case of the second structure (double sized unit cells) there are 28 doubly degenerate bands. From these 11 are completely filled and one is half filled. The physically most interesting metallic band is a π-band (its Bloch function is composed of the 2p AO of N and the 3p AO of S, both of π symmetry) with a width of 4.12 eV (ε^{CO}_{min} = -9.854 eV, ε^{CO}_{max} = -5.734 eV) and the Fermi level lies at

ε_F = -7.821 eV. Unfortunately, there is no direct experimental information about the width of this band in $(SN)_x$. Therefore, the effective electronic mass (1.7 m_e) and the density of states at the Fermi level (0.14/(eV spin molecule)) has been calculated [17]. Both agree quite with the corresponding experimental values (2 m_e from the analysis of polarized reflectivity spectra of single crystalline $(SN)_x$ [22] and 0.18/(eV spin molecule) from the contribution of the linear temperature dependent term to the specific heat [23], respectively.) Further according to a Mulliken type population analysis of the wave function 0.40 e charge is transferred in the ground state from the S to the N atom which agrees again reasonable well with experiment (transferred charge 0.3-0.4 e from X-ray photo emissionspectra [24]).

It is worth mentioning that according to a minimal basis *ab initio* SCF LCAO CO calculation of the linear $(SN)_x$ chain [8] the width of the halfly filled band is ~10 eV with a corresponding effective mass of 0.72 m_e and with a density of states of 0.06 (eV spin molecule). This shows - in accordance with the general experience - that the properties of a partially filled band are strongly dependent on the basis set+ and therefore, one has to go beyond the minimal basis approach.

As it was mentioned above, recentyl 5-10 mol. per cent hydrogen was found in $(SN)_x$ [10]. Most probably the H atoms bind to the N atoms. In this way they change the hybridization state of the N atoms and with this the number of π electrons in the partially filled band of $(SN)_x$ (in an ∖S⫽N∖ unit there are 3 π electrons, while in a ∖S⁄N(H)∖ unit 4). To find out the shift of the Fermi level and the change in the density of states caused by the hydrogen binding, a calculation is in progress in the coherent potential approximation (CPA). This method uses as input the density of states ($\varrho^A(\varepsilon)$) [25] of one of the systems and the $\varepsilon(k)$ curves of a pure polyA and polyB chain (in our case of an $(SN)_x$ and of an $(\overset{}{S}\underset{H}{N})_x$ chain). This density of states was calculated with the aid of a program using Delhalle's method [26]. Their values based on an *ab initio* SCF LCAO CO calculation of the $(SN)_x$ and $(S\underset{H}{N})_x$ chains (using a minimal basis [27] set and the same geometry for $(S\underset{H}{N})_x$ as for $(SN)_x$ [19] with a N-H distance of 1.00 Å) are shown in Table I. We can see from the Table that the width of the completely filled valence band of $(S\underset{H}{N})_x$ (3.5 eV) is only about one third of the width of halfly filled valence band of $(SN)_x$ (10.9 eV). It is unclear that this effect is completely due to the too small basis set (see the remarks above by the double ζ calculation of $(SN)_x$) or also the lack of the treatment of the correlation plays a significant role (as probably is the case). The results of this CPA calculation will be published elsewhere [27].

+ Most probably in this case also correlation effects are much more important than in cases of completely filled or empty bands.

Table I.

The density of states values of the $(SN)_x$ and $(\underset{H}{SN})_x$ chains based on a minimal basis set *ab initio* SCF LCAO CO calculation (in (eV spin molecule)$^{-1}$ units) [27]

$(SN)_x$ chain		$(\underset{H}{SN})_x$ chain	
$\varepsilon^{CO}_{val.}$	$\rho^{A}(\varepsilon)$	$\varepsilon^{CO}_{val.}$	$\rho^{B}(\varepsilon)$
-8.24	0.60	-7.86	0.28
-8.14	0.93	-7.77	0.83
-8.03	0.54	-7.70	0.61
-7.92	0.38	-7.66	0.55
-7.02	0.18	-7.62	0.50
-5.47	0.12	-7.18	0.30
-4.57	0.10	-6.71	0.12
-3.12	0.06	-4.79	0.39
-1.79	0.10	-4.76	1.18
1.33	0.14	-4.71	2.75
2.12	0.32	-4.68	3.16
2.45	0.49	-4.54	1.54
2.56	0.69	-4.38	3.76
2.67	1.03		
$\varepsilon^{min.}_{val.} = -8.24$ eV		$\varepsilon^{min.}_{val.} = -7.86$ eV	
$\varepsilon^{max.}_{val.} = 2.67$ eV		$\varepsilon^{max.}_{val.} = -4.38$ eV	
$\varepsilon_{Fermi} = -1.90$ eV		$\varepsilon_{Fermi} = -4.38$ eV (completely filled band)	

3. THE TCNQ-TTF SYSTEM

Using the MINDO/3 parametrization [15] the band structures of the neutral stacked poly(TCNQ) and poly(TTF) were recalculated using a newly developed MINDO/3 CO program [28]. For these calculations which were performed -similarly to the previous investigations [14]- in the first neighbor's interactions approximation, the same geometry of the stacked chains was applied as it was found in the mixed TCNQ-TTF crystal [29]. According to this geometry the perpendicular interplane distance is 3.17 Å in the TCNQ stack and 3.47 Å in the TTF case, respectively. To reach self consistency about 30 iteration steps were needed to fulfil simoultaneously the applied

$$\left| p(o)^{(n+1)}_{r,s} - p(o)^{(n)}_{r,s} \right| \leq 5.10^{-4} \quad , \tag{1a}$$

$$\left| p(1)^{(n+1)}_{r,s} - p(1)^{n}_{r,s} \right| \leq 5.10^{4} \tag{1b}$$

SCF criteria. For the numerical integrations necessary to calculate the generalized bond orders $p(q)_{r,s}$ (q=0,1) [30] five different values of k were used in the o - π/a interval (according to our previous experience 5 values of k give already a consistent band structure). The expressions of the applied MINDO crystal orbital method are also given in reference [30] and therefore we do not repeat them here.

In Table II we give the two highest filled and lowest unfilled energy bands of poly(TCNQ) and poly(TTF) at R=3.17 Å and R=3.47 Å stacking distances, respectively (more detailed band structures taking into account also other stacking distances, will be published elsewhere [28]).

<u>Table II.</u>

The physically most intersting energy bands (in eV-s) of poly(TCNQ) and poly(TTF) (R=3.47 Å).

	poly(TCNQ) R=3.17 Å				poly(TTF) R=3.47 Å			
Band level	ε^{MO}	$\varepsilon^{CO}_{min}(k_{min}a)$	$\varepsilon^{CO}_{max}(k_{max}a)$	$\Delta\varepsilon$	ε^{MO}	$\varepsilon^{CO}_{min}(k_{min}a)$	$\varepsilon^{CO}_{max}(k_{max}a)$	$\Delta\varepsilon$
$n^{*}+2$	0.976	0.693(o)	0.778(π)	0.085	0.539	0.529(o)	0.549(π)	0.016
conduction	-1.492	-1.975(o)	-1.419(π)	<u>0.556</u>	-0.118	-0.140(o)	-0.075(π)	0.065
valence	-7.576	-7.837(π)	-7.743(o)	0.095	-7.411	-7.581(π)	-7.057(o)	<u>0.524</u>
$n^{*}-1$	-9.362	-9.578(π)	-9.534(o)	0.044	-8.593	-8.597(π/4)	-8.511(π)	0.086

We see in the Table that in agreement with the previous CNDO/2 CO calculations and the the MINDO/2 CO calculation for poly(TCNQ) [14] again the physically most interesting conduction band of poly(TCNQ) and the valence band of poly(TTF) (underlined values in the Table) are rather broad. Further the MINDO/3 band width of 0.556 eV obtained for the conduction band of poly(TCNQ) agrees quite well with the previous MINDO/2 value of 0.628 eV [14]. Also the previously estimated band width of 0.5 eV for the valence band of poly(TTF) (taking the half of the corresponding (CNDO/2 band width [14]) agrees very well with the here calculated MINDO/3 value of 0.524 eV.

Having also a working MINDO/3 MO program we have tried [28] to calculated the charge transfer between a TTF and TCNQ molecule taking the pair of these molecules as a supermolecule. In a relative geometry which corresponded to the "lateral position" of the two molecules (see Fig. 4 of [13]) which seemed to be the most advantegeous one for the charge transfer we did not find any charge transfer. This surprising result cannot be explained by the shortcomings of the MINDO/3 method because this method - as it is well known - gives a smaller gap between the HOMO and LEMO levels than the *ab initio* HF method. Our result is, however, in complete agreement with that of the *ab initio* supermolecule calculation of Cavallone and Clementi [13], who did not find either charge transfer in this relative geometrical position (see Table V part b of [13]). On the other hand they have found a rather strong charge transfer at the "head on" relative position (see Fig. 3 and the part a of Table V in [13]) showing the possibility of charge transfer in this somewhat unexpected direction. To investigate this problem in more detail we execute MINDO/3 MO supermolecule calculations also for relative positions in which the two molecules are in different planes (but keeping still the experimental geometry [29] of the mixed crystal).

4. FURTHER OUTLOOK

In the case of the $(SN)_x$ system we plan to perform *ab initio* SCF LCAO CO calculations in the double ζ level also for the different two-dimensional segments of the three-dimensional structure and for the 3-D structure itself. (The development of the necessary programs is in progress). In this way the effects on the band structure of the interchain interactions can be studied. One can perform also CPA calculations with these 2- and 3-D band structures to obtain a better knowledge of the possible effects of the hydrogen impurities. These investigations can be completed by calculations in which also other possible positions of the H atoms would be taken into account.

A further aspect of the $(SN)_x$ problem is the treatment of the correlation in the ground state. As first approximation one could apply for the long range correlation an electron gas method [31] (treating in this way only the electron in the halfly filled band), while for the short range correlation a pair correlation method (like the CEPA method [32]) or the cluster expansion method of Cizek [33] could be applied in a suitably modified form.

In the case of the TCNQ-TTF system as next step one should aim for ab initio SCF LCAO band structure calculations both for the neutral and for the suitably charged poly(TCNQ) and poly(TTF) chains (using the experimentally found ~0.6 e value of the charge transfer per molecule pair [5]). To keep the charge neutrality in these calculations the counter charges on the partner chain could be taken at first in the monopole approximation. For a better treatment of the interactions between the charged chains the newly developed mutually consistent field (MCF) method [34] could be applied which would take care automatically also of the polarization.

The band structures of the charged poly(TCNQ) and poly(TTF) stacks could then be recalculated also with the aid of the "open shell" SCF LCAO CO method [16] which would take care on the fact that these charged chains are built up from not very strongly interacting (stacked units) with an odd (partial) number of electrons.

For the treatment of the long range correlation in the partially filled bands one could use again the technique mentioned at $(SN)_x$ [31] and for the short range correlation again the CEPA method [32] or the cluster expansion method [33] could be applied.

One could study also such analogues of $(SN)_x$ as $(ON)_x$ $(SeN)_x$, $(SP)_x$ etc. chains and the other members of the TCNQ donor class of compounds (like TCNQ-NMP, TCNQ-K, TCNQ with the Se analogue of TTF, etc.) We do not intend, however, to discuss these existing and hypothetical systems here.

One can hope that obtaining with the help of the planned investigations a more thorough knowledge about the electronic structure of the highly conducting polymers one will be able to interpret their different interesting physical properties on a more sound basis. In this way the band theory of polymers would give a significant contribution to this field.

ACKNOWLEDGMENTS

The author should like to express his sincere gratitude to the NATO Scientific Affairs Division and to the University of Namur whose sponsorship made it possible to organize the ASI on the "Electronic Structure and Properties of Polymers".

He is further very much indebted to Drs. S. Suhai, T.C. Collins and M. Seel for the stimulating fruitful discussions.

+ Actually it will be more easy according to all probability to apply these methods in these stacked cases than in the case of the chemically bound $(SN)_x$.

REFERENCES

1. For a review see: H.P. Geserich and L. Pintschovius, Adv. Solid State Phys. 16 (1976) 65.
2. L.B. Coleman, M.J. Cohen, D.J. Sandman, F.G. Yamagishi, A.F. Garito and A.J. Heeger, Solid State Comm. 12 (1973) 1125.
3. H. Fröhlich, Proc. Roy. Soc. A223 (1954) 296; J. Bardeen, Solid State Comm. 13 (1973) 357; D. Allender, J.W. Bray and J. Bardeen, Phys. Rev. B9 (1974) 119.
4. R.E. Peierls, "Quantum Theory of Solids" (Clarendon Press, Oxford, 1935).
5. See the following review papers: I.F. Schegolev, Phys. Stat. Solidi 12 (1972) 9; H.R. Zeller, Adv. Solid State Phys. 13 (1973) 31; Z.G. Soos, Ann. Rev. Phys. Chem. 25 (1974) 12; A.J. Heeger and A.F. Garito in"Low-dimensional Cooperative Phenomena" H.J. Keller, Ed., (Plenum Press, New York-London, 1975) p. 89; G. Thomas et al, Phys. Rev. B11 (1976) 5105; M.J. Cohen, L.B. Coleman, A.F. Garito and A.J. Heeger, Phys. Rev. B13 (1976) 5111.
6. K. Krogmann, Angew. Chemie 81 (1969) 10; K. Krogmann in "One-dimensional conductors" H.G. Schuster, Ed., (Springer Verlag, Berlin) p. 12.
7. H. Kamimura, A.J. Grant, F. Levy, A.D. Yoffe, G.D. Pitt, Solid State Comm. 17 (1975) 49; D.E. Parry, J.M. Thomas, J. Phys. C8 (1975) L45; A. Zunger, J. Chem. Phys. 63 (1975) 4854; S. Suhai and M. Kertész, J. Phys. C9 (1976) L347; V.T. Rajan and L.M. Falicov, Phys. Rev. B12 (1975) 1240.
8. C. Merkel and J. Ladik, Phys. Lett. 56A (1976) 395; M.Kertész, J. Koller, A. Azman and S. Suhai, Phys. Lett. 55A (1975) 107.
9. W.E. Rudge and P.M. Grant, Phys. Rev. Lett. 35 (1975) 1799.
10. B. Györffy and S. Faulkner (personal communication).
11. M. Ratner, J.R. Sabin and E.E. Ball, Mol. Phys. 26 (1973) 1177; D.A. Lowitz, J. Chem. Phys. 12 (1967) 4698; S. Hiroma, J. Kuroda and J. Akamutu, Bull. Chem. Soc. Japan 15 (1971) 9; H.T. Jonkmann and J. Kommandeur, Chem. Phys. Lett. 15 (1972) 496; F. Herman, A.R. Williams and K.H. Johnson, Phys. Rev. Lett. 33 (1974) 94; J. Ladik, A. Karpfen, G. Stollhoff and P. Fulde, Chem. Phys. 7 (1975) 267.
12. H.T. Junkmann, G.A. Van der Welde and W.C. Nieuwport, Chem. Phys. Lett. 25 (1974) 62.
13. F. Cavallone and E. Clementi, J. Chem. Phys. 63 (1975) 4304.
14. A. Karpfen, J. Ladik, G. Stollhoff and P. Fulde, Chem. Phys. 8 (1975); Chem. Phys. Lett. 31 (1975) 291; J. Ladik, Int. J. Quant. Chem. S9 (1975) 563.
15. R.C. Bingham, M.J.S. Dewar and D.H. Lo, J. Aner. Chem. Soc. 97 (1975) 1285.
16. G. Del Re, J. Ladik and G. Biczó, Phys. Rev. 155 (1967) 997; J.-M. André. L. Gouverneur and G. Leroy, Int. J. Quant. Chem.

1 (1967) 427 and 451; J. Ladik in "Electronic Structure of Polymers and Molecular Crystals", J.-M. André and J. Ladik, Eds., (Plenum Press, New York-London, 1975) p. 23.
17. S. Suhai and J. Ladik, Solid State Comm. 22 (1977) 227.
18. B. Roos and P. Siegbahn, Theor. Chim. Acta (Berlin) 17 (1970) 209.
19. M. Boudeulle and P. Michelle, Acta Cryst. A28 (1972) S199.
20. C.M. Mikulski, P.J. Russo, M.S. Saran, A.G. Madoiarmid, A.F. Garito, A.J. Heeger, J. Aner. Chem. Soc. 97 (1975) 6358.
21. See J. Ladik in "Electronic Structure and Properties of Polymers", J.-M. André, J. Delhalle and J. Ladik, Eds., (Reidel Publ. Co., Dordrecht, 1978)
22. P.M. Grant, R.L. Greene and G.B. Street, Phys. Rev. Lett. 35 (1975) 1740.
23. R.L. Greene, P.M. Grant and G.B. Street, Phys. Rev. Lett. 34 (1975) 89.
24. P. Mengel, P.M. Grant, W.E. Rudge and B.H. Schechtmann, Phys. Rev. Lett. 35 (1975) 1803.
25. P. Soven, Phys. Rev. 156 (1967) 809; B. Velický, S. Kirkpatrick and H. Ehrenreich, Phys. Rev. 175 (1968) 747; S. Kirkpatrick, B. Velický and H. Ehrenreich, Phys. Rev. B1 (1970) 3250.
26. J. Delhalle and S. Delhalle, Am. Soc. Brux. 89 (1975) 403.
27. M. Seel, T.C. Collins, D.K. Rai, S. Suhai, R.D. Singh and J. Ladik, Chem. Phys. Lett. (submitted).
28. R.D. Singh, J. Ladik and S. Suhai, Chem. Phys. Lett. (submitted)
29. T.E. Phillips, T.J. Kistenmacher, J. P. Ferraris and D.O. Cowan, Chem. Comm. (1973) 471; T.J. Kistenmacher, T.E. Phillips and D.O. Cowan, Acta Cryst. 33 (1974) 763.
30. J. Ladik in "Electronic Structure of Polymers and Molecular Crystals", J.-M. André and J. Ladik, Eds., (Plenum Press, New York-London, 1975) p. 663.
31. See for instance: D.N. Leroy and G.E. Brown, Phys. Rev. B12 (1975) 2138.
32. W. Meyer, J. Chem. Phys. 58 (1973) 1017.
33. J. Čížek, Adv. Chem. Phys. 14 (1969) 35.
34. P. Otto and J. Ladik, Chem. Phys. 8 (1975) 192; ibid 19 (1977) 209.

QUANTUM MECHANICAL TREATMENT OF TRANSPORT PROPERTIES OF SEMICONDUCTORS: POSSIBLE APPLICATION TO POLYMERS

P. Csavinszky

Department of Physics, University of Maine
Orono, Maine 04473, U.S.A.

ABSTRACT. Charge and energy transport comprises a very large part of the physics of semiconductors. The aim of the present set of lectures is a discussion of some of the fundamental concepts of transport theory.

I. INTRODUCTION

In recent years, considerable theoretical effort has gone into the formulation of a quantum theory of transport [1]. It has been found, however, that the semiclassical transport theory, relying on the quantum theory of solids, can be considered as a good approximation.

Our first goal is the description of a system of electrons by means of a distribution function, both at zero and at finite temperatures [2]. This accomplished, we shall be concerned with the behavior of an electron gas under the action of an external (electric and/or magnetic) field and a temperature gradient. This will lead us to our second goal, the finding of the distribution function under steady-state conditions [3]. The theoretical framework is developed in Sections I - VI. Some facts and speculations relevant to organic and polymeric semiconductors are given in Sections VII to IX.

II. EQUILIBRIUM DISTRIBUTION (THE FERMI-DIRAC DISTRIBUTION)

To develop the FD distribution function, we must first examine the quantum theory of the free electron gas. Before we

J.-M. André et al. (eds.), Quantum Theory of Polymers, 289-312.

can discuss the electron gas at $T \neq 0°K$, we must examine it in its ground state, i.e. at $T = 0°K$.

(a) $T = 0°K$

Let us consider N electrons, confined to a volume V. Making the independent electron approximation, i.e. assuming that the electrons do not interact with one another, we shall refer to such an electron gas as the Fermi gas. (The term Fermi liquid refers to a Fermi gas with interactions [4]). The ground state of the N-electron system can be found by first finding the energy levels of a single electron confined to the volume V. This done, we can create the ground state of the Fermi gas by successively filling the one-electron levels in a manner consistent with the Pauli principle. (From here on the term "state" will refer to the N-electron system, and the term "level" to the one-electron system).

A single electron can be described by a wave function $\psi(\vec{r})$ [standing for $\psi(x, y, z)$], and by the specification of the orientation of its spin. The wave function $\psi(\vec{r})$, associated with the energy level ε, is the solution of the time-independent Schrödinger equation

$$-\frac{\hbar^2}{2m}\nabla_r^2\,\psi(\vec{r}) = \varepsilon\,\psi(\vec{r}) \quad \left(\hbar = \frac{h}{2\pi}\right), \tag{1}$$

where the subscript r on ∇^2 refers to the fact that the Laplacian is defined in coordinate-space.

Assuming that the electron under consideration is confined to a cube of side $L = V^{1/3}$, we solve (1) by the Born-von Kármán (or, as it is also called, periodic) boundary condition which is stated by

$$\psi(x, y, z + L) = \psi(x, y, z) \tag{2a}$$

$$\psi(x, y + L, z) = \psi(x, y, z) \tag{2b}$$

$$\psi(x + L, y, z) = \psi(x, y, z). \tag{2c}$$

One can verify, by differentiation, that the wave function

$$\psi_{\vec{k}}(\vec{r}) = \frac{1}{\sqrt{V}}\, e^{i\,\vec{k}\cdot\vec{r}}, \tag{3}$$

which belongs to the energy level

$$\varepsilon(\vec{k}) = \frac{\hbar^2\,\vec{k}\cdot\vec{k}}{2m}, \tag{4}$$

does satisfy (1). One can also verify that $\psi_{\vec{k}}(\vec{r})$ is normalized to unity, i.e.

$$\int_{\substack{\text{over all} \\ \text{of r-space allowed}}} |\psi_{\vec{k}}(\vec{r})|^2 d\vec{r} = 1, \tag{5}$$

where $d\vec{r}$ stands for dxdydz.

The vector $\vec{k}$ is called the wave vector. Its significance lies in the fact that the wave function $\psi_{\vec{k}}(\vec{r})$ is an eigenstate of the momentum operator

$$\vec{p} = \frac{\hbar}{i} \vec{\nabla}_r \equiv \frac{\hbar}{i} \frac{\partial}{\partial \vec{r}}, \tag{6}$$

with the eigenvalue

$$p = \hbar k . \tag{7}$$

Invoking now the boundary conditions in (2a) to (2c), we find that only certain discrete values of k_x, k_y, k_z, the components of $\vec{k}$, are possible. To see that this is so, let us apply (2a) to (3). We find that

$$\frac{1}{\sqrt{V}} e^{i(k_x x + k_y y + k_z z)} = \frac{1}{\sqrt{V}} e^{i[k_x x + k_y y + k_z (z + L)]}. \tag{8}$$

Simplification of (8) results in

$$e^{ik_z L} = 1 , \tag{9}$$

which can only be satisfied if

$$k_z L = 2\pi n_z , \tag{10a}$$

where n_z is an integer.

Similarly, application of (2b) and (2c) to (3) leads to

$$k_y L = 2\pi n_y \tag{10b}$$

$$k_x L = 2\pi n_x , \tag{10c}$$

where n_y and n_x are integers.

The three-dimensional space, with Cartesian axes k_x, k_y, and k_z, is known as the k-space. Returning to the magnitude k of the wave vector $\vec{k}$, given by

$$k^2 = k_x^{\,2} + k_y^{\,2} + k_z^{\,2} , \tag{11}$$

we see that the allowed values of k are quantized. Substitution of (11) into (4), with consideration of (10a) to (10c), leads to the energy levels

$$\varepsilon(\vec{k}) = \frac{\hbar^2}{2m} \left(\frac{2\pi}{L}\right)^2 (n_x^2 + n_y^2 + n_z^2) . \tag{12}$$

Let us ask now how many allowed values of k are contained in a region of k-space of volume Ω that is very large compared to the volume $(2\pi/L)^3$ which contains two electrons with opposite spins. We know that such a region of k-space contains a vast number of allowed points. To an excellent approximation, the number of allowed values of k is just the volume Ω of k-space divided by the volume of k-space per point. The latter quantity is $(2\pi/L)^3$ if we ignore spin. From this argument, we conclude that the region of k-space of volume Ω will contain

$$\nu = \frac{\Omega}{(2\pi/L)^3} = \frac{\Omega\, V}{8\pi^3} \tag{13}$$

allowed values of k. Equivalently, the number of allowed k-values per unit volume of k-space, called the k-space density of levels, is just

$$d = \frac{\nu}{\Omega} = \frac{V}{8\pi^3} . \tag{14}$$

Let us pass now to the ground state of the N-electron system, recalling that the one-electron energy levels are specified by the wave vectors $\vec{k}$, and by the projections of the electron's spin along an arbitrary axis. (The projections can take either of the two values $\hbar/2$ or $-\hbar/2$). For this reason, two one-electron levels are associated with each allowed value of k.

When N is a very large number, the occupied region of k-space is indistinguishable from a sphere. This sphere is called the Fermi sphere, and its radius, k_F, is called the Fermi wave vector. The ground state of the (noninteracting) N-electron system is thus formed by occupying all one-electron levels with $k \leq k_F$, while all one-electron levels with $k > k_F$ are unoccupied. The surface of the Fermi sphere, separating the occupied levels from the unoccupied ones, is called the Fermi surface. (It is of fundamental importance in the theory of metals).

The momentum $p_F = \hbar k_F$ is known as the Fermi momentum, the velocity $v_F = p_F/m$ is referred to as the Fermi velocity. In a similar manner, the energy $\varepsilon_F = \hbar^2 k_F{}^2/(2m)$ is called the Fermi energy, and, a related concept, the Fermi temperature is defined by $T_F = \varepsilon_F/k_B$, where k_B denotes the Boltzmann constant.

(b) $T \neq 0°K$

We shall consider now the N-particle system in thermal equilibrium, at temperature T. Next, we assume that E_α^N is the energy of the αth stationary state of this system. According to statistical mechanics [5], the weight $P_N(E)$, which must be assigned to a state of energy E, is given by

$$P_N(E) = \frac{e^{-E/k_BT}}{\sum e^{-E_\alpha^N/k_BT}} , \tag{15}$$

where the summation extends over all states α.

The denominator in (15) is known as the partition function. It is related to the Helmholz free energy (defined by

$$F = U - T\,S, \tag{16}$$

where U is the internal energy, and S is the entropy) by

$$\sum e^{-E_\alpha^N/k_BT} = e^{-F_N/k_BT}. \tag{17}$$

Using (17), we can rewrite (15) as

$$P_N(E) = e^{-(E - F_N)/k_BT} . \tag{18}$$

Since our particles are electrons, the Pauli principle must be satisfied when an N-electron state is constructed by the filling of the N different one-electron levels. Assuming that the number of possible one-electron levels is larger than N, many states of the N-electron system can be constructed. This means that we can specify a particular state of the N-electron system by listing which of the N one-electron levels are filled. The quantity we want to know is the probability f_i^N, that there is an electron in the one-electron level i when the N-electron system is in thermal equilibrium. (The one-electron level i, itself, can be specified by the electron's wave vector $\vec{k}$, and by the projection s of its spin along some axis).

The probability f_i^N is the sum of the independent probabilities of finding the N-electron system in any one of those possible states in which the ith level is occupied. In view of (15), we then have

$$f_i^N = \sum P_N(E_\alpha^{\ N}) \ , \tag{19}$$

where the summation covers all states α in which there is an electron in the one-electron level i.

The exclusion principle permits us only two possibilities. Level i is either occupied, or it is unoccupied. Accordingly, (19) can also be written as

$$f_i^N = 1 - \sum P_N(E_\gamma^{\ N}), \tag{20}$$

where the summation extends over all N-electron states γ in which there is no electron in the one-electron level i.

To proceed, we observe that, from an (N + 1)-electron state in which there is an electron in the one-electron level i, we can construct an N-electron state in which there is no electron in the level i, by simply removing an electron from the ith level of the (N + 1)-electron system and leaving the occupation of all the other one-electron levels unaltered. Evidently, the energy of the N-electron state created in such a manner differs from the energy of the (N + 1)-electron state only by the energy ε_i of the one-electron level i whose occupation is different in the two systems. Thus, the set of energies of all N-electron states with the one-electron level i unoccupied is the same as the set of energies of all (N + 1)-electron states with the level i occupied, provided that each energy in the latter set is reduced by ε_i. We can, therefore, rewrite (20) as

$$f_i^N = 1 - \sum P_N \ (E_\alpha^{N + 1} - \varepsilon_i), \tag{21}$$

where the summation extends over all (N + 1)-electron states α in which there is an electron in the one-electron level i.

Using (18), we can write P_N in (21) as

$$\begin{aligned} P_N(E_\alpha^{N + 1} - \varepsilon_i) &= e^{-(E_\alpha^{\ N + 1} - \varepsilon_i - F_N)/k_BT} \\ &= e^{(\varepsilon_i - F_{N + 1} + F_N)/k_BT} \ e^{-(E_\alpha^{\ N + 1} - F_{N + 1})/k_BT} \\ &= e^{(\varepsilon_i - \mu)/k_BT} \ P_{N + 1} \ (E_\alpha^{\ N + 1}) \ , \end{aligned} \tag{22}$$

where μ, known as the chemical potential, is given (at temperature

T) by

$$\mu = F_{N+1} - F_N . \tag{23}$$

Substituting (22) into (21), we find that

$$f_i^N = 1 - e^{(\varepsilon_i - \mu)/k_B T} \sum P_{N+1}(E_\alpha^{N+1}) , \tag{24}$$

where, we repeat again, the summation is over all (N + 1)-electron states α in which there is an electron in the one-electron level i. Using (19), with N replaced by N + 1, we can cast (24) into the form

$$f_i^N = 1 - e^{(\varepsilon_i - \mu)/k_B T} \; f_i^{N+1} . \tag{25}$$

The relation in (25) is an exact relation. It states the connection between the probability of the one-electron level i being occupied (at temperature T) in an N-electron system and that in an (N + 1)-electron system.

Our chief interest is in situations when N is a very large number. (In a metal N is of the order of 10^{22} cm^{-3}). In this case, it can be shown [6] that the addition of an extra electron to the N-electron system alters the probability of occupation of a given level, such as the ith, by the order of 1/N. We may, therefore, to a very good approximation, replace f_i^{N+1} by F_i^N in (25) and solve for f_i^N. The result is

$$f_i^N = \frac{1}{e^{(\varepsilon_i - \mu)/k_B T} + 1} , \tag{26}$$

which is known as the Fermi-Dirac distribution function.

In what follows, we shall drop the superscript N, i.e. the explicit reference to the N-dependence of f_i. [Such a dependence is carried through the chemical potential, as can be seen from (23)].

We now want to verify that the FD distribution function is consistent with the ground state properties (T = 0°K) of the N-electron system. Recalling that in the ground state of the system the energy levels are occupied up to and including the Fermi level ε_F, we conclude that the ground state distribution function must be such that

$$f_{k,s} = \begin{cases} 1 & \varepsilon(k) \le \varepsilon_F \\ 0 & \varepsilon(k) > \varepsilon_F \end{cases} . \tag{27}$$

On the other hand, the limiting form of the FD distribution, as seen from (26), is

$$\lim_{T\to 0} f_{k,s} = \begin{cases} 1 & \varepsilon(k) \leq \mu \\ 0 & \varepsilon(k) > \mu \,. \end{cases} \tag{28}$$

For (27) and (28) to be consistent, it is necessary that we have

$$\lim_{T\to 0} \mu = \varepsilon_F \,. \tag{29}$$

III. NONEQUILIBRIUM DISTRIBUTION

When an electron gas is under the action of an external field, the probability that a given quantum state $\vec{k}$ is occupied depends not only on the energy $\varepsilon(\vec{k})$ of that state but also on the wave vector $\vec{k}$ itself [7]. To see that this is so, let us look at the equation of motion of the wave vector,

$$\hbar \frac{d\vec{k}}{dt} = \vec{F}, \tag{30}$$

where $\vec{F}$ is the force acting on the electron. Integration of (30) tells us that, an electron in a state $\vec{k}$ at a given time t, will find itself in a state $\vec{k} + \delta\vec{k}$ at a later time $t + \delta t$.

Recalling, from (26), that the probability of occupation of level ε depends on $\varepsilon - \mu$, rather than on ε alone, we conclude that the presence of a temperature gradient, which makes μ a function of position, will also make the distribution function a function of position. For this reason, we seek to describe the system of N electrons by means of a modified distribution function $f(\vec{k}, \vec{r}, t)$.

We shall assume that the modified distribution function is defined in such a way that the number of electrons in the six-dimensional volume element $d\vec{k}\, d\vec{q} = dk_x dk_y dk_z\, dxdydz$, at time t, is given by

$$\frac{1}{4\pi^3} f(\vec{k}, \vec{r}, t) d\vec{k}\, d\vec{q} \,. \tag{31}$$

This quantity is just the phase-space density of levels, $(4\pi^3)^{-1} f(\vec{k}, \vec{r}, t)$, multiplied by the volume element of phase-space, $d\vec{k}\, d\vec{q}$. As displayed in (31), an allowed value of k is associated with two electrons of opposite spins. At equilibrium, $f(\vec{k}, \vec{r}, t)$ depends only on $\varepsilon(\vec{k})$ and reduces to the FD distribution

$$f[\varepsilon(\vec{k})] = \frac{1}{e^{[\varepsilon(\vec{k}) - \mu]/k_B T} + 1} . \tag{32}$$

In principle, the electronic properties of a conductor are completely specified once $f(\vec{k}, \vec{r}, t)$ is known [8]. To convince ourselves that this is so, let us consider the contribution to the electric current by an electron in state $\vec{k}$. This is given by

$$\vec{I}_{\vec{k}} = - e_o \vec{v}_{\vec{k}} , \tag{33}$$

where e_o is the magnitude of the electronic charge, and

$$\begin{aligned} \vec{v}_{\vec{k}} &= \frac{1}{\hbar} \vec{\nabla}_k \, \varepsilon(\vec{k}) \\ &\equiv \frac{1}{\hbar} (\hat{i} \frac{\partial}{\partial k_x} + \hat{j} \frac{\partial}{\partial k_y} + \hat{k} \frac{\partial}{\partial k_z}) \, \varepsilon(\vec{k}) \end{aligned} \tag{34}$$

is the velocity of the electron in question with $\hat{i}$, $\hat{j}$, $\hat{k}$ standing for unit vectors along the axes of k-space. In view of (34), the current density $\vec{J}$, at position $\vec{r}$ at time t, is obtained by simply summing the individual current contributions $\vec{I}_{\vec{k}}$ of all the electrons of the system. Such a summation is achieved not only by integrating the electron velocities $\vec{v}_{\vec{k}}$ over all possible values of $\vec{k}$, but also by weighing each velocity by the distribution function $f(\vec{k}, \vec{r}, t)$. Thus, the total current density at $\vec{r}$ at time t is given by

$$\vec{J}(\vec{r}, t) = - \frac{e_o}{4\pi^3} \int_{\substack{\text{over all of} \\ \text{k-space}}} \vec{v}_{\vec{k}} \, f(\vec{k}, \vec{r}, t) d\vec{k} . \tag{35}$$

In view of the above illustration, the central problem in the theory of transport processes is the finding of the modified distribution function $f(\vec{k}, \vec{r}, t)$ under specified conditions. This task is accomplished in two steps. First, we must derive an equation that $f(\vec{k}, \vec{r}, t)$ must satisfy. Second, we must then find solutions of this equation subject to specific conditions (involving external fields and temperature gradients).

The first step is relatively simple in the semiclassical treatment to which we adhere. The second step, however, presents formal mathematical difficulties under all realistic physical conditions.

IV. THE BOLTZMANN EQUATION

Let us consider those electrons which, at time t, are within the volume element $d\vec{k}\, d\vec{q}$. Such electrons may, as time goes on, leave the volume element as a result of three processes.

(1) An electron may leave the r-space element $d\vec{q}$, centered about $\vec{r}$, by virtue of the fact that its velocity $\vec{v}$ may carry it out of the volume element.

(2) An electron may leave the k-space element $d\vec{k}$, centered about $\vec{k}$, by virtue of the fact that its wave vector $\vec{k}$ changes under the action of a force $\vec{F}$ resulting from an externally applied field.

(3) An electron may also change its $\vec{k}$ vector as a result of a scattering event (by a lattice vibration, for instance), and so it may be scattered out of the k-space element $d\vec{k}$.

An electron, which at time t is located in $d\vec{k}\, d\vec{q}$, centered about $\vec{k}$ and $\vec{r}$, will at time t + dt be located in an equal volume, centered about $\vec{r} + (d\vec{r}/dt)dt$ and $\vec{k} + (d\vec{k}/dt)dt$.

Thus, we find that the (total) time rate of change of the modified distribution function $f(\vec{k}, \vec{r}, t)$ is governed by the formula

$$\frac{Df}{Dt} \equiv \frac{df}{dt} = \left(\frac{\partial f}{\partial t}\right)_d + \left(\frac{\partial f}{\partial t}\right)_c , \tag{36}$$

where $(\partial f/\partial t)_d$ and $(\partial f/\partial t)_c$ are called the drift term and the collision term, respectively. The drift term is obtained by forming the difference $[f(\vec{k}, \vec{r}, t) - f(\vec{k} + \dot{\vec{k}}\, dt, \vec{r} + \dot{\vec{r}}\, dt, t + dt)]$, and performing a Taylor series expansion in the variables $d\vec{k}$, $d\vec{r}$, and dt. Retaining only linear terms in these variables, we find that (36) becomes

$$\frac{df}{dt} = -\dot{\vec{k}}\cdot\vec{\nabla}_k f - \vec{v}\cdot\vec{\nabla}_r f - \frac{\partial f}{\partial t} + \left(\frac{\partial f}{\partial t}\right)_c . \tag{37}$$

The equation above is called the Boltzmann transport equation. In what follows, we shall consider only steady-state conditions characterized by $\partial f/\partial t = 0$. Such a situation prevails when charge transport proceeds under the influence of time-independent forces ($dF/dt = 0$). In the steady-state, Boltzmann's equation is reduced to

$$\left(\frac{\partial f}{\partial t}\right)_c = \dot{\vec{k}}\cdot\nabla_k f + \vec{v}\cdot\nabla_r f \,. \tag{38}$$

The difficulties encountered in the solution of (38) become apparent as soon as the collision term is written in a more explicit form. For this purpose, let us introduce the quantity $S(\vec{k}, \vec{k}')$, denoting the probability per unit time for the scattering of an electron from level $\vec{k}$ into a level $\vec{k}'$. The problem of actually calculating $S(\vec{k}, \vec{k}')$ for different scattering processes is one in its own right and it is done by quantum mechanical treatment of scattering models.

Assuming an explicit expression for $S(\vec{k}, \vec{k}')$, we may write the collision term as

$$\begin{aligned}\left(\frac{\partial f}{\partial t}\right)_c = \int \{&S(\vec{k}', \vec{k})\, f(\vec{k}')\, [1 - f(\vec{k})] \\ &- S(\vec{k}, \vec{k}')\, f(\vec{k})\, [1 - f(\vec{k}')]\} d\vec{k}' .\end{aligned} \tag{39}$$

The idea behind this expression is the following: the first term in the integrand in (39) equals the number of electrons that are scattered from the volume element $d\vec{k}'$ into the volume element $d\vec{k}$ per unit time. The factor $S(\vec{k}', \vec{k})$ gives the a priori probability of such a scattering event. The factor $f(\vec{k}')$ stands for the probability that an electron initially occupies the state $\vec{k}'$, while the factor $[1 - f(\vec{k})]$ represents the probability that the state $\vec{k}$, into which the electron is assumed to be scattered, is unoccupied. In view of the above, the product $f(\vec{k}')[1 - f(\vec{k})]$ stands for the simultaneous probability of occupancy of the state $\vec{k}'$ and availability of state $\vec{k}$, a prerequisite for the scattering event. In a similar manner, the second term in the integrand in (39) equals the number of electrons that are scattered per unit time out of the volume element $d\vec{k}$ and into the volume element $d\vec{k}'$.

At equilibrium, when $f(\vec{k}, \vec{r}, t) = f(\varepsilon)$, we may write

$$S(\vec{k}', \vec{k})\, f(\varepsilon')\, [1 - f(\varepsilon)] = S(\vec{k}, \vec{k}')\, f(\varepsilon)[1 - f(\varepsilon')], \tag{40}$$

which simply means that scattering out of $d\vec{k}'$ and into $d\vec{k}$ proceeds at the same rate as scattering into $d\vec{k}'$ from $d\vec{k}$.

It is generally assumed that the scattering probabilities, $S(\vec{k}', \vec{k})$ and $S(\vec{k}, \vec{k}')$, do not depend on the applied electric or magnetic fields. If so, we can use (40) to rewrite (39). This is done by substituting $S(\vec{k}', \vec{k})$ from (40) into (39). The result is

$$\left(\frac{\partial f}{\partial t}\right)_c = \int S(\vec{k}, \vec{k}')\, f(\varepsilon)[1 - f(\varepsilon')] \times \left\{\frac{f(\vec{k}')[1 - f(\vec{k})]}{f(\varepsilon')[1 - f(\varepsilon)]} - \frac{f(\vec{k})[1 - f(\vec{k}')]}{f(\varepsilon)[1 - f(\varepsilon')]}\right\} d\vec{k}' . \quad (41)$$

V. THE BLOCH EQUATION

To proceed, we assume that the modified distribution function $f(\vec{k}, \vec{r})$ can be written as a sum of two terms. For the first term, we choose the equilibrium FD distribution function. As to the second term, we assume that it represents the deviation of the distribution from the equilibrium one. In this framework, we have

$$f(\vec{k}, \vec{r}) = f(\varepsilon) + f_1(\vec{k}, \vec{r}), \quad (42)$$

and the determination of $f(\vec{k}, \vec{r})$ is now reduced to the determination of $f_1(\vec{k}, \vec{r})$. In what follows, we shall find it convenient to write $f_1(\vec{k}, \vec{r})$ as

$$f_1(\vec{k}, \vec{r}) = -\phi(\vec{k}, \vec{r})\, \frac{\partial f(\varepsilon)}{\partial \varepsilon} , \quad (43)$$

a form which leaves the generality of $f_1(\vec{k}, \vec{r})$ unimpaired since the function $\phi(\vec{k}, \vec{r})$ remains unrestricted. In view of (43), the determination of $f(\vec{k}, \vec{r})$ is now reduced to the determination of $\phi(\vec{k}, \vec{r})$.

For reasonable external fields, we may assume that the deviation of the steady-state distribution from the equilibrium distribution is rather small. (For "hot electron" effects, such as one encounters in semiconductors subjected to very strong electric fields, this assumption is not sufficient. A distribution function for this case, has, however, been also established [8]).

Substituting (42) into (41), and maintaining terms only through first-order in f_1, we find, after some algebra, that

$$\left(\frac{\partial f}{\partial t}\right)_c = \int S(\vec{k}, \vec{k}')\, f(\varepsilon)\, [1 - f(\varepsilon')] \times \left\{- \frac{f_1(\vec{k}, \vec{r})}{f(\varepsilon)\, [1 - f(\varepsilon)]} + \frac{f_1(k, r)}{f(\varepsilon')\, [1 - f(\varepsilon')]}\right\} d\vec{k}' . \quad (44)$$

Making use of the identities

$$f(\varepsilon)\, [1 - f(\varepsilon)] = - k_B T\, \frac{\partial f(\varepsilon)}{\partial \varepsilon} , \quad (45a)$$

$$f(\varepsilon')\,[1 - f(\varepsilon')] = -k_B T \frac{\partial f(\varepsilon')}{\partial \varepsilon'} , \tag{45b}$$

we can bring (44) to the form

$$\left(\frac{\partial f}{\partial t}\right)_c = \frac{1}{k_B T} \int S(\vec{k}, \vec{k}')\, f(\varepsilon)\,[1 - f(\varepsilon')] \times \{ \phi(\vec{k}', \vec{r}) - \phi(\vec{k}, \vec{r})\}\, d\vec{k}' . \tag{46}$$

We shall refer to the integral in (46) as the collision integral.

Choosing for $\vec{F}$, in (30), the Lorentz force expressed with (34),

$$\vec{F} = -e_o[\vec{E} + \frac{1}{c\hbar} (\vec{\nabla}_k \varepsilon) \times \vec{H}],$$

we can express $\dot{\vec{k}}$ and substitute it into (38). At the same time, we can express $\vec{v}$ from (34) and substitute it also into (38). The result, upon consideration of (46), is

$$\begin{aligned}\left(\frac{\partial f}{\partial t}\right)_c &= -\frac{e_o}{\hbar}[\vec{E} + \frac{1}{c\hbar}(\vec{\nabla}_k\, \varepsilon) \times \vec{H}]\cdot\vec{\nabla}_k f(\vec{k}, \vec{r}) \\ &\quad + \frac{1}{\hbar}(\vec{\nabla}_k\, \varepsilon)\cdot\vec{\nabla}_r\, f(\vec{k}, \vec{r}) \\ &= \frac{1}{k_B T} \int S(\vec{k}, \vec{k}')\, f(\varepsilon)\,[1 - f(\varepsilon')] \\ &\quad \times [\phi(\vec{k}') - \phi(\vec{k})]d\vec{k}' .\end{aligned} \tag{47}$$

Equation (47) is the linearized form of the Boltzmann equation, and it is known as the Bloch equation. The Bloch equation is an integral equation for the unknown function ϕ and, in general, its solution cannot be obtained in a closed form.

There are two techniques for solving Bloch's equation. The first one is based on a variational principle, formulated by Kohler [9]. It is a powerful method, often resorted to when the second technique, the relaxation-time approximation is not applicable.

VI. THE RELAXATION-TIME APPROXIMATION

The relaxation-time τ is defined by

$$\left(\frac{\partial f}{\partial t}\right)_c = -\frac{f_1}{\tau} . \tag{48}$$

It is a time constant, which characterizes the return of a

disturbed distribution to the equilibrium one. To understand this point clearly, we rewrite (48) as

$$\frac{d}{dt}(f_0 + f_1) = -\frac{f_1}{\tau}, \tag{49}$$

where we have written f_0 for $f(\varepsilon)$. This quantity does not depend on time, so $\dot{f}_0 = 0$. Rearranging, (49) becomes

$$\frac{df_1}{f_1} = -\frac{dt}{\tau}. \tag{50}$$

Integration of (50) leads to

$$f_1 = \gamma\, e^{-\frac{t}{\tau}}, \tag{51}$$

where γ is a constant. In view of (51), we now see that the return of the disturbed distribution to the equilibrium one is exponential.

We have stated without proof that the collision integral, as given in (46), may be replaced by (48). We shall now develop the solution of the Bloch equation on the basis of this assumption.

As a first step, we wish to rewrite the left-hand side of (47) and then equate it to the right-hand side of (48). Defining the quantity

$$\vec{P} = -e_o\vec{E} - [\varepsilon - \eta\,(x, y, z)]\vec{\nabla}_r \ln T\,(x, y, z), \tag{52}$$

and using the relation

$$\begin{aligned} \vec{\nabla}_r f &\simeq \vec{\nabla}_r f_0 \\ &= \left(\frac{\partial f_0}{\partial T}\vec{\nabla}_r T\right) = -\frac{\partial f_0}{\partial \varepsilon}\left[\vec{\nabla}_r \eta + (\varepsilon - \eta)\vec{\nabla}_r \ln T\right], \end{aligned} \tag{53}$$

where, for notational simplicity, the x, y, z-dependence on η and T has been dropped, the result is

$$\vec{P}\cdot\vec{v}\,\frac{\partial f_0}{\partial \varepsilon} - \frac{e_o}{c\hbar}(\vec{v} \times \vec{H})\cdot\vec{\nabla}_k f = \frac{\phi}{\tau}\frac{\partial f_0}{\partial \varepsilon}. \tag{54}$$

The correctness of (54) is not obvious at first sight. It rests on the identity in (53), which can be verified by direct differentiation. In steady-state, the Fermi energy η is constant throughout the material. For this reason, (53) reduces to

$$\vec{\nabla}_r f_0 = -\frac{\partial f_0}{\partial \varepsilon}(\varepsilon - \eta)\vec{\nabla}_r \ln T. \tag{55}$$

Next, let us consider the left-hand side of (47), namely

$$-\frac{e_o}{\hbar}[\vec{E} + \frac{1}{c\hbar}(\vec{\nabla}_k \varepsilon) \times \vec{H}]\cdot\vec{\nabla}_k f(\vec{k}, \vec{r}) + \frac{1}{\hbar}\vec{\nabla}_k \varepsilon\cdot\vec{\nabla}_r f(\vec{k}, \vec{r}) \,. \tag{56}$$

The quantity in (56), using (34), can be written as

$$\frac{1}{\hbar}[-e_o\vec{E} - \frac{e_o}{c}\vec{v} \times \vec{H}]\cdot\vec{\nabla}_k f(\vec{k}, \vec{r}) + \vec{v}\cdot\vec{\nabla}_r f(\vec{k}, \vec{r}) \,. \tag{57}$$

Substituting for $\vec{\nabla}_r f$ from (53), the quantity in (57) is written as

$$\frac{1}{\hbar}[-e_o\vec{E} - \frac{e_o}{c}\vec{v} \times \vec{H}]\cdot\vec{\nabla}_k f(\vec{k}, \vec{r}) - \frac{\partial f_o}{\partial \varepsilon}(\varepsilon - \eta)\vec{v}\cdot\vec{\nabla}_r \ln T \,. \tag{58}$$

Writing out (58) in detail, we obtain

$$-\frac{e_o}{c\hbar}(\vec{v} \times \vec{H})\cdot\vec{\nabla}_k f - \frac{e_o}{\hbar}E\cdot\vec{\nabla}_k f_0 - (\varepsilon - \eta)\frac{\partial f_o}{\partial \varepsilon}\vec{v}\cdot\vec{\nabla}_r \ln T \,, \tag{59}$$

where, in the term involving the electric field $\vec{E}$, f has been replaced by f_0.

To proceed, we have to work out $\vec{\nabla}_k f_0$. Considering that

$$\frac{\partial}{\partial k_x} f_0 = \frac{\partial f_0}{\partial \varepsilon} \frac{\partial \varepsilon}{\partial k_x} \,, \tag{60}$$

with similar equations for the derivatives involving y and z, we find that

$$\vec{\nabla}_k f_0 = \frac{\partial f_0}{\partial \varepsilon} \vec{\nabla}_k \varepsilon \,. \tag{61}$$

Substituting (61) into (59), and using (34), we obtain

$$[-e_o \vec{E} - (\varepsilon - \eta)\vec{\nabla}_r \ln T]\cdot\vec{v}\,\frac{\partial f_0}{\partial\varepsilon} - \frac{e_o}{c\hbar}(\vec{v}\times\vec{H})\cdot\vec{\nabla}_k f\ . \tag{62}$$

The quantity in (62) is recognized, upon consideration of the definition of $\vec{P}$ in (52), as the left-hand side of (54). The right-hand side of (54) easily follows upon substituting (43) into (48). This completes the verification of (54).

The next step consists in the modification of the second term on the left-hand side of (54). This term, using (42), can be written as

$$\begin{aligned} -\frac{e_o}{c\hbar}(\vec{v}\times\vec{H})\cdot\vec{\nabla}_k f &= -\frac{e_o}{c\hbar}(\vec{v}\times\vec{H})\cdot\vec{\nabla}_k f_0 \\ &\quad -\frac{e_o}{c\hbar}(\vec{v}\times\vec{H})\cdot\vec{\nabla}_k f_1\ . \end{aligned} \tag{63}$$

Considering (61) and (34), the first term on the right-hand side of (63) becomes

$$-\frac{e_o}{c\hbar}(\vec{v}\times\vec{H})\cdot\vec{\nabla}_k f_0 = -\frac{e_o}{c}(\vec{v}\times\vec{H})\cdot\vec{v}\,\frac{\partial f_0}{\partial\varepsilon}\ . \tag{64}$$

The quantity in (64) is equal to zero. The reason is that $(\vec{v}\times\vec{H})$ is perpendicular to $\vec{v}$. With (63) and (64), we may now write (54) as

$$\vec{P}\cdot\vec{v}\,\frac{\partial f_0}{\partial\varepsilon} - \frac{e_o}{c\hbar}(\vec{v}\times\vec{H})\cdot\vec{\nabla}_k f_1 = \frac{\phi}{\tau}\frac{\partial f_0}{\partial\varepsilon}\ . \tag{65}$$

The second term on the left-hand side of (65) contains a scalar triple product. Using the identity

$$(\vec{B}\times\vec{C})\cdot\vec{A} = (\vec{C}\times\vec{A})\cdot\vec{B} = (\vec{A}\times\vec{B})\cdot\vec{C}\ , \tag{66}$$

we find that this term can be brought to the form

$$-\frac{e_o}{c\hbar}(\vec{v}\times\vec{H})\cdot\vec{\nabla}_k f_1 = -\frac{e_o}{c\hbar}\,\vec{H}\cdot(\vec{\nabla}_k f_1 \times \vec{v}), \tag{67}$$

which, using (34) and defining the operator

$$\vec{\Omega} = \vec{\nabla}_k\,\varepsilon \times \vec{\nabla}_k, \tag{68}$$

can also be written as

$$-\frac{e_o}{c\hbar}(\vec{v}\times\vec{H})\cdot\vec{\nabla}_k f_1 = \frac{e_o}{c\hbar^2}[\vec{H}\cdot(\vec{\Omega} f_1)] \; . \tag{69}$$

Using (69), the linearized transport equation in (65) becomes

$$\vec{P}\cdot\vec{v}\;\frac{\partial f_0}{\partial\varepsilon} + \frac{e_o}{c\hbar^2}\;\vec{H}\cdot(\vec{\Omega} f_1) = \frac{\phi}{\tau}\,\frac{\partial f_0}{\partial\varepsilon} \; . \tag{70}$$

To proceed, the identity

$$\vec{\Omega}\, g(\varepsilon) = 0 \; , \tag{71}$$

where $g(\varepsilon)$ is a function only of the energy, will be used. The correctness of (71) may be verified by making the simplest choice of

$$g(\varepsilon) = \varepsilon. \tag{72}$$

In this case, with (68) and (72), we have

$$\vec{\nabla}_k\varepsilon \times \vec{\nabla}_k\varepsilon = |\vec{\nabla}_k\varepsilon|^2\,\sin(\vec{\nabla}_k\varepsilon,\,\vec{\nabla}_k\varepsilon) = 0, \tag{73}$$

which verifies (71).

We now cast the second term on the left-hand side of (70) into the form

$$\frac{e_o}{c\hbar^2}\,\vec{H}\cdot(\vec{\Omega} f_1) = -\frac{e_o}{c\hbar^2}\,\vec{H}\cdot(\Omega\phi)\,\frac{\partial f_0}{\partial\varepsilon} \; . \tag{74}$$

Equation (74) is not immediately obvious. To check on its correctness, let us work it out in detail. Using (43), we find that

$$\begin{aligned}\frac{e}{c\hbar^2}\,\vec{H}\cdot(\vec{\Omega} f_1) &= \frac{e_o}{c\hbar^2}[\vec{H}\cdot\vec{\Omega}(-\phi\,\frac{\partial f_0}{\partial\varepsilon})] \\ &= -\frac{e_o}{c\hbar^2}\{H\cdot[(\Omega\phi)\,\frac{\partial f_o}{\partial\varepsilon} + \phi(\vec{\Omega}\,\frac{\partial f_0}{\partial\varepsilon})]\} \; .\end{aligned} \tag{75}$$

Considering (71), with the choice of

$$g(\varepsilon) = \frac{\partial f_0}{\partial\varepsilon} \; , \tag{76}$$

we see that the second term in the square brackets in (75) is zero.

This verifies (74).

To state our result, we may write (70), with (74), as

$$\vec{P}\cdot\vec{v}\,\frac{\partial f_0}{\partial \varepsilon} - \frac{e_o}{c\,\hbar^2}\,\vec{H}\cdot(\vec{\Omega}\phi)\,\frac{\partial f_0}{\partial \varepsilon} = \frac{\phi}{\tau}\,\frac{\partial f_0}{\partial \varepsilon}\ . \qquad (77)$$

In words, the relaxation-time approximation permitted us to reduce the Bloch integral equation to the simpler differential equation

$$\phi = \tau(\vec{P}\cdot\vec{v}) - \frac{e_o\,\tau}{c\,\hbar^2}\,\vec{H}\cdot(\vec{\Omega}\phi)\ , \qquad (78)$$

[(78) results from (77) upon canceling the factor $\partial f_0/\partial\varepsilon$ in each term].

Equation (78) is the basic equation in those transport problems where a meaningful relaxation-time can be defined.

VII. ORGANIC SEMICONDUCTORS

Many of these materials are characterized [10] by a low measured value of the carrier mobility μ. This poses a problem, namely that the de Broglie wavelength λ of the carrier may become larger than its mean free path ℓ. Since the charge carrier is "smeared out" over a distance λ, the concept of its motion over a path $\ell < \lambda$ is meaningless.

Experimental results exist which are not compatible with the idea of the band model. Such is the case in monocrystals of ferrocene [11]. The limit of applicability of the band model can be determined approximately [12]. In order to be able to consider a definite mechanism of scattering, described by a relaxation time τ, the uncertainty principle requires that

$$\tau > \frac{\hbar}{\Delta E}\ ,$$

where ΔE is the width of the conduction band. If we estimate ΔE from

$$\Delta E = \frac{\hbar^2\,k_o^{\,2}}{2\,m}\ ,$$

where, for a linear crystal of lattice constant a we have

$$k_o \simeq \frac{\pi}{a}\ ,$$

we find that

$$\mu = \frac{e_o}{m}\tau > \frac{4}{\pi}\ \frac{e_o a^2}{h} \simeq 1\ \frac{cm^2}{V\ sec},$$

for the usual values of a. Consequently, if experiments tell us that $\mu > 1\ cm^2(V\ sec)^{-1}$, we may apply the concept of charge transfer in an energy band. At lower mobilities, a different form of charge transfer, such as a "hopping motion" of electrons and holes, must be considered [13]. This state of affairs clearly points to the necessity of experimental mobility data before one should attempt a theoretical formulation of charge transport in a particular material.

Low mobilities point to strong interactions of the charge carriers with the lattice modes. If the width of the energy band is very narrow, the charge carriers have "large" effective masses. In this case, an electron, for instance, may be treated as a nearly localized wave packet. Conduction then occurs via a phonon assisted "hopping" of the electrons from one lattice site to another and not by band conduction. Consequently, a transport theory approach based on band conduction is likely to be a correct approach when one deals with organic semiconductors in which the charge carriers have "small" effective masses.

The mechanism of electron-phonon interaction, a major factor that limits charge transport, is vastly more complex in organic semiconductors than in inorganic ones. In inorganic materials, a lattice point is merely a geometrical point endowed with a mass. In organic materials, the "lattice point" is a molecule endowed by a structure capable of exhibiting intramolecular vibrations. While in an inorganic crystal the rotation of a lattice point is meaningless, the librational motion of a molecule in an organic crystal is of great importance since, as it is now known, such a motion is responsible for the Raman spectrum of the crystal [14].

The pressure and temperature dependences of physical constants of organic crystals are also expected to be markedly different from those of inorganic crystals. The cohesion of organic materials is due mainly to dispersion forces, such as the quadrupole-quadrupole interaction. Such forces depend on the distance between molecules by a much stronger power law than the essentially electrostatic interactions which are responsible for the covalent bonding of inorganic semiconductors. It has, indeed, been found that the widths of the electron and hole bands in the anthracene crystal change by approximately a factor of 6 upon the application of a 160 Kilobar pressure [15].

The concept of an "impurity" is also more complicated in organic semiconductors than in inorganic ones. In inorganic materials, one can usually treat a charged impurity as a point charge, and a neutral impurity as a H-like atom. In organic crystals, the internal structure of the impurity molecule, charged or uncharged, must also be considered. In most studies, aromatic hydrocarbon crystals have been used which included another aromatic hydrocarbon molecule as an impurity [16]. In such cases, for the clarification of the energy transfer process, the highest electronic energy level of the impurity is of importance. Radicals as impurities are of special importance since they have a half-occupied molecular orbital as their highest energy level. For instance, the 1-hydronaphtyl radical,

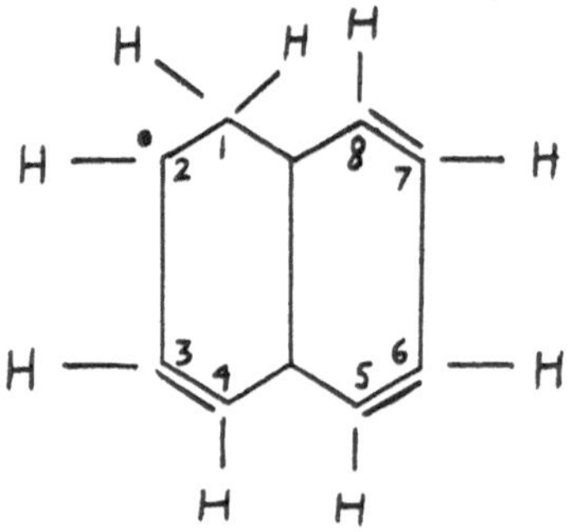

,

in naphtalene and anthracene is analogous to à donor state in an ordinary semiconductor, while the 1-naphtyl radical,

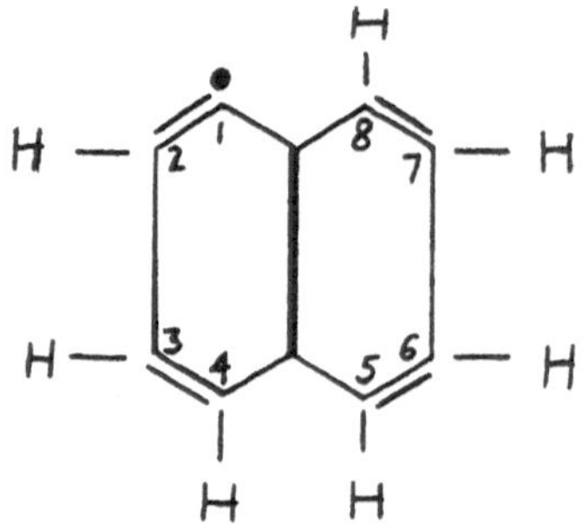

,

is analogous to an acceptor state.

Impurities might also act as traps. It has been established in photo-excitation studies that anthraquinone in an anthracene single crystal acts as an electron trap, while naphtacene traps holes [17]. For this reason, one cannot expect "general" theories of impurity scattering. Each impurity, in each host crystal, poses a separate problem.

The coupling of an impurity molecule to the host lattice may be of special significance. In ordinary semiconductors, the introduction of an impurity into the host lattice leads to local modes. Essentially, this is just a mass effect. In organic semiconductors, however, we may have to worry about the coupling of the intramolecular modes to the lattice modes.

In inorganic semiconductors, one can usually be satisfied by the use of a constant effective mass. In narrow band organic semiconductors, on the other hand, the possibility of an energy dependent m* might have to be entertained. The reason is, that in low molecular weight organic semiconductors the width of the conduction band is of the order of k_BT at room temperature [18]. Such a state of affairs could lead to a variety of complications: the Fermi level, the density of states, the transport theory approach, might all need a careful revision.

VIII. SEMICONDUCTING POLYMERS (POLYMERIC SEMICONDUCTORS)

The mechanism of conduction in polymeric organic semiconductors is more complex than that in low molecular weight crystals. The reason for this lies in the more complicated structure of polymers.

In linear polymers, regions of polyconjugation may be separated by regions not containing double bonds. An electron, in a region of polyconjugation, may be assumed to be described by the quantum mechanical model of a particle in a one-dimensional (linear) box. The same statement should also be true for the excited states of an electron in such a region. Since, in a realistic polymer, the "linear boxes" are separated by barriers, charge transport by tunneling, with or without phonon assistance, appears to be a plausible charge transfer process [19]. This process, for electrons in an excited state, appears to be an even more probable charge transport mechanism because the barrier to be surmounted between regions of polyconjugation is now reduced in height. It would appear, that the formulation of a realistic quantum mechanical model of charge transport by tunneling requires a fairly reliable knowledge of the structure of the polymer. Such a knowledge should include information about the lengths and distribution of the regions of polyconjugation, together with a corresponding knowledge of the barrier regions where no bond conjugation occurs.

A further structural quantity of interest is the side group. As the electron is imagined to move along the main chain, it polarizes the side groups which, in turn, affect the movement of the electron [20].

Structural defects, (whatever their own microscopic structure), are very important in polymeric semiconductors. Experiments on irradiated and thermally tested polyethylene show that the unpaired spins recorded in the EPR spectra cannot be current carriers [21]. The reason is, that the concentration of current carriers, recorded from electrical measurements, is lower than the concentration of unpaired spins by several orders of magnitude. For this reason, the EPR signal is assumed to arise from structural defects not appearing clearly in the electrical conductivity. The interpretation of the electrical conductivity, by itself, is not free from problems which arise, mainly, by the indeterminacy of the structure of polymeric semiconductors. To quote an example, studies on pyrolized polyacrylonitrile show that the specimen investigated consists of a large number of highly conducting subvolumes [22]. Each such region consists of $\geq 5 \times 10^5$ C atoms. In better conducting samples of this material, the regions of polyconjugation increase in dimensions and may contain as many as $\gtrsim 5 \times 10^8$ C atoms.

The interpretation of measured mobility data is also more uncertain in polymeric semiconductors than in the wide band gap inorganic materials. To see why, let us look at the technique of measuring the drift velocity of excited carriers in an electric field [23]. In this technique, nonequilibrium carriers are generated by photons and injected into the sample. They move in the field between two electrodes and then the current in the sample ceases. If the sample contains deep traps, (whatever their detailed microscopic structure), with a trapping time greater than the transfer time, the carriers fallen into the traps will not reach the second electrode. As a result, a serious error might be made in trying to deduce the number of charge carriers from the measured mobility. A similar error might arise when free radicals,with unpaired electrons, are present in the sample. The reason is, that the unpaired electrons might participate in the electrical conductivity by a jump process, for instance [24].

In linear polymers, involving the benzene ring, for instance, in a region of polyconjugation, the problem of charge transport should be even more complicated. The reason is, that the quantum mechanical model of a particle in a box must then be augmented to involve a particle on a circle. A further complication could arise when the barrier, regions, not containing conjugated double bonds, have, for instance, isolated benzene rings. In this case, charge transfer by tunneling from one region of polyconjugation to another one might be assisted or hindered by the presence of the conjugated ring playing the role of a possible intermediate state.

IX. SEMICONDUCTING BIOPOLYMERS

Jordan [25] and Szent-Györgyi [26] have suggested that proteins may posses the nature of semiconductors. For this class of materials, the interpretation of conductivity data is even more complicated than for polymeric semiconductors. The reason is, that it is difficult to distinguish the ionic conductivity from the electronic conductivity.

ACKNOWLEDGEMENT

The author is indebted to Pat Byard for the demanding typing of this manuscript.

REFERENCES

[1]. L. P. Kadanoff and G. Baym, Quantum Statistical Mechanics (Benjamin, New York, 1962).
[2]. Our treatment of the problem is essentially that of N. W. Ashcroft and N. D. Mermin, Solid State Physics (Holt, Rinehart and Winston, New York, 1976), Chpt. 2.
[3]. Our treatment of the problem is essentially that of F. J. Blatt, Physics of Electronic Conduction in Solids (McGraw-Hill, New York, 1968), Chpt. 5.
[4]. For a brief introduction to the Fermi liquid see C. Kittel, Introduction to Solid State Physics (Wiley, New York, 1976), 5th ed., Chpt. 10.
[5]. F. W. Sears, Thermodynamics (Addison-Wesley, Reading, 1953), 2nd ed., Chpt. 14.
[6]. See Ref. [2].
[7]. See Ref. [4]., Chpt. 8.
[8]. B. R. Nag, Theory of Electrical Transport in Semiconductors (Pergamon, New York, 1972), Chpt. 7.
[9]. M. Kohler, Ann. Physik 40, 601 (1942).
[10]. For a compilation, see F. Gutman and L. E. Lyons, Organic Semiconductors (Wiley, New York, 1967), Section 4.4.
[11]. See L. I. Boguslavskii and A. V. Vannikov, Organic Semiconductors and Biopolymers (Plenum, New York, 1970), p. 85 ff.
[12]. J. Yamashita and T. Kurosawa, J. Phys. Soc. Jpn. 15, 802 (1960).
[13]. For inorganic semiconductors, theories of "hopping" have been given by A. Miller and A. Abrahams, Phys. Rev. 120, 745 (1960); N. F. Mott and W. D. Twose, Advan. Phys. 10, 107 (1961).
[14]. See (an article by M. Ito in) A. B. Zahlan, Excitons, Magnons, and Phonons in Molecular Crystals (Cambridge University Press, Cambridge, 1968).
[15]. See Ref. [11], p. 10 ff.
[16]. See (an article by N. Itoh and T. Chong in) K. Masuda and M. Silver, Energy and Charge Transfer in Organic Semiconductors (Plenum, New York, 1974).
[17]. See (an article by U. Itoh and K. Takeishi in) Ref. [16].

[18]. See, for instance, band structure calculations on the best studied material, anthracene, by O. H. LeBlanc, J. Chem. Phys. 35, 1275 (1961); J. L. Katz, S. A. Rice, S. Choi, and J. Jortner, J. Chem. Phys. 39, 1683 (1963); R. Silbey, J. Jortner, S. A. Rice, and M. T. Vala, Jr., J. Chem. Phys. 42, 733 (1965).
[19]. See Ref. [11], p. 69 ff.
[20]. Recent mobility measurements on substituted durene crystals bring out the importance of this point. Furthermore, Z. Burshtein and D. F. Williams, J. Chem. Phys. 66, 2746 (1977), have also established that conduction in substituted durene occurs definitely by the band mechanism.
[21]. See Ref. [11], p. 37 ff.
[22]. See Ref. [21].
[23]. See Ref. [11], p. 79 ff.
[24]. See Ref. [11], p. 58.
[25]. P. Jordan, Naturwiss 26, 693 (1938).
[26]. A. Szent-Györgyi, Nature 148, 158 (1941).

QUANTUM THEORETICAL TREATMENT OF ELECTRONIC EFFECTS ON THE MECHANICAL PROPERTIES OF SEMICONDUCTORS: POSSIBLE APPLICATION TO POLYMERS

P. Csavinszky

Department of Physics, University of Maine
Orono, Maine 04473, U.S.A.

ABSTRACT. The electronic contribution to the elastic constants of semiconductors represents an example of how the mechanical properties of a crystalline solid are influenced by its energy band structure. The aim of the present set of lectures is a discussion of some of the fundamental concepts involved in this phenomenon.

I. INTRODUCTION

The energy levels in the conduction (or valence) band of a semiconductor, such as Ge and Si, depend on the state of strain of the crystal. For this reason, if some of the energy levels are occupied by electrons (or holes), the electronic free energy of the semiconductor will also depend on the state of strain of the crystal. The energy levels of importance are those which are close to the minimum or maximum energy of their band. In view of the fact that energy band extrema usually occur at points of k-space which are endowed with special symmetry properties, the electronic contribution to the elastic constants is tied also to the special symmetries of the energy bands of the semiconductor. The electronic effect in the elastic properties of semiconductors is also of interest for transport phenomena since it can be correlated with the propagation of elastic and thermal waves in semiconductor crystals.

To understand the origin of the effect, let us consider an unstrained n-type semiconductor crystal, such as Ge or Si.

J.-M. André et al. (eds.), Quantum Theory of Polymers, 313-333.

Depending on doping concentration and temperature, some of the energy levels in the conduction band will be occupied by mobile electrons. Consequently, the crystal possesses a certain amount of electronic free energy and zero amount of strain energy. Let us imagine now that the semiconductor is strained. In this case, the mobile electrons will redistribute themselves among the energy levels in such a way that their electronic free energy in the strained crystal is minimized. To satisfy energy conservation, the change in the electronic free energy must show up as a contribution to the now nonzero strain energy function of the crystal. In view of the fact that the strain energy function depends on the elastic constants of the pure crystal, a change in this quantity must then amount to a change in the elastic constants relative to the pure crystal. Measurements of the speed of propagation of ultrasonic waves in specific directions in doped semiconductor crystals do, indeed, confirm the existence of the electronic contribution to the elastic constants.

The paper is structured as follows. In Section II, the theory of the electronic effect is illustrated for both n-type and p-type Ge and Si. In Section III, some facts and speculations relevant to organic and polymeric semiconductors are advanced.

II. THEORY OF THE ELECTRONIC CONTRIBUTION TO THE ELASTIC CONSTANTS.

(A) The effect of strain on the band structure.

The theory of the electronic contribution to the elastic constants of semiconductors requires a knowledge of some important physical constants of these materials. One of the needed quantities is the effective mass of the charge carrier, while another one is the deformation potential constant associated with a particular energy band. Such parameters are accurately known for Ge and Si and, for this reason, these materials are ideal for the study of the electronic contribution to the elastic constants.

For n-type Ge and Si, for instance, the effect of strain on the band structure can be interpreted within the framework of the deformation-potential approximation [1] which states that all electronic levels in a particular valley change by the same amount under the action of strain. To put it another way, the valley shifts rigidly under the action of strain. For this reason, we can describe the effect of strain on valley (i) by giving the change of $E^{(i)}$, the energy of the lowest electronic level in this valley.

The most general linear dependence of level $E^{(i)}$ on the strain tensor $\underline{\varepsilon}$ is of the form [2]

$$E^{(i)} = \underline{\Xi}^{(i)} : \underline{\varepsilon} \, , \tag{1}$$

where $\underline{\Xi}^{(i)}$ is a second-rank tensor, called the deformation-potential tensor, whose components are referred to as the deformation-potential constants.

To find the form the tensor $\underline{\Xi}^{(i)}$ must have, we recall that the valleys in a multiband semiconductor can be transformed into one another by subjecting the unstrained crystal to rotational transformations of the required symmetry. In Ge, for instance, the four equivalent valleys lie along the [111] directions, which are axes of three-fold symmetry. In Si, however, the six equivalent valleys are along the [100] directions, which are axes of four-fold symmetry. Consequently, both in Ge and Si, a second-rank tensor property of valley (i), such as the shift of $E^{(i)}$ with strain, must be invariant under a rotation of $2\pi/3$ and $2\pi/4$, respectively. As shown by Keyes, this requirement is met by the tensor

$$\underline{\Xi}^{(i)} = \Xi_d \, \underline{I} + \Xi_u \, \overrightarrow{a^{(i)}} \, \overrightarrow{a^{(i)}} \, , \tag{2}$$

where $\underline{I}$ stands for the second-rank unit tensor, and $\overrightarrow{a^{(i)}}$ is a unit vector that locates valley (i). Such a unit vector has components $a_1^{(i)}$, $a_2^{(i)}$, and $a_3^{(i)}$ along the x, y, and z coordinate axes and is directed along the [111] axes in Ge, and along the [100] axes in Si. Considering the first

$$\overrightarrow{a^{(i)}}$$

in

$$\overrightarrow{a^{(i)}} \, \overrightarrow{a^{(i)}}$$

as a column vector, and the second

$$\overrightarrow{a^{(i)}}$$

as a row vector, we may also write their matrix product as

$$\overrightarrow{a^{(i)}} \, \overrightarrow{a^{(i)}} = \begin{bmatrix} a_1^{(i)} \, a_1^{(i)} & a_1^{(i)} \, a_2^{(i)} & a_1^{(i)} \, a_3^{(i)} \\ a_2^{(i)} \, a_1^{(i)} & a_2^{(i)} \, a_2^{(i)} & a_2^{(i)} \, a_3^{(i)} \\ a_3^{(i)} \, a_1^{(i)} & a_3^{(i)} \, a_2^{(i)} & a_3^{(i)} \, a_3^{(i)} \end{bmatrix} . \tag{3}$$

We shall discuss now n-type Ge and n-type Si separately, as has been done by Keyes [2], and by Einspruch and Csavinszky [3], respectively.

(B) The case of n-type Ge

Let us introduce the dimensionless quantity

$$w^{(i)} = \frac{E^{(i)}}{k_B T} , \tag{4}$$

which might be called the reduced lowest electronic level of valley (i). Upon consideration of (1) and (2), (4) can be written as

$$w^{(i)} = (k_B T)^{-1} [\Xi_d \underline{I} : \underline{\varepsilon} + \Xi_u \underline{a^{(i)} a^{(i)}} : \underline{\varepsilon}], \tag{5}$$

or as

$$w^{(i)} = (k_B T)^{-1} [\Xi_d \sum_{k,l} I_{kl} \varepsilon_{kl} + \Xi_u \sum_{k,l} (a^{(i)} a^{(i)})_{kl} \varepsilon_{kl}], \tag{6}$$

where $(a^{(i)} a^{(i)})_{kl}$ refer to the components of the tensor

$$\vec{a}^{(i)} \vec{a}^{(i)} = \underline{a^{(i)} a^{(i)}} ,$$

defined in (3).

Performing the summations, (6) assumes the form

$$\begin{aligned} w^{(i)} = (k_B T)^{-1} \{ & \Xi_d [\varepsilon_{11} + \varepsilon_{22} + \varepsilon_{33}] \\ & + \Xi_u [(a^{(i)} a^{(i)})_{11} \varepsilon_{11} + (a^{(i)} a^{(i)})_{21} \varepsilon_{21} \\ & + (a^{(i)} a^{(i)})_{31} \varepsilon_{31} + (a^{(i)} a^{(i)})_{12} \varepsilon_{12} \\ & + (a^{(i)} a^{(i)})_{22} \varepsilon_{22} + (a^{(i)} a^{(i)})_{32} \varepsilon_{32} \\ & + (a^{(i)} a^{(i)})_{13} \varepsilon_{13} + (a^{(i)} a^{(i)})_{23} \varepsilon_{23} \\ & + (a^{(i)} a^{(i)})_{33} \varepsilon_{33}]\} . \end{aligned} \tag{7}$$

Recalling that $\overrightarrow{a^{(i)}}$ must be a unit vector, we must have

$$[a^{(i)}]^2 = \overrightarrow{a^{(i)}} \cdot \overrightarrow{a^{(i)}} = 1 , \tag{8}$$

which means that

$$[a_1^{(i)}]^2 + [a_2^{(i)}]^2 + [a_3^{(i)}]^2 = 1 . \tag{9}$$

Let us consider now the four valleys of the conduction band structure one by one. Designating the valley in the [111] direction as (i) = (1), we see that (9) can be satisfied with the components

$$a_1^{(1)} = \frac{1}{\sqrt{3}} , \quad a_2^{(1)} = \frac{1}{\sqrt{3}} , \quad a_3^{(1)} = \frac{1}{\sqrt{3}} , \tag{10}$$

which permits us to write (3) as

$$\overrightarrow{a^{(1)}}\,\overrightarrow{a^{(1)}} = \begin{bmatrix} \frac{1}{3} & \frac{1}{3} & \frac{1}{3} \\ \frac{1}{3} & \frac{1}{3} & \frac{1}{3} \\ \frac{1}{3} & \frac{1}{3} & \frac{1}{3} \end{bmatrix} . \tag{11}$$

Similarly, by designating the valleys in the $[\bar{1}11]$, $[\bar{1}\bar{1}1]$, and $[1\bar{1}1]$ directions as (i) = (2), (3), and (4), the components of the respective

$\overrightarrow{a^{(i)}}$'s ,

satisfying (9), are given by

$$a_1^{(2)} = -\frac{1}{\sqrt{3}} , \quad a_2^{(2)} = \frac{1}{\sqrt{3}} , \quad a_3^{(2)} = \frac{1}{\sqrt{3}} \tag{12}$$

$$a_1^{(3)} = -\frac{1}{\sqrt{3}} , \quad a_2^{(3)} = -\frac{1}{\sqrt{3}} , \quad a_3^{(3)} = \frac{1}{\sqrt{3}} \tag{13}$$

$$a_1^{(4)} = \frac{1}{\sqrt{3}} , \quad a_2^{(4)} = -\frac{1}{\sqrt{3}} , \quad a_3^{(4)} = \frac{1}{\sqrt{3}} . \tag{14}$$

With the vector components displayed in (12) to (14), the tensors corresponding to (3) are

$$\overrightarrow{a^{(2)}}\,\overrightarrow{a^{(2)}} = \begin{bmatrix} \frac{1}{3} & -\frac{1}{3} & -\frac{1}{3} \\ -\frac{1}{3} & \frac{1}{3} & \frac{1}{3} \\ -\frac{1}{3} & \frac{1}{3} & \frac{1}{3} \end{bmatrix} \tag{15}$$

$$\overrightarrow{a^{(3)}}\,\overrightarrow{a^{(3)}} = \begin{bmatrix} \frac{1}{3} & \frac{1}{3} & -\frac{1}{3} \\ \frac{1}{3} & \frac{1}{3} & -\frac{1}{3} \\ -\frac{1}{3} & -\frac{1}{3} & \frac{1}{3} \end{bmatrix} \tag{16}$$

$$\overrightarrow{a^{(4)}}\,\overrightarrow{a^{(4)}} = \begin{bmatrix} \frac{1}{3} & -\frac{1}{3} & \frac{1}{3} \\ -\frac{1}{3} & \frac{1}{3} & -\frac{1}{3} \\ \frac{1}{3} & -\frac{1}{3} & \frac{1}{3} \end{bmatrix}. \tag{17}$$

Let us write out now (7), with (11), (15), (16), and (17), for the four valleys. The result is

$$\begin{aligned} w^{(1)} &= (k_B T)^{-1} \left[(\Xi_d + \tfrac{1}{3}\,\Xi_u)(\varepsilon_{11} + \varepsilon_{22} + \varepsilon_{33}) \right. \\ &\quad \left. + \tfrac{1}{3}\,\Xi_u\,(\varepsilon_{12} + \varepsilon_{13} + \varepsilon_{23} + \varepsilon_{21} + \varepsilon_{31} + \varepsilon_{32}) \right] \end{aligned} \tag{18}$$

$$\begin{aligned} w^{(2)} &= (k_B T)^{-1} \left[(\Xi_d + \tfrac{1}{3}\,\Xi_u)(\varepsilon_{11} + \varepsilon_{22} + \varepsilon_{33}) \right. \\ &\quad \left. + \tfrac{1}{3}\,\Xi_u\,(-\varepsilon_{12} - \varepsilon_{13} + \varepsilon_{23} - \varepsilon_{21} - \varepsilon_{31} + \varepsilon_{32}) \right] \end{aligned} \tag{19}$$

$$w^{(3)} = (k_B T)^{-1}[(\Xi_d + \frac{1}{3}\Xi_u)(\varepsilon_{11} + \varepsilon_{22} + \varepsilon_{33}) + \frac{1}{3}\Xi_u(\varepsilon_{12} - \varepsilon_{13} - \varepsilon_{23} + \varepsilon_{21} - \varepsilon_{31} - \varepsilon_{32})] \tag{20}$$

$$w^{(4)} = (k_B T)^{-1}[(\Xi_d + \frac{1}{3}\Xi_u)(\varepsilon_{11} + \varepsilon_{22} + \varepsilon_{33}) + \frac{1}{3}\Xi_u(-\varepsilon_{12} + \varepsilon_{13} - \varepsilon_{23} - \varepsilon_{21} + \varepsilon_{31} - \varepsilon_{32})]. \tag{21}$$

Next, we want to compute $\bar{w}$, the average value of the scaled changes of the $E^{(i)}$'s. This quantity is defined by

$$\bar{w} = \frac{1}{4}\sum_{i=1}^{4} w^{(i)}. \tag{22}$$

Upon substitution of (18) to (21) into (22), we have

$$\bar{w} = (k_B T)^{-1}[(\Xi_d + \frac{1}{3}\Xi_u)(\varepsilon_{11} + \varepsilon_{22} + \varepsilon_{33})] . \tag{23}$$

A quantity of further interest is $\overline{w^2}$, defined by

$$\overline{w^2} = \frac{1}{4}\sum_{i=1}^{4} [w^{(i)}]^2. \tag{24}$$

Performing some algebra, with the aid of (18) to (21), we find that, upon consideration of $\varepsilon_{12} = \varepsilon_{21}$, $\varepsilon_{13} = \varepsilon_{31}$, $\varepsilon_{23} = \varepsilon_{32}$,

$$\overline{w^2} = (k_B T)^{-2}[(\Xi_d + \frac{1}{3}\Xi_u)(\varepsilon_{11} + \varepsilon_{22} + \varepsilon_{33})]^2 + \frac{4}{9}(k_B T)^{-2}\,\Xi_u^2\,(\varepsilon_{12}^2 + \varepsilon_{13}^2 + \varepsilon_{23}^2). \tag{25}$$

Using (23) and (25), we can now form

$$\overline{w^2} - \bar{w}^2 = (4/9)(k_B T)^{-2}\,\Xi_u^2\,(\varepsilon_{12}^2 + \varepsilon_{13}^2 + \varepsilon_{23}^2), \tag{26}$$

a quantity of which use will be made later.

In the case of degenerate n-type Ge, a case which prevails when there are more than about 2 x 10^{17} monovalent donors per cm^3, the electrons remain in the conduction band at all temperatures and form a degenerate electron gas. The application of a pure shear strain to such a Ge crystal changes the total electronic energy only to second-order in the strain. As we shall see, this amounts to a change in the elastic constants.

To proceed, let us consider the density of electronic states E of valley (i). This quantity is given by

$$N^{(i)}(E) = \frac{4\pi}{h^3}(2m^*)^{3/2}(E - E^{(i)})^{\frac{1}{2}}, \tag{27}$$

where m* is an average effective mass of an electron, defined by

$$m^* = (m_t^{\,2}\, m_\ell)^{1/3}. \tag{28}$$

The effective mass m* is called the density-of-states effective mass, with m_t and m_ℓ standing for the transverse and longitudinal effective masses associated with an energy ellipsoid.

The concentration of electrons in valley (i) is given by

$$n^{(i)} = \int_{E^{(i)}}^{\infty} N^{(i)}(E)\, f_o(E)\, dE, \tag{29}$$

where $f_o(E)$ denotes the Fermi-Dirac distribution function, namely

$$f_o(E) = \frac{1}{e^{(E - \zeta)/k_B T} + 1}, \tag{30}$$

with ζ standing for the Fermi energy. Introducing the dimensionless quantity

$$\eta = \frac{\zeta}{k_B T}, \tag{31}$$

which may be called the reduced Fermi level, we can rewrite (29), with the aid of (27) and (30), as

$$n^{(i)} = 2\pi^{-\frac{1}{2}} N_c F_{\frac{1}{2}}(\eta - w^{(i)}). \tag{32}$$

In (32), $F_{\frac{1}{2}}$ stands for the Fermi-Dirac integral, given by

$$F_\nu(\eta) = \int_o^{\infty} \frac{x^\nu\, dx}{e^{x-\eta} + 1}, \tag{33}$$

and N_c is a constant, defined by

$$N_c = 2(2\pi m^* k_B T\, h^{-2})^{3/2} . \tag{34}$$

The Fermi level ζ is determined from the condition that, in the case of degeneracy, the sum of the concentrations of electrons in the four valleys must add up to the total electron concentration N. Thus, we must have

$$N = \sum_{i=1}^{4} n^{(i)} , \tag{35}$$

which, upon consideration of (32), can be written as

$$N = 2\pi^{-\frac{1}{2}} N_c \sum_{i=1}^{4} F_{\frac{1}{2}}(\eta - w^{(i)}) . \tag{36}$$

To proceed, we assume that strain changes the Fermi level ζ_o of the unstrained crystal. It is convenient to work with the reduced Fermi level η, defined in (31), and write

$$\eta = \eta_o + \delta, \tag{37}$$

where, in analogy to (31), the quantity $\eta_o = \zeta_o/k_B T$ refers to the reduced Fermi level in the unstrained lattice and δ stands for the change due to strain. For small strains, the Fermi-Dirac functions can be expanded. After expansion, to terms quadratic in the strain, (36) becomes

$$N = 2\pi^{-\frac{1}{2}} N_c \sum_{i=1}^{4} [F_{\frac{1}{2}}(\eta_o) + (\delta - w^{(i)}) F_{\frac{1}{2}}{}' + \frac{1}{2} (\delta - w^{(i)})^2 F_{\frac{1}{2}}{}''], \tag{38}$$

where the derivatives of the Fermi-Dirac function $F_{\frac{1}{2}}$ are defined by

$$F_{\frac{1}{2}}{}' = \left. \frac{d F_{\frac{1}{2}}(\eta)}{d\eta} \right|_{\eta = \eta_o} , \tag{39}$$

and

$$F_{\frac{1}{2}}{}'' = \left. \frac{d^2 F_{\frac{1}{2}}(\eta)}{d \eta^2} \right|_{\eta = \eta_o} . \tag{40}$$

Carrying out the summation in (38), we obtain

$$N = 2\pi^{-\frac{1}{2}}(4N_c)[F_{\frac{1}{2}}(\eta_o) + (\delta - \bar{w})\, F_{\frac{1}{2}}' + \frac{1}{2}(\delta^2 - 2\delta\bar{w} + \overline{w^2})\, F_{\frac{1}{2}}''] , \tag{41}$$

where the quantities $\bar{w}$ and $\overline{w^2}$ have already been defined in (23) and (25).

The total electron concentration N, as given in (41), refers to the strained crystal. The same quantity, in the unstrained crystal, is given by

$$N = 2\pi^{-\frac{1}{2}}(4N_c)\, F_{\frac{1}{2}}(\eta_o), \tag{42}$$

where we have taken into account that the number of electrons in the conduction band is conserved when the strain ceases.

By equating Eqs. (41) and (42), we can solve for δ. To terms second order in the $w^{(i)}$, the result is

$$\delta = \bar{w} - \frac{1}{2}(F_{\frac{1}{2}}'' / F_{\frac{1}{2}}')(\overline{w^2} - \bar{w}^2). \tag{43}$$

Next, let us consider the total free energy per unit volume associated with the electrons in valley (i). This quantity is given by

$$A_e^{(i)} = n^{(i)}\,\zeta + k_B T \int_{E^{(i)}}^{\infty} N^{(i)}(E)\,[1 - \ln f_o(E)]\, dE, \tag{44}$$

which, upon consideration of (27) and (30), can be brought to the form

$$A_e^{(i)} = 2\pi^{-\frac{1}{2}} N_c\, k_B T\, [\eta\, F_{\frac{1}{2}}(\eta - w^{(i)}) - \frac{2}{3} F_{3/2}(\eta - w^{(i)})]. \tag{45}$$

As given in (45), $A_e^{(i)}$ refers only to valley (i). The quantity of interest is, however, the sum of the free energy densities over all four valleys, namely

$$A_e = \sum_{i=1}^{4} A_e^{(i)} . \tag{46}$$

Considering (37) and (45), the free energy in (46) can be expanded, To terms quadratic in δ, the expansion of (46) leads to the expression

$$
\begin{aligned}
A_e = 2\pi^{-\frac{1}{2}} k_B T(4N_c)[(\eta_o + \delta) F_{\frac{1}{2}}(\eta_o) \\
+ \eta_o(\delta - \bar{w}) F_{\frac{1}{2}}' + (\delta^2 - \delta\bar{w}) F_{\frac{1}{2}}' \\
+ \frac{1}{2}\eta_o(\delta^2 - 2\delta\bar{w} + \overline{w^2}) F_{\frac{1}{2}}'' - \frac{2}{3} F_{3/2}(\eta_o) \\
- \frac{2}{3}(\delta - \bar{w}) F_{3/2}' - \frac{1}{3}(\delta^2 - 2\delta\bar{w} + \overline{w^2}) F_{3/2}''],
\end{aligned} \tag{47}
$$

where the primes on the function $F_{3/2}$ denote derivatives in a manner described in (39) and (40) for $F_{1/2}$.

Substituting for N_c and δ from (42) and (43), we find that (47) becomes, to terms quadratic in the strain components,

$$
\begin{aligned}
A_e = N k_B T[\eta_o - \frac{2}{3}(F_{3/2}(\eta_o)/F_{1/2}(\eta_o))] + N k_B T \bar{w} \\
+ N k_B T (\overline{w^2} - \bar{w}^2)[-\frac{1}{2}(F_{1/2}'/F_{1/2}(\eta_o))].
\end{aligned} \tag{48}
$$

The first term in (48) represents the free energy of the degenerate electron gas in the unstrained Ge crystal. The second term in (48) is linear in the strain, as can bee seen from (22) and (7). This term, as has been shown by Keyes, gives rise to a dilatation of the crystal. The third term in (48) contains the factor

$$(\overline{w^2} - \bar{w}^2),$$

and, therefore, it is quadratic in the strain. This is the term which is responsible for the change in the elastic constants.

To see that this is so, let us consider the elastic strain energy W_g, given by

$$
\begin{aligned}
W_g = \frac{1}{2}[B(\varepsilon_{11} + \varepsilon_{22} + \varepsilon_{33})^2 \\
+ \frac{4}{3} C'(\varepsilon_{11}^2 + \varepsilon_{22}^2 + \varepsilon_{33}^2 - \varepsilon_{11}\varepsilon_{22} - \varepsilon_{22}\varepsilon_{33} - \varepsilon_{11}\varepsilon_{33}) \\
+ 4 c_{44}(\varepsilon_{23}^2 + \varepsilon_{13}^2 + \varepsilon_{12}^2)],
\end{aligned} \tag{49}
$$

In (49), the constant B is called the bulk modulus, which is related to the ordinary elastic constants by

$$B = (1/3)\ (c_{11} + 2c_{12}), \tag{50}$$

while the constant C' is referred to as the shear constant, and it is related to the same quantities by

$$C' = (1/2)(c_{11} - c_{12}). \tag{51}$$

Substituting (26) into (48), we find that the last term of (48) becomes

$$\begin{aligned} \tilde{A}_e &= -\frac{1}{2} N\, k_B T\, (\overline{w^2} - \bar{w}^2)[F_{\frac{1}{2}}{}'/F_{\frac{1}{2}}(\eta_o)] \\ &= -\frac{2\, N\, \Xi_u^{\ 2}}{9\, k_B T}\, [F_{\frac{1}{2}}{}'/F_{\frac{1}{2}}(\eta_o)](\varepsilon_{12}^2 + \varepsilon_{13}^2 + \varepsilon_{23}^2). \end{aligned} \tag{52}$$

Comparison of (52) with (49) shows that the term in the elastic constant c_{44} has the same dependence on the strain components. Consequently, it is c_{44} that is modified by the electronic free energy. To find the change in c_{44}, let us write the change in elastic energy upon doping as

$$\delta W_g = W_g^{\ d} - W_g^{\ p}, \tag{53}$$

where the subscripts d and p refer to the doped and pure crystals, respectively. Using (49), we can write (53) as

$$\begin{aligned} \delta W_g &= \frac{1}{2}\,\delta B(\varepsilon_{11} + \varepsilon_{22} + \varepsilon_{33})^2 \\ &+ \frac{2}{3}\,\delta C'(\varepsilon_{11}^2 + \varepsilon_{22}^2 + \varepsilon_{33}^2 - \varepsilon_{11}\varepsilon_{22} - \varepsilon_{22}\varepsilon_{33} - \varepsilon_{11}\varepsilon_{33}) \\ &+ 2\delta c_{44}(\varepsilon_{23}^2 + \varepsilon_{13}^2 + \varepsilon_{12}^2), \end{aligned} \tag{54}$$

where we have introduced the quantities

$$\begin{aligned} \delta B &= B^d - B^p \\ \delta C' &= C'^d - C'^p \\ \delta c_{44} &= c_{44}^d - c_{44}^p \ . \end{aligned} \tag{55}$$

Let us denote now the third term of (54) by

$$\delta \tilde{W}_g = 2\delta c_{44}(\varepsilon_{23}^2 + \varepsilon_{13}^2 + \varepsilon_{12}^2) \ . \tag{56}$$

Conservation of energy requires that (56) be equal to (52). The reason is, that both

$$\delta\tilde{W}_g$$

and

$$\tilde{A}_e$$

depend on the strain components in an identical manner. Writing then

$$\tilde{A}_e = \delta\tilde{W}_g, \tag{57}$$

we find that the change in c_{44} is given by the formula

$$\delta c_{44} = - \frac{N\,\Xi_u^2}{9\,k_B T} \left[\frac{F_{\frac{1}{2}}'}{F_{\frac{1}{2}}(\eta_o)}\right]. \tag{58}$$

Equation (58) is our central result. The change in the elastic constant c_{44} is negative, because part of the free energy needed to strain the crystal is recovered by the transfer of electrons from valleys that are raised in energy by the strain to valleys that are lowered in energy by the strain. Measurements [4] indicate that the change in c_{44} is significant since 0.06% As donors (2.8×10^{19} cm^{-3}) produces a 5.5% reduction in c_{44}.

In the following, we shall find it convenient to introduce the dimensionless quantity

$$\frac{\delta c_{44}}{c_{44}} = - \frac{N\,\Xi_u^2}{9\,k_B T\,c_{44}} \left[\frac{F_{\frac{1}{2}}'}{F_{\frac{1}{2}}\,(\eta_o)}\right] . \tag{59}$$

Our next goal is the investigation of the temperature dependence of $\delta c_{44}/c_{44}$. This is done by introducing the concept of the degeneracy temperature, defined by [2]

$$T_D = \left(\frac{3}{32\,\pi}\right)^{2/3} \frac{h^2\,N^{2/3}}{2\,m^*\,k_B} \ . \tag{60}$$

To proceed, we substitute (34) into (42) and express T from the resulting equation. We obtain

$$T = \left(\frac{1}{16\,\pi}\right)^{2/3} \frac{h^2 N^{2/3}}{2\, m^* k_B} F_{\frac{1}{2}}(\eta_o)^{-2/3} . \tag{61}$$

Multiplying the right-hand side of (61) by $(3/2)^{2/3}(3/2)^{-2/3}$, and considering (60), we can write T as

$$T = T_D \left[\frac{3\, F_{\frac{1}{2}}(\eta_o)}{2}\right]^{-2/3} . \tag{62}$$

The next step is the substitution of (62) into (59). We find, after some algebra, and upon consideration of (60), that

$$\frac{\delta c_{44}}{c_{44}} = \left(\frac{\delta c_{44}}{c_{44}}\right)_o L_{44}(\eta_o), \tag{63}$$

where the quantities on the right-hand side of (63) are defined by

$$\left(\frac{\delta c_{44}}{c_{44}}\right)_o = -\frac{4}{3}\left(\frac{4\pi}{3}\right)^{2/3} \frac{m^*\, \Xi_u^2\, N^{1/3}}{h^2\, c_{44}} , \tag{64}$$

and

$$L_{44}(\eta_o) = \frac{2}{3}\left[\frac{3\, F_{\frac{1}{2}}(\eta_o)}{2}\right]^{2/3} \left[\frac{F'_{\frac{1}{2}}}{F_{\frac{1}{2}}(\eta_o)}\right] . \tag{65}$$

The meaning of (63) is now easily seen. The first factor on the right-hand side describes $\delta c_{44}/c_{44}$ at the absolute zero of temperature, while the second factor on the right-hand side establishes the temperature dependence of this quantity [$L_{44}(\eta_o)$ approaches unity as T appraches zero].

(C) The case of n-type Si.

As far as the theory of the electronic contribution to the elastic constants is concerned, the conduction band of Si differs from that of Ge only in the number and orientation of the equivalent energy ellipsoids.

Locating the equivalent valleys in Si by the unit vectors

$$\vec{a^{(1)}} = \begin{bmatrix}1\\0\\0\end{bmatrix} \qquad \vec{a^{(2)}} = \begin{bmatrix}-1\\0\\0\end{bmatrix}$$

$$\vec{a^{(3)}} = \begin{bmatrix}0\\1\\0\end{bmatrix} \qquad \vec{a^{(4)}} = \begin{bmatrix}0\\-1\\0\end{bmatrix} \tag{66}$$

$$\vec{a^{(5)}} = \begin{bmatrix}0\\0\\1\end{bmatrix} \qquad \vec{a^{(6)}} = \begin{bmatrix}0\\0\\-1\end{bmatrix} ,$$

Einspruch and Csavinszky [3] find that it is the elastic constant C' that is changed by the electronic contribution. Their treatment, proceeding along that of the previous section, has resulted in the formula

$$\delta C' = -2(2\pi)^{2/3}\left(\frac{m^* \Xi_u^2 N^{1/3}}{h^2}\right) L_2(T/T_D) . \tag{67}$$

It is seen from (67) that, the formula for $\delta C'$, is, again, the product of temperature-independent and temperature-dependent factors since

$$L_2(T/T_D) = \frac{2}{3}\,\frac{T_D}{T}\,\frac{F_{\frac{1}{2}}'}{F_{\frac{1}{2}}(\eta_o)} . \tag{68}$$

(D) The case of p-type Ge and Si.

The structure of the valence band in Ge and Si is considerably more complicated than that of the conduction band. In these materials, the valence band maximum, in the absence of spin-orbit coupling, is three-fold degenerate (sixfold including spin) at k = 0. Spin-orbit splitting partially removes the degeneracy, and the sixfold degenerate level splits into a fourfold level and into a twofold level. The twofold level is called the split-off band, and the fourfold level comprises the heavy-hole band and the light-hole band. The structure of these bands is not spherical since the dependence of the corresponding holes on the wave vector components is given by

$$E_\pm(\vec{k}) = Ak^2 \pm [B^2k^4 + C^2(k_x^2k_y^2 + k_y^2k_z^2 + k_z^2k_x^2)]^{\frac{1}{2}}, \tag{69}$$

where A, B, and C are parameters [5], expressed in units of $\hbar^2/(2m_o)$, where m_o is the electron mass.

In (69), the upper band (+) is called the light-hole band, while the lower band (-) is called the heavy-hole band. Both of these bands are associated with "warped" energy surfaces.

The description of the heavy-hole and light-hole bands is much simplified if the spherical approximation is made. This approximation consists in replacing in (69) the original set of parameters A, B, C by a new set A', B', C', defined by

$$\begin{aligned} A' &= A \\ B' &= B \\ C' &= 0 \ . \end{aligned} \tag{70}$$

Within this approximation, the heavy-hole and light-hole masses are given by

$$m_H = \frac{1}{A' - B'} \; m_o \tag{71}$$

and

$$m_L = \frac{1}{A' + B'} \; m_o \ . \tag{72}$$

The theory of the electronic contribution to the elastic constants of p-type Ge has been given by Keyes [2]. He used the spherical approximation, and further simplified the problem by neglecting the split-off band and the light-hole band. The rationale for the former is the large magnitude of the spin-orbit splitting, while the rationale for the latter is the fact that in p-type Ge most of the holes are in the large mass band.

Omitting details of this very approximate calculation, Keyes' result, for degenerate p-type Ge, is expressed by

$$\frac{\delta C'}{C'} = - \frac{1}{5} \left(\frac{8\pi}{3}\right)^{2/3} \frac{m_H \, \Xi_s'^2 \, N^{1/3}}{h^2 \, C'} \ , \tag{73}$$

where C' is an elastic shear constant, Ξ_s' is a deformation potential constant, and N is the hole concentration.

Keyes' theory for p-type Ge has been extended by Csavinszky and Einspruch [6] to degenerate p-type Si. The extension consists in the inclusion into the theory the light-hole band and the split-off band. These steps are necessary for two reasons. First, the heavy-hole mass to light-hole mass ratio in Si is only 3.06, while in Ge it is 6.36. Second, the split-off band in Si is only 0.044 eV removed from the heavy-hole and light-hole bands at k = 0, while in Ge this distance (in k-space) is 0.28 eV. These material parameters mean that, in degenerate p-type Si, all three valence bands are appreciably populated.

Omitting calculational details again, the result is

$$\frac{\delta C'}{C'} = -\frac{1}{5}\left(\frac{8\pi}{3}\right)^{2/3} \frac{\Xi_s'^{\,2}\,(m_H^{3/2} + m_L^{3/2})^{2/3}\, N^{1/3}}{h^2\, C'} , \tag{74}$$

when only the heavy-hole and light-hole bands are considered. When all three valence bands are considered the result is

$$\frac{\delta C'}{C'} = -\frac{1}{5}\left(\frac{8\pi}{3}\right)^{2/3} \frac{\Xi_s'^{\,2}\, m_v \left[1 - \frac{1}{15}\frac{\lambda}{\zeta}\left(\frac{m_s}{m_v}\right)^{3/2}\right] N^{1/3}}{h^2\, C} , \tag{75}$$

where

$$m_v = (m_H^{3/2} + m_L^{3/2} + m_s^{3/2})^{2/3} , \tag{76}$$

with λ, ζ, m_s standing for the split-off distance (in k-space), the Fermi energy, and the split-off-hole mass, respectively.

Both in p-type Ge and in p-type Si, the electronic effect is a result of the shifts of the various valence bands with strain which is followed by a repopulation of holes. There is thus, again, a change in the electronic free energy of holes which results in a contribution to the elastic constant C' of these materials.

To close this section, it is mentioned that a recent treatment of the electronic effect by Cardona et.al. [7] went beyond the spherical approximation and was based essentially on (69).

(E) The electronic contribution to the third-order elastic constants.

Up to the present section, we have considered only terms quadratic in the strain components in the various expansions. Carrying the expansions one step further, the retention of terms cubic in the strain components leads to an additional term to the electronic free energy [given in (48)].

Keyes [2] has found that, for n-type Ge, the additional term is of the form

$$A_e' = N k_B T \,(\overline{w^3} + 2\,\bar{w}^3 - 3\,\bar{w}\,\overline{w^2})\,[\frac{1}{6}\frac{F_{\frac{1}{2}}''}{F_{\frac{1}{2}}(\eta_o)}], \tag{77}$$

where

$$\overline{w^3} = \frac{1}{4}\sum_{i=1}^{4}[w^{(i)}]^3. \tag{78}$$

When the factor involving the averages in (77) is evaluated, it is found that

$$\overline{w^3} + 2\bar{w}^3 - 3\bar{w}\,\overline{w^2} = \frac{16}{9}\left(\frac{\Xi_u}{k_B T}\right)^3 \varepsilon_{23}\varepsilon_{31}\varepsilon_{12}\,. \tag{79}$$

A comparison (not given here) of (77) and (79) with the strain energy function, involving also cubic terms in the strain components, shows that it is c_{456} which is affected by the electronic contribution. Keyes finds that the change in this third-order elastic constant is given by

$$\delta c_{456} = \frac{4}{27}\,\frac{N\,\Xi_u^3}{(k_B T)^2}\,[\frac{F_{\frac{1}{2}}''}{F_{\frac{1}{2}}(\eta_o)}]. \tag{80}$$

When the temperature dependence of c_{456} is referred to the degeneracy temperature, in a manner described before, (80) can be written as

$$\delta c_{456} = (\delta c_{456})_o\, L_{456}(\eta_o), \tag{81}$$

where $(\delta c_{456})_o$ is the change in c_{456} at 0°K, and L_{456} is a temperature-dependent factor.

Measurements by Hall [8] on degenerate n-type Ge verified the correctness of (80).

(F) Electronic effects in thermal properties.

The thermal properties of a crystal are related to its mechanical properties via the elastic constants of the material. When the elastic constants of a semiconductor are changed by the electronic contribution, we may also assume that its thermal properties will also be changed to some extent.

An important quantity in the description of the thermal

properties of a lattice is the Debye temperature, Θ_D, which can be related to the elastic constants. It is, therefore, this quantity where we may expect an electronic effect. Measurements by Bryant and Keesom [9], on heavily-doped n-type Ge, have, indeed, substantiated this expectation. These workers find that, in the range of 10^{17} to 10^{19} donors/cm^3, the reduction in Θ_D amount to several °K.

Another quantity of interest, in the description of the thermal properties of a lattice, is the thermal expansion coefficient, β, which can be related to the Debye temperature. An electronic effect in β was, indeed, predicted [10] and found [11] in this quantity both in n-type and p-type Ge. In both cases, the electronic effect manifests itself as an increase in the thermal expansion coefficient.

III. ORGANIC SEMICONDUCTORS AND POLYMERIC SEMICONDUCTORS

In Section II, we have proceeded along two assumptions. First, we have taken conduction by the band mechanism for certain. Second, we have considered rather specific band structures (multi-valley for the n-type materials, and degenerate for the p-type materials).

The first assumption is by no means universally valid for molecular crystals. Recent work [12] on carrier mobility in durene, however, definitely points to the existence of such a charge transport mechanism.

The second assumption will probably not be fulfilled in case of molecular crystals. Some of these materials may, however, have lowest and excited state energy bands which are close to each other. In this case, both of these bands should be appreciably populated by charge carriers at an appropriate temperature. Application of strain to such a material, containing charge carriers in the bands, may then shift the bands to a different amount. The band shifts would then be followed by a repopulation of the (one particle) energy levels with a concomitant change in the electronic free energy of the "carrier gas". This change in free energy might be expected to show up as an electronic contribution to the elastic constants of the crystal. A calculation carried out along these lines for n-type Ge by Csavinszky [13] has, indeed, confirmed this viewpoint.

As far as polymers are concerned, the hope for finding an electronic effect in the elastic constants of these materials is currently less than that for molecular crystals. The chief reason is, that measurements of elastic constants require sizeable single crystals of a substance. It should, however, be mentioned that substituted polydiacetylene single crystals, representing fully-conjugated organic polymers, have recently been obtained [14] as large-dimension single crystals.

ACKNOWLEDGEMENT

The author is indebted to Pat Byard for the demanding typing of this manuscript.

REFERENCES

[1]. J. Bardeen, Phys. Rev. 75, 1777 (1949); W. Shockley and J. Bardeen, Phys. Rev. 77, 407 (1950); ibid. 80, 72 (1950).

[2]. R. W. Keyes, IBM J. Res. Develop. 5, 266 (1961).

[3]. N. Einspruch and P. Csavinszky, Appl. Phys. Letters 2, 1 (1963).

[4]. For Ge, see L. Bruner and R. W. Keyes, Phys. Rev. Letters 7, 55 (1961). For Si, see Ref. [3].

[5]. J. M. Luttinger and W. Kohn, Phys. Rev. 97, 969 (1955).

[6]. P. Csavinszky and N. G. Einspruch, Phys. Rev. 132, 2434 (1963); P. Csavinszky, J. Appl. Phys. 36, 3723 (1965); P. Csavinszky, Phys. Rev. B 10, 1539 (1974).

[7]. C. K. Kim, M. Cardona, and S. Rodriguez, Phys. Rev. B 13, 5429 (1976).

[8]. J. J. Hall, Phys. Rev. 137, A960 (1965).

[9]. C. Bryant and P. H. Keesom, Phys. Rev. 124, 698 (1961).

[10]. T. A. Kontorova, Fiz. Tverd. Tela 4, 3328 (1962) [Soviet Phys.-Solid State 4, 2435 (1963)].

[11]. V. V. Zhdanova, Fiz. Tverd. Tela 5, 3341 (1963) [Soviet Phys.-Solid State 5, 2450 (1964)]; V. V. Zhdanova and J. Kontorova, Fiz. Tverd. Tela 7, 333 (1965) [Soviet Phys.-Solid State 7, 2685 (1966)].

[12]. Z. Burshtein and D. F. Williams, J. Chem. Phys. 66, 2746 (1977).

REFERENCES (continued)

[13]. P. Csavinszky, J. Appl. Phys. 37, 1967 (1966).

[14]. R. R. Chance and R. H. Baughman, J. Chem. Phys. 64 3889 (1976).

THEORETICAL INVESTIGATION OF THE TRANSPORT PROPERTIES OF POLYMERS AND ORGANIC MOLECULAR CRYSTALS

S. Suhai

Lehrstuhl für Theoretische Chemie
Friedrich Alexander Universität Erlangen - Nürnberg
852 Erlangen, BRD

ABSTRACT. The basic concepts involved in the calculation of various transport properties of polymers and condensed organic systems are outlined in some detail. Special attention is paid to the unique electronic properties of these solids which make it necessary in some cases to reconsider the conventional calculation methods of electron transport theory. Applications are presented to some biopolymers and charge transfer molecular crystals of the TCNQ type.

1. INTRODUCTION

The interest in the solid state chemistry and physics of polymers and molecular crystals has grown considerably during the past years. The transport properties of these materials have been very extensively studied from the experimental side and some of them have been found to display specific electric and magnetic properties which may be of great importance from the point of view of practical applications. Others are known to play a fundamental role in life processes. Since the electronic structures of many representative models of the above systems have also been calculated [1] it seems worth while to calculate with their help some of the transport quantities and to relate them to the corresponding experimental ones. Besides testing the quality of the electronic indices through this procedure, we may hope to get important information about the mechanism of electron transport in these solids.

The electrical conductivity of these materials is varying over a very broad range. Most of them are semiconductors over the whole temperature region of interest (biopolymers,

J.-M. André et al. (eds.), Quantum Theory of Polymers, 335-366.

alkali-TCNQ compounds, etc.), others are semiconductors at lower temperatures but evolve to extremely well conducting organic 'metals' at higher temperatures (NMP-TCNQ, TTF-TCNQ, etc.), or the sulfur-nitride polymer is for instance superconducting at very low temperatures and behaves like a metal above 0.26°K. We have to say, unfortunately, that up to the present only the electronic structure of the semiconducting phase of these systems is more or less reliably described from the theoretical point of view. The mechanism of the semiconductor - to - metal transition in the TCNQ compounds or the stability of the metallic phase against correlation effects in $(SN)_x$ are still open problems and consequently there are no adequate energy band structures for the calculation of the transport properties of these solids in the metallic phase. For this reason we will only very shortly outline how the transport coefficients should be calculated in metallic systems, but in the detailed derivations and applications we will have always semiconductors in mind.

The present lecture intends to present in a simple way only the most fundamental concepts of the transport theory of polymers and to show how one can practically apply these concepts in concrete situations. In Section 2 we set up the Boltzmann transport equation, establish the different phenomenological transport coefficients, relate them to the experimentally measured quantities and show how they can be obtained by the solution of the transport equation. In Section 3 we present different methods for the calculation of the scattering matrix elements of the electron-phonon interaction and finally in Section 4 we apply the methods discussed in the previous Sections for the calculation of the electrical conductivity, thermopower, Hall coefficient, etc. in polymers having very different energy band structures.

2. FORMAL TRANSPORT THEORY

In calculating the various transport quantities of polymers we will confine our attention to steady-state situations in the linear regime where the macroscopic response of the electrons is linearly related to the applied perturbations. In this case the Boltzmann equation (BE) which is essentially a classical transport equation based on the quasi-classical equations of motion for an electron moving in a single energy band, provides a good description of all the phenomena discussed. One should not forget , however, that there are situations where this approach fails [2] (very high magnetic fields, interband transitions due to high energy photons, etc.) and in these cases the more general Kubo formalism [3] has to be applied. We will not base our treatment of the transport phenomena on this formalism because its great generality makes specific calculations very difficult even in cases which are easily handled using the BE. There are excellent textbooks giving a comprehensive account of the material treated in this Section. We refer the reader interested for further details to the works of Wilson [4], Ziman [5] and Smith et al. [5].

2.1 The Boltzmann Equation

In the transport theory of solids we are concerned with average effects produced by many electrons. Each of them is supposed to follow a trajectory in the six-dimensional phase space $(\vec{k}, \vec{r})$ which is determined by the dynamical equations [5]

$$\vec{v}(\vec{k}) = \hbar^{-1} \nabla_{\vec{k}} E(\vec{k}), \quad \hbar\dot{\vec{k}} = -e\,(\vec{\epsilon}' + c^{-1}[\vec{v}(\vec{k}) \times \vec{B}]) \tag{2.1}$$

where $E(\vec{k})$ is the band energy, $\vec{\epsilon}'$ and $\vec{B}$ are the electric and magnetic field vectors, respectively. The gas of conduction electrons (or holes) can be characterized by a distribution function $f(\vec{k},\vec{r},t)$ which is equal to the probability of finding an electron with wave vector $\vec{k}$ and a given spin orientation in the unit volume element around $\vec{r}$ at time t. The $\vec{r}$ dependence can be omitted when the solid is homogeneous (in the absence of internal inhomogeneities this means isothermal conduction). For the problems we will consider, $f(\vec{k},\vec{r},t)$ will be the same for both spins. Hence, since the density of states in $\vec{k}$ space per unit volume of $\vec{r}$ space is $1/(2\pi)^3$, the density of electrons in $(\vec{k},\vec{r})$ space is $(4\pi^3)^{-1} \cdot f(\vec{k},\vec{r},t)$.

The electric current density $\vec{J}$ and the heat current density $\vec{U}$ (the flux of energy measured relative to the Fermi energy η) can be defined in terms of $f(\vec{k},\vec{r},t)$:

$$\vec{J}=-(e/4\pi^3)\int f(\vec{k},\vec{r},t)\vec{v}(\vec{k})d^3\vec{k}, \quad \vec{U}=(4\pi^3)^{-1}\int f(\vec{k},\vec{r},t)[E(\vec{k})-\eta]\vec{v}(\vec{k})d^3\vec{k} \tag{2.2}$$

In thermal equilibrium (no electric field, no thermal gradient) the distribution function is the well known Fermi-Dirac function: $f^o(\vec{k})=\{\exp[(1/k_BT)(E(\vec{k})-\eta)]+1\}^{-1}$, which can be very well approximated, however, in most polymers and molecular crystals by the classical Maxwell-Boltzmann distribution (non-degenerate semiconductors). When the thermal equilibrium is upset we will assume that the disturbed distribution is still sufficiently close to the equilibrium such that

$$f(\vec{k},\vec{r})=f^o(\vec{k},\vec{r})+f^1(\vec{k},\vec{r}), \text{ where } f^1 \ll f. \tag{2.3}$$

In the transport calculations we will linearize in the external disturbances represented by $f^1(\vec{k},\vec{r})$, i.e. we will perform a linear approximation in the field strengths $\vec{\epsilon}'$ and $\vec{B}$ (small fields). Since the equilibrium distribution does not carry any net flow, we have to calculate the true distribution at least to first order in the driving forces. From statistical mechanics we know that the phase points move in $(\vec{k},\vec{r})$ space like an incompressible fluid, thus the total time derivative of $f(\vec{k},\vec{r},t)$,

$$(\partial f/\partial t)_{drift}=\partial f/\partial t+\vec{v}(\vec{k})\nabla_{\vec{r}}f+\dot{\vec{k}}\nabla_{\vec{k}}f \tag{2.4}$$

ought to be zero. Since, however, there are always imperfections in a real solid which scatter the electrons, the above drift term will not be zero but, instead, equal to the negative rate at which f increases as a result of these collision processes. Writing down this ballance equation we get the general BE:

$$\partial f/\partial t+\vec{v}\nabla_{\vec{r}}f+\dot{\vec{k}}\nabla_{\vec{k}}f+(\partial f/\partial t)_{coll.}=0 \tag{2.5}$$

2.2 The Collision Term

To derive the appropriate form of $(\partial f/\partial t)_{coll.}$ we will suppose that the effect of the various collision events is to induce transitions from an occupied state $\vec{k}$ to an empty one $\vec{k}'$. Let $P(\vec{k},\vec{k}')$ be the probability per unit time that an electron known to be occupying the state $\vec{k}$, will be scattered into the unoccupied state $\vec{k}'$. Then using Fermi statistics, we have:

$$(\partial f/\partial t)_{coll.}=(1/4\pi^3)\int d^3\vec{k}'\{f(\vec{k}')[1-f(\vec{k})]P(\vec{k}',\vec{k})-f(\vec{k})[1-f(\vec{k}')]P(\vec{k},\vec{k}')\} \tag{2.6}$$

In this equation the variables $\vec{r}$ and t have been suppressed and the integration is over the interior of the first Brillouin zone. The evaluation of the transition probabilities $P(\vec{k},\vec{k}')$ forms generally the most tedious part of any transport calculation. We will take up this question later, for the moment we will suppose that $P(\vec{k},\vec{k}')$ is known for the scattering of interest in our solid. In thermal equilibrium the collisions alone do not change the total density of representative points in the phase space and (2.6) vanishes. Thus we arrive at the principle of detailed balancing:

$$f^o(\vec{k}')[1-f^o(\vec{k})]P(\vec{k}',\vec{k})=f^o(\vec{k})[1-f^o(\vec{k}')]P(\vec{k},\vec{k}') \tag{2.7}$$

In studying now the effect of applied fields and/or of spatial gradients of the thermodynamic parameters we will make the basic assumption that $P(\vec{k},\vec{k}')$ is uneffected by these perturbations and therefore continues to satisfy the condition (2.7) (which is not true for high field effects, phonon drag, etc.).

2.3 The Linearized Boltzmann Equation

First we note, that provided Eq.(2.7) remains valid, the application of a magnetic field has no effect on the uniform steady-state distribution since $f^o(\vec{k})$ is independent of both $\vec{r}$ and t, and from (2.5) and (2.1) we have:

$$-(e/\hbar c)[\vec{v}(\vec{k})x\vec{B}]df^o(\vec{k})/dE(\vec{k})\cdot\nabla_{\vec{k}}E(\vec{k})=-(e/c)[\vec{v}(\vec{k})x\vec{B}]\cdot\vec{v}(\vec{k})\cdot df^o(\vec{k})/dE(\vec{k})=0 \tag{2.8}$$

This shows that the Lorentz force is not a thermodynamical force, the magnetic field itself does not give rise to any net flow. The carriers are just constantly turning around their cyclotron orbits and this does not change the distribution function. As stated above, we assume the field strengths and spatial gradients to be first order quantities while $f^o(\vec{k},\vec{r})$ and $f^1(\vec{k},\vec{r})$ in Eq.(2.3) will be supposed to be of zeroth and first order, respectively. Then in zeroth order the BE reduces to

$$-(e/\hbar c)[\vec{v}(\vec{k})\times\vec{B}]\cdot\nabla_{\vec{k}}f^o(\vec{k})=-(\partial f^o/\partial t)_{coll}, \tag{2.9}$$

and it can be seen that (2.9) is satisfied by the Fermi-Dirac distribution even if T and η are functions of $\vec{r}$, provided $P(\vec{k},\vec{k}')$ satisfies Eq. (2.7).

Collecting now the first order terms on both sides of Eq. (2.5) we obtain the linearized BE on which all our subsequent discussion will be based:

$$\begin{aligned}&\vec{v}(\vec{k})\cdot\nabla_{\vec{r}}f^o(\vec{k})-(e/\hbar)\vec{\mathcal{E}}'\cdot\nabla_{\vec{k}}f^o(\vec{k})-(e/\hbar c)[\vec{v}(\vec{k})\times\vec{B}]\cdot\nabla_{\vec{k}}f^1(\vec{k})=\\&=-(1/4\pi^3)\int d^3\vec{k}'\{f^1(\vec{k})[(1-f^o(\vec{k}))P(\vec{k}',\vec{k})+f^o(\vec{k})P(\vec{k},\vec{k}')]-\\&-f^1(\vec{k}')[f^o(\vec{k})P(\vec{k}',\vec{k})+(1-f^o(\vec{k}))P(\vec{k},\vec{k}')]\}.\end{aligned} \tag{2.10}$$

It is customary to define the non-equilibrium part of our distribution function in the form

$$f^1(\vec{k},\vec{r})=-\phi(\vec{k},\vec{r})df^o/dE(\vec{k})=\phi(\vec{k},\vec{r})(1/k_BT)f^o(\vec{k})[1-f^o(\vec{k})]. \tag{2.11}$$

The collision term on the right hand side of Eq. (2.10) takes the following simple form by substituting (2.11):

$$-(\partial f^1/\partial t)_{coll.}=(1/4\pi^3k_BT)\int d^3\vec{k}V(\vec{k},\vec{k}')[\phi(\vec{k})-\phi(\vec{k}')], \tag{2.12}$$

where

$$V(\vec{k},\vec{k}')=f^o(\vec{k})[1-f^o(\vec{k}')]P(\vec{k},\vec{k}') \tag{2.13}$$

is the equilibrium transition rate between the states $\vec{k}$ and $\vec{k}'$, the actual number of transitions per unit time, when the system is in overall thermodynamic equilibrium. For later purposes we note that (2.12) can be written in the very simple form

$$-(\partial f^1/\partial t)_{coll.}=\hat{P}\,\phi(\vec{k},\vec{r}), \tag{2.14}$$

where $\hat{P}$ is the integral operator of the scattering, having a symmetrical and positive kernel $V(\vec{k},\vec{k}')$. Similarly, the substitution of (2.11) on the left hand side of Eq. (2.10) results in a very transparent from of the linearized BE:

$$X(\vec{k},\vec{r})+\hat{M}\,\phi(\vec{k},\vec{r})=\hat{P}\,\phi(\vec{k},\vec{r}), \tag{2.15}$$

where

$$X(\vec{k},\vec{r})=\left[-e\vec{\mathcal{E}}'-\nabla\eta+(E(\vec{k})-\eta)T\nabla(1/T)\right]\vec{v}(\vec{k})\,df^{o}/dE(\vec{k}), \tag{2.16}$$

and for the magnetic operator $\hat{M}$ we have (noting that $\nabla_{\vec{k}}(df^{o}/dE)$ is parallel to $\vec{v}(\vec{k})$):

$$\hat{M}\phi(\vec{k},\vec{r})=(df^{o}/dE(\vec{k}))(e/\hbar c)\left[\vec{v}(\vec{k})x\vec{B}\right]\nabla_{\vec{k}}\phi(\vec{k},\vec{r}). \tag{2.17}$$

Since $\hat{M}$ is not associated with any flow, we may formally regard this drift term as a sort of collision term, as if the electrons were constantly 'colliding' with the magnetic field.

2.4 The Phenomenological Transport Coefficients

Before seeking for concrete solutions of the linearized BE in specific systems we will derive some general relationships between the different transport coefficients. First, the formal solution of the integro-differential equation (2.15) can be written in the form

$$\phi=(\hat{P}-\hat{M})^{-1}X. \tag{2.18}$$

Substituting this expression into Eq. (2.11) and using (2.16) we obtain the first order correction of the distribution function:

$$f^{1}=\frac{df^{o}}{dE}(\hat{P}-\hat{M})^{-1}\left[e\vec{\mathcal{E}}'\cdot\vec{v}(\vec{k})+\nabla\eta\cdot\vec{v}(\vec{k})-(E(\vec{k})-\eta)T\nabla(1/T)\cdot\vec{v}(\vec{k})\right]\frac{df^{o}}{dE} \tag{2.19}$$

The first two terms of the square bracket can be drawn together by noting that the electromotive force in a crystal with non-uniform chemical potential is given by $\vec{\mathcal{E}}=\vec{\mathcal{E}}'+(1/e)\nabla\eta$. Substituting now the distribution $f^{o}+f^{1}$ into the expressions (2.2) of the electric and heat current densities and observing that f^{o} gives no contribution to either $\vec{J}$ or $\vec{U}$ since $E(\vec{k})$ and $\vec{v}(\vec{k})$ have, respectively, even and odd parity, we can write down the transport equations in the canonical form of linear relationships between flows and associated thermodynamic forces:

$$\vec{J}=K^{EE}\vec{\mathcal{E}}+T\cdot K^{ET}\cdot\nabla(1/T) \tag{2.20a}$$

$$\vec{U}=K^{TE}\vec{\mathcal{E}}+T\cdot K^{TT}\cdot\nabla(1/T), \tag{2.20b}$$

where the elements of the transport coefficient tensors may be determined to be:

$$K^{EE}_{ij}=-(e^{2}/4\pi^{3})\int d^{3}\vec{k}'(df^{o}/dE)v_{i}(\hat{P}-\hat{M})^{-1}v_{j}(df^{o}/dE) \tag{2.21a}$$

$$K^{ET}_{ij}=(e/4\pi^3)\int d\vec{k}'(df^o/dE)v_i(\hat{P}-\hat{M})^{-1}v_j[E(\vec{k})-\eta](df^o/dE) \tag{2.21b}$$

$$K^{TE}_{ij}=(e/4\pi^3)\int d\vec{k}'(df^o/dE)[E(\vec{k})-\eta]v_i(\hat{P}-\hat{M})^{-1}v_j(df^o/dE) \tag{2.21c}$$

$$K^{TT}_{ij}=-(1/4\pi^3)\int d^3\vec{k}'(df^o/dE)[E(\vec{k})-\eta]v_i(\hat{P}-\hat{M})^{-1}v_j[E(\vec{k})-\eta](df^o/dE) \tag{2.21d}$$

These equations supply the general solution of the linear transport problem. Before applying them to concrete physical situations we want to show that the symmetry properties of the operator $(\hat{P}-\hat{M})^{-1}$ immediately lead to the well known Onsager relations derived originally by arguments of the irreverisible thermodynamics. Let the functions $\phi(\vec{k})$ and $\Psi(\vec{k})$ be two arbitrary real periodic functions of $\vec{k}$ with odd parity (this can be assumed without any loss of generality since any part of these functions with even parity would not contribute to either $\vec{J}$ or $\vec{U}$). By using the definitions (2.12), (2.14) and (2.17) for the operators $\hat{P}$ and $\hat{M}$ one can easily prove with some algebra that $\hat{P}$ is symmetric and $\hat{M}$ is antisymmetric within the specified set of basis functions:

$$\int\phi(\vec{k})\hat{P}\,\Psi(\vec{k})d\vec{k}=\int\Psi(\vec{k})\hat{P}\,\phi(\vec{k})d^3\vec{k} \tag{2.22a}$$

$$\int\phi(\vec{k})\hat{M}\,\Psi(\vec{k})d\vec{k}=-\int\Psi(\vec{k})\hat{M}\,\phi(\vec{k})d^3\vec{k}. \tag{2.22b}$$

Furthermore, since $V(\vec{k},\vec{k}')>0$, the collision operator turns out to be positive definite which poperty will be very important later in applying the variational procedure to the solution of the BE. Using Eqs. (2.22a-b) we find for the inverse operator $(\hat{P}-\hat{M})^{-1}$:

$$\int\phi(\vec{k})(\hat{P}-\hat{M})^{-1}\Psi(k)d^3\vec{k}=\int\Psi(\vec{k})(\hat{P}+\hat{M})^{-1}\phi(\vec{k})d^3\vec{k}. \tag{2.23}$$

Eq. (2.23) allows us to directly verify the Onsager ralations on the basis of the microscopic treatment. Since $\hat{M}$ is linear in $\vec{B}$, $-\hat{M}$ is just $\hat{M}$ calculated for a system in which $\vec{B}$ is replaced by $-\vec{B}$ (see Eq. (2.17)). By a little inspection we get the following relationships from Eqs. (2.21 a-d):

$$K^{EE}_{ij}(\vec{B}) = K^{EE}_{ji}(-\vec{B}) \tag{2.24a}$$

$$K^{TT}_{ij}(\vec{B}) = K^{TT}_{ji}(-\vec{B}) \tag{2.24b}$$

$$K^{TE}_{ij}(\vec{B}) = K^{ET}_{ji}(-\vec{B}) \tag{2.24c}$$

As obtained from the canonical transport equations, these relations are not yet expressed in terms of the actually measured coefficients, but the change to the usual transport quantities is very easy. The experimental definitions are the following:

a) Isothermal el. cond.(σ): $\vec{J}=\sigma\vec{E}$, under the condition $\nabla T=0$.

b) Thermal conductivity (κ): $\vec{U}=\kappa(-\nabla T)$, under the condition $\vec{J}=0$.
c) Seebeck coefficient (Q): $\vec{E}=Q\cdot\nabla T$, under the condition $\vec{J}=0$.
d) Peltier coefficient (π): $\vec{U}=\pi\cdot\vec{J}$, under the condition $\nabla T=0$.

Applying now these definitions and Eqs. (2.20 a-b) we can easily establish that

$$\sigma = K^{EE} \tag{2.25a}$$

$$\kappa = (1/T)\left[K^{TT}-K^{TE}(K^{EE})^{-1}K^{ET}\right] \tag{2.25b}$$

$$Q = (1/T)(K^{EE})^{-1}K^{ET} \tag{2.25c}$$

$$\Pi = K^{TE}(K^{EE})^{-1} \tag{2.25d}$$

Finally, using the relations (2.24a-c) we can derive the symmetry properties of these transport coefficients:

$$\sigma_{ij}(\vec{B}) = \sigma_{ji}(-\vec{B}) \tag{2.26a}$$

$$\kappa_{ij}(\vec{B}) = \kappa_{ji}(-\vec{B}) \tag{2.26b}$$

$$Q_{ij}(\vec{B}) = (1/T)\Pi_{ji}(-\vec{B}) \tag{2.26c}$$

To proceed further with the study of electron transport we have to take into account the detailed form of the operator $(\hat{P}-\hat{M})^{-1}$ and calculate with it the quantities (2.21a-d). We will accomplish this work in two steps. First, we shall define a relaxation time for the scattering process in consideration and investigate various transport properties in metallic and semiconducting systems with this assumption. Lateron, we will see that the necessary conditions for the existence of such a relaxation time are not fulfilled in a large class of organic solids, therefore we will consider in a second step the solution of the linearized BE by the variational method for this case.

2.5 Solution of the Boltzmann Equation with a Relaxation Time

As we have seen in Point 2.2, the collision term of the BE involves a complicated integral over the unknown distribution function. In many cases, however, we can make certain assumptions which greatly simplify the problem. The most widely used simplification in the relaxation time approximation. Let us see how this comes about.

The collisions suffered by the electrons involve a change in their momenta and energies. From the point of view of electrical conductivity the main thing is the change of $\vec{k}$ as this determines the loss of forward drift. There are actually many cases in which the energy change during scattering is completely negligible besides the average energy of the conduction electrons. As we shall see later, this elasticity of the scattering is a necessary condition for the existence of a relaxation time. Obvious examples of elastic scattering are the collisions of electrons with structural defects. The electron-phonon interaction in most metals is also elastic at room temperatures as well as the scattering of conduction electrons in semiconductors with long wavelength phonons at all temperatures. Now let us consider a time dependent but spatially homogeneous situation in the absence of applied fields. We write again the distribution function in the form $f=f^o+f^1$, where f^o is the thermal equilibrium distribution and f^1 is a small perturbation. From (2.5) we have:

$$\partial f^1/\partial t = -(\partial f^1/\partial t)_{coll.} \qquad (2.27)$$

We would expect that our system with any given perturbation $f^1=f-f^o$ at t=o should decay away to the distribution f^o with a time constant characteristic of the collision processes. In the relaxation time approximation we will assume therefore, that

$$(\partial f^1/\partial t)_{coll.} = f^1(\vec{k})/\tau , \qquad (2.28)$$

which yields the solution of Eq.(2.27) in the form $f^1(\vec{k},t)=f^1(\vec{k},o)e^{-t/\tau}$. The actual τ, if it exists at all, can be determined, of course, only be the investigation of the scattering process itself. For the moment we will assume only that it does exist and depends only on the electron energy: $\tau=\tau(E)$. Using Eqs. (2.11) and (2.14) we get from (2.28) that the scattering operator is also a scalar function of the energy: $\hat{P}=(df^o/dE)\ \tau(E)^{-1}$. To proceed further with the solution of the linearized BE, we will assume the magnetic field to be sufficiently low that the electrons undergo collisions much before they can complete a cyclotron orbit. Then we can regard the solution for the case $\vec{B}$=o as slightly perturbed by the addition of the magnetic field and we can write down a power series expansion of $(\hat{P}-\hat{M})^{-1}$ in the form

$$(\hat{P}-\hat{M})^{-1} = \hat{P}^{-1}\sum_n(\hat{P}^{-1}\hat{M})^n \qquad (2.29)$$

Introducing the operator $\hat{\Omega}=(df^o/dE)^{-1}\hat{M}=(e/\hbar c)[\vec{v}(\vec{k})x\vec{B}]\nabla_{\vec{k}}$ and taking into account that the energy dependent $\hat{P}$ commutes with $\hat{\Omega}$ we obtain the Jones-Zener expansion [7]:

$$(\hat{P}-\hat{M})^{-1} = \frac{\tau}{df^o/dE}\left\{1+\tau\hat{\Omega}+(\tau\hat{\Omega})^2+\dots\right\}. \qquad (2.30)$$

It only remains to substitute this expansion into the transport coefficients and we get the solutions of the transport problem in increasing powers of $\vec{B}$. Before doing this let us consider in which circumstances a meaningful relaxation time can be defined. From Eqs. (2.12) and (2.28) we have

$$\frac{1}{\tau} = + \frac{1}{4\pi^3 k_B T(df^o/dE)}\int d^3\vec{k}'\, V(\vec{k},\vec{k}')\left[1-\frac{\phi(\vec{k}')}{\phi(\vec{k})}\right] \qquad (2.31)$$

Using (2.16) and (2.18) on the other hand we get:

$$\phi=(\hat{P}-\hat{M})^{-1}X=\frac{\tau}{(df^o/dE)}\{1+\tau\hat{\Omega}+(\tau\hat{\Omega})^2+\ldots\}X. \tag{2.32}$$

The assumption that τ depends only on $E(\vec{k})$ involves that the constant energy surfaces are spherical, otherwise there would be no reason to expect τ to be independent on the orientation of $\vec{k}$. Consequently, we can write $\vec{v}(\vec{k})=(\hbar\vec{k})/m^*$, where m^* is the effective mass. Furthermore, introducing the notation $\vec{Y}=-e\vec{\mathcal{E}}+(E(\vec{k})-\eta)T\nabla(1/T)$, from (2.16) we get $X=\vec{Y}\vec{v}(\vec{k})(df^o/dE)$. Having these quantities in mind, a typical term of (2.32) has the form:

$$\frac{\tau\hat{\Omega}X}{(df^o/dE}=\frac{e\tau}{\hbar c}[\vec{v}(\vec{k})x\vec{B}]\nabla_{\vec{k}}\vec{Y}\cdot\frac{\hbar\vec{k}}{m^*}=\frac{e\tau}{m^*c}[\vec{B}x\vec{Y}]\cdot\vec{v}(\vec{k}). \tag{2.33}$$

Similarly, we obtain

$$\frac{(\tau\hat{\Omega})^2X}{(df^o/dE)}=\left(\frac{e\tau}{m^*c}\right)^2\left[\vec{B}x[\vec{B}x\vec{Y}]\right]\cdot\vec{v}(\vec{k}). \tag{2.34}$$

Thus Eq. (2.32) becomes:

$$\phi=\vec{Z}\vec{v}(\vec{k}),\ \text{where}\ \vec{Z}=\tau\left\{\vec{Y}+\frac{e\tau}{m^*c}[\vec{B}x\vec{Y}]+\left(\frac{e\tau}{m^*c}\right)^2\left[\vec{B}x[\vec{B}x\vec{Y}]\right]+\ldots\right\} \tag{2.35}$$

Let us return now to our earlier assumption, that the scattering processes are elastic, i.e. $V(\vec{k},\vec{k}')$ in (2.31) contains a factor $\delta\{E(\vec{k})-E(\vec{k}')\}$. In this case the ratio $\phi(\vec{k}')/\phi(\vec{k})$ reduces with (2.35) to $\vec{Z}\vec{k}'/\vec{Z}\vec{k}$. Noting that in the spherically symmetric case $V(\vec{k},\vec{k}')$ must have a cylindrical symmetry in $\vec{k}'$ about $\vec{k}$, we can see that from $\vec{Z}\vec{k}'$ only the projection $k'\cdot\cos\vartheta\cdot\vec{k}/k$ of $\vec{k}'$ onto $\vec{k}$ gives non-zero contribution to the integral (ϑ is the angle between $\vec{k}$ and $\vec{k}'$). Since $V(\vec{k},\vec{k}')\neq 0$ only for $k=k'$, we obtain for the relaxation time in this case from (2.31):

$$\tau^{-1}=(4\pi^3k_BTdf^o/dE)^{-1}\int d^3\vec{k}'V(\vec{k},\vec{k}')(1-\cos\vartheta). \tag{2.36}$$

A similar result can be obtained without the cosine factor for the so called velocity-randomizing collisions for which $V(\vec{k},\vec{k}')=V(\vec{k},-\vec{k}')$. Except for the two cases discussed above, elastic and velocity randomizing collisions in a spherically symmetric band, the introduction of a single energy dependent relaxation time is always an uncertain approximation. It is true, however, that in many cases the formal use of Eq. (2.36) and a subsequent averaging of τ^{-1} over an energy surface yealds meaningful results when substituted into the transport expressions [6]. At this point two short remarks are in order. First, even if a correct relaxation time cannot be calculated, its approximate form may provide a very useful guide in the variational solution of the Boltzmann equation [5]. At second, it should be stressed that the possible range of validity of the Jones-Zener method is much wider than the relaxation time model. In fact, Eq. (2.29) suggests that it can be used for any scattering mechanism. Suppose, the operator

$\hat{P}$ can be inverted by means of a variational calculation. Then using the power series expansion of $(\hat{P}-\hat{M})^{-1}$ we can simply account for the small perturbation due to the magnetic field.

2.6 Solution of the Boltzmann Equation by the Variational Principle

As we shall see later, in many organic solids (periodic DNA models, TCNQ crystals, etc.) the energy band widths are of some hundredths of an eV, thus they are comparable with the energies of the acoustic phonon modes by which they are most effectively scattered. For these cases the relaxation time approximation cannot be rigorously applied, since the assumption about the elasticity of the scattering does not hold. From our previous discussion we know, however, that the BE is a linear inhomogeneous integral equation of the from $X(\vec{k})=\hat{P}\phi(\vec{k})$, where $X(\vec{k})$ is a known function of the external fields and the scattering operator $\hat{P}$ is symmetric and positive definite. Fortunately, there is a general mathematical procedure how to derive such an equation from the variation of a certain integral and how to get the unknown function ϕ from a variational principle. This method has been applied by Kohler [8] and Sondheimer [9] for the calculation of the electrical conductivity of solids. We will show here the expression of the electrical conductivity in the case $\nabla T=o$ and $\vec{B}=o$, but we note that the range of applicability of this method is much more general [10] (see also Chapters 7 and 12 of Ziman's book [5]). Let us note first that Eq. (2.15) implies $\langle\phi,X\rangle=\langle\phi,\hat{P}\phi\rangle$. The variational principle states now that of all the functions satisfying this equation, the solution of the original integral equation gives $\langle\phi,\hat{P}\phi\rangle$ its maximum value. An equivalent statement is, that the solution of the integral equation gives to

$$\frac{\langle\phi,\hat{P}\phi\rangle}{|\langle\phi,X\rangle|^2} \tag{2.37}$$

its minimum value (for the proof of these theorems see Chapt. 7 of Ref. [5]). In practice we have to define a trial function with a certain number of arbitrary parameters and we have to vary them until the functional (2.37) becomes an extremum.

There is, however, an interesting possibility for a short cut to calculate the electrical conductivity directly with the help of the expression (2.37). We know namely from statistical mechanics that the entropy of an assembly of fermions is given in the equilibrium by

$$S=-k_B\int\left\{f(\vec{k})\ln f(\vec{k})+[1-f(\vec{k})]\ln[1-f(\vec{k})]\right\}d^3\vec{k}\,. \tag{2.38}$$

Extending the validity of this formula to the case when $f-f^o$ is small and differentiating with respect to time, we get in the linear approximation for the entropy production:

$$\dot{S} = -(1/T)\int \phi(\vec{k})\dot{f}(\vec{k})d^3\vec{k}. \tag{2.39}$$

In steady state we may suppose that the balance of entropy is maintained between two equal and opposite changes due to $\dot{f}_{coll.}$ and $\dot{f}_{field}$, for which we get respectively:

$$\dot{S}_{coll.} = (1/T)\langle\phi, \hat{P}\phi\rangle \text{ and } \dot{S}_{field} = -(1/T)\langle\phi, X\rangle = -(1/T)\vec{\epsilon}\cdot\vec{J}. \tag{2.40}$$

The macroscopically observed entropy production (Joule heat) is $\dot{S}_{macr.} = \dot{S}_{coll.} = -\dot{S}_{field} = \rho\cdot J^2/T$, where the resistivity is defined by $\rho = \sigma^{-1}$. Substituting $\dot{S}_{coll.}$ from Eq. (2.40) and noting that from Eqs. (2.2) and (2.11) $J = \langle\phi, X(\epsilon=1)\rangle$, where $X(\epsilon=1)$ represents the left hand side of the BE in unit electric field, we obtain for the resistivity

$$\rho = \frac{\langle\phi, \hat{P}\phi\rangle}{|\langle\phi, X(\epsilon=1)\rangle|^2}. \tag{2.41}$$

From Eq. (2.37) we know, however, that this ratio has to be a minimum for the true solution ϕ, thus it provides an upper bound for the resistivity (or a lower bound for the conductivity) with any approximate trial distribution function.

3. THE ELECTRON-PHONON INTERACTION

Our aim in the present Section will be to derive the matrix elements of the electron-lattice vibration interaction which is at normal temperatures the main electron scattering process in our systems. As stated in the introduction we will have semiconducting polymers in mind and will confine ourselves to the derivation of the actual form of these matrix elements for physically different situations in polymers. Concerning the general problematics of the electron-lattice interactions we refer the interested reader to Refs. [5], [11], [12].

The organic solids in consideration consist mainly of large planar molecules and their characteristic feature is that the intermolecular interactions are by an order of magnitude smaller than the intramolecular ones. Consequently, one can distinguish to a fairly good approximation between two different phonon systems, the effect of which on the electron transport is entirely different. The first type of lattice vibrations involves the motion of the molecular units as entities themselves, these are the so called acoustic-type vibrations with characteristic phonon energies of a hundredth of an eV. These phonons will be identified as the most active scattering objects in these solids at normal temperatures.The coupling of the delocalized electrons with the high frequency intramolecular vibrational modes, on the other hand, is expected to be much weaker due to the largeness of the corresponding phonon energy (0.2-0.3eV). If, however, the electronic energy bands are narrow enough, the average Bloch velocity of the conduction electrons becomes small, so that the time a charge carrier will spend in the neighbourhood of a given molecule will be comparable with the period of the intramolecular oscillations. The presence of the extra electron (or hole) will give rise in this case to the formation of a 'phonon-polaron' state due to the polarization of the intramolecular oscillators. As we will outline in Point 3.3 this essentially static effect introduces a vibrational overlap factor into the energy band structures, but does not influence otherwise the electron transport.

The physically decisive quantity in the description of the electron-acoustic phonon interaction is the ratio of the one electron bandwidth ΔE to the mean thermal energy k_BT. If $\Delta E/k_BT \gg 1$, the carriers will populate only the band edge points, consequently they can be well described with an effective mass m^* corresponding to the band edge, and the most important is that they can be scattered only by long-wavelength phonons (which have negligibly small energy). This scattering process can be excellently treated by the relatively simple deformation potential method [13] which we will sketch in the next point. If, however, ΔE is comparable to k_BT, and this is the case in a very large class of organic solids, the deformation-potential approximation breaks down, since points differing from the band edges will be also populated with non-vanishing probability. In this case we have to take into account scattering events between arbitrary points of the Brillouin zones (normal and Umklapp processes) including, of course, transitions due to short-wavelength phonons. The derivation of the corresponding scattering matrix elements will be the subject of Point 3.2.

3.1 The Deformation Potential Approximation

We will derive the matrix elements of the interaction between charge carriers and long-wavelength longitudinal acoustic vibrations but it should be noted that the method in its more general form is applicable also to shear strains [14]. The origin of the deformation potential is the following: if we expand a crystal corresponding to a fractional change in volume given by the dilatation Δ, we obtain bands slightly differring from those corresponding to the normal atomic distances. The shifts in the energy of the band extrema will in general be linear in the dilatation Δ. The proportionality factor D is called the deformation potential constant: $\delta V = D \cdot \Delta$. In the case of a long-wavelength lattice vibration the dilatation varies with position and it can be assumed, that there is an effective potential seen by the electrons, which also varies with position: $V(\vec{r}) = D \cdot \Delta(r)$. To treat the electron-lattice interaction it is convenient to expand the displacements $\vec{x}_j$ of the lattice site at $\vec{R}_j$ in terms of normal coordinates:

$$\vec{x}_j = \sum_{\vec{q}} \vec{u}_{\vec{q}} \exp\left\{ i\vec{q}\vec{R}j \right\} \tag{3.1}$$

Since we will be interested always in longitudinal vibrations of quasi 1D systems (linear chains using the classification of McCubbin et al. [15]) we will omit mostly the vector notation in the subsequent discussions if we refer to the component of a vector parallel to the chain axis. The phonon amplitudes u_q can be written in terms of phonon creation an annihilation operators [5]:

$$u_q = [\hbar/2MG\omega(q)]^{1/2} (a_q + a^+_{-q}), \tag{3.2}$$

where M is the mass of the lattice site, G is the number of sites and $\omega(q)$ is the phonon frequency. Since the dilatation is given by $\Delta = \mathrm{div}(x)$ the potential seen by an electron at the position r due to the instantaneous displacements of the lattice with amplitude u_q is:

$$V(r) = D \sum_q [\hbar/2MG\omega(q)]^{1/2}\, iq(a_q + a^+_{-q}) \exp\{iqr\}. \tag{3.3}$$

Assuming nearly-free electrons with plane wave states $\Psi(k) = G^{-1/2}\exp\{ikr\}$ and with energies $E(k) = \hbar^2 k^2/2m^*$, and introducing the electron creation and annihilation operators c_k^+ and c_k, the Hamiltonian of the electron-phonon interaction becomes:

$$H_{el,qh} = \sum_{k,q}\sum V_q (a_q + a^+_{-q}) c^+_{k+q}\, c_k, \tag{3.4}$$

where the matrix element V_q is given by $V_q = iD[\hbar/2MG\omega(q)]^{1/2}q$. From Eq. (3.4) it can be seen, that the interaction couples only states in which a single electron has changed its wavenumber. Furthermore, if the starting phonon state is characterized by $|n_q, n_{-q}\rangle$, we have

$$(a_q + a^+_{-q})|n_q, n_{-q}\rangle = n_q^{1/2}|n_q - 1, n_{-q}\rangle + (n_{-q}+1)^{1/2}|n_q, n_{-q}+1\rangle. \tag{3.5}$$

Using now the Golden Rule of time dependent perturbation theory [16] we get for the probability of scattering per unit time from the state $\Psi(k)$ to $\Psi(k+q)$:

$$P(k,k+q) = \frac{2\pi}{\hbar}|V_q|^2\{n_q \delta[E(k+q) - E(k) - \hbar\omega(q)] + (n_q+1)\delta[E(k+q) - E(k) + \hbar\omega(-q)]\} \tag{3.6}$$

We shall substitute this expression in the next Section to calculate the electrical conductivity of wide band polymers using the relaxation time defined by Eq. (2.36).

..2 Electron-Acoustic Phonon Interaction in Narrow-Band Semiconductors

As we have discussed above in the case of solids whose energy bandwidths are of the order of the thermal energy we cannot restrict ourselves to scatterings with small phonon momenta and also the single effective mass approximation breaks down since in a narrow tight-binding band the effective mass will be a sensitive function of the electron wavenumber, thus we have to take into account the k dependence of the matrix elements in the entire band. A further difficulty which we have to account for is the strongly localized nature of the wave functions in organic solids (this is the origin of the narrowness of the bands as well). In the standard procedures for the calculation of the electron-phonon interaction matrix-elements, however, only the change of the local potentials due to the lattice motion is taken into account while the wave-functions themselves are constrained at the original lattice sites. As pointed out by different authors [17 - 19], this leads to physically unreasonable spurious scattering terms in the tight-binding limit. To overcome this difficulty Whitfield [18] proposed the use of a 'deformed Bloch' representation in which the local displacements are built into the starting wave function and the perturbing Hamiltonian is expressed directly in terms of the relative displacements. While this method is physically more correct, the complicated structure of the resulting matrix elements is less appropriate for practical applications. Another solution of this problem has been suggested by Friedman [19] who expressed the interaction constants directly with the gradients of the characteristic molecular overlap integrals (within the framework of the simple tight-binding LCMO theory). In this Point we will follow this latter proposal and derive the matrix elements of the electron-acoustic phonon scattering using the approximations of the SCF LCAO crystal orbital method [20] and time dependent perturbation theory.

The total wave function of the complete (electron-phonon) system can be written using the above method as a linear combination of the instantaneous local atomic wave functions:

$$\Psi_m(\vec{r},\ldots\vec{R}_{s'}\ldots,t)=G^{-1/2}\sum_{s=1}^{G}\sum_{l=1}^{g} d_{sm}(\ldots\vec{R}_{s'}\ldots,t)\,c_{lm}\phi_l^s(\vec{r}-\vec{R}_s-\vec{R}_{sl}), \quad (3.7)$$

where the quantum number m will be identified later as a band index (we will not consider interband scattering), $\vec{R}_s$ is the instantaneous lattice position ($\vec{R}_s=\vec{R}_s^0+\vec{x}_s$), $\vec{R}_{sl}$ is the position of the atomic orbital ϕ_l^s at site s, G and g are the number of lattice sites and that of the atomic orbitals per site, respectively, and finally d_{sm} and c_{lm} are expansion coefficients whose physical meaning will be clear shortly. As before, we will restrict our interest for one-dimensional systems and will use the component of the vectors along this dimension whenever possible. The Hamiltonian of the system is written as

$$H=H_e+H_L, \quad (3.8)$$

where the electronic part is an effective one-electron Hamiltonian, $H_e=H_e^{eff}$ which will be specified later for different approximations, and the lattice is treated in the harmonic approximation by

$$H_{L}=-\frac{\hbar^2}{2M}\sum_{h=1}^{G}\nabla^2_{x_h}+V_{L}(\ldots R_s,\ldots) \tag{3.9}$$

where M is the mass of the lattice site again. The total wave function obeys now the time dependent Schrödinger equation:

$$i\hbar\partial\Psi_m/\partial t = H\Psi_m . \tag{3.10}$$

Substituting Ψ_m from (3.7), multiplying from the left by ϕ_i^{t*} and integrating over $\vec{r}$ we get:

$$i\hbar\sum_s\sum_l(\partial d_{sm}/\partial t)c_{lm}\langle\phi_i^t|\phi_l^s\rangle = \sum_s\sum_l d_{sm}c_{lm}\langle\phi_i^t|H_e^{eff}|\phi_l^s\rangle +$$
$$+\sum_s\sum_l(H_L d_{sm})c_{lm}\langle\phi_i^t|\phi_l^s\rangle-\frac{\hbar^2}{M}\sum_s\sum_l\sum_h(\nabla_{x_h}d_{sm})c_{lm}\langle\phi_i^t|\nabla_{x_h}|\phi_l^s\rangle -$$
$$-\frac{\hbar^2}{2M}\sum_s\sum_l\sum_h d_{sm}c_{lm}\langle\phi_i^t|\nabla^2_{x_h}|\phi_l^s\rangle. \tag{3.11}$$

The following approximations may be helpful to bring this equation to a more promising form: (1) It can be simply verified, that the last term on the right hand side is proportional to $(m_{el}/M)\cdot E^{el}_{kin}$, which can be discarded as mall. (2) We may expand the electronic matrix elements about their equilibrium values according to the relative displacements to first order:

$$H_{il}^{ts}=\langle\phi_i^t|H_e^{eff}|\phi_l^s\rangle=(H_{il}^{ts})_o+(x_t-x_s)(\nabla H_{il}^{ts})_o \tag{3.12}$$

(i.e. only one-phonon processes will be allowed). (3) We will introduce at this point the ZDO approximations for convenience (that is we will set $\langle\phi_i^t|\phi_l^s\rangle = S_{il}^{ts}=\delta_{il}\delta_{ts}$), since the actual applications of this method which we will report in the next Section were based all on π-electron methods. It should be stressed, however, that this procedure could be equally well applied also within the framework of an <u>ab intio</u> crystal orbital scheme (we will discuss later the numerical problems of such a calculation). Having these approximations in mind and setting for a moment all lattice displacements and velocities equal to zero, we get the Schrödinger equation of the system with complete translational symmetry. It can be easily verified by some algebra, that the solutions of this zeroth-order equations can be written into the usual Bloch-form:

$$d_{tm}^{k,\{n\}}c_{lm}(k) = e^{ikR_t^o}\chi^{\{n\}}e^{-(it/\hbar)[E_m(k)+E^{\{n\}}]}c_{lm}(k) \tag{3.13}$$

where $\{n\}$ represents the totality of vibrational quantum numbers, and the vibrational wave function $\chi^{\{n\}}$ is a product of harmonic oscillator eigenfunctions, for which

$$H_L\cdot\chi^{\{n\}} = E^{\{n\}}\cdot\chi^{\{n\}}. \tag{3.14}$$

Upon substitution of (3.13) into the zeroth order form

$(x_s=0,\ (\hbar/iM)\nabla_s=0)$ of Eq. (3.11) we get the matrix eigenvalue equation:

$$\sum_s\sum_l e^{ikR_s^o}(H_{il}^{o,s})_o c_{lm}(k) = E_m(k)c_{im}(k),\ (i=1,\dots g), \tag{3.15}$$

which is identical with the eigenvalue problem solved by determining the band structure of our solid [20]. The LCAO coefficients $c_{lm}(k)$ completely determine the lattice Bloch-functions having the form

$$\Psi_m(k) = G^{-1/2}\sum_s\sum_l e^{ikR_s^o}c_{lm}(k)\phi_l^s \tag{3.16}$$

From the preceding discussion it is evident that the effective one-electron Hamiltonian in Eq. (3.8) has to be the same which is applied to the band structure calculation. In an earlier application of this method [21] a simple Hückel-type H_e^{eff} has been used, in the SCF ab initio or Pariser-Parr-Pople (PPP) procedures, to be discussed later, we have to build a Slater-determinant-type many electron wave function from the Bloch-orbitals (3.16) and the effective electronic Hamiltonian will have the following form:

$$H_e^{eff} = H^N + \sum_k\sum_{m=1}^{\bar{m}}(2J_{km}-K_{km}), \tag{3.17}$$

where $\bar{m}$ is the number of the filled bands, H^N is the one-electron part of the Hamiltonian (kinetic energy+core potentials), and the Coulomb and exchange operators are defined respectively by

$$J_{km}(1)|\phi_l^s(1)\rangle = \langle\Psi_{km}(2)|r_{12}^{-1}|\Psi_{km}(2)\rangle|\phi_l^s(1)\rangle \tag{3.18}$$

$$K_{km}(1)|\phi_l^s(1)\rangle = \langle\Psi_{km}(2)|r_{12}^{-1}|\phi_l^s(2)\rangle|\Psi_{km}(1)\rangle. \tag{3.19}$$

The next step in our procedure in now to take into account the two terms on the right hand side of Eq. (3.11) which we have neglected by looking for a zeroth-order solution. These arise from the part proportional to $(x_t-x_s)$of the first term on the right hand side of (3.11) and from the originally third term on the same right hand side. These two terms are identified now as the sources of the electron-phonon scattering. Their physical meaning is also rather apparent: the first one results from the variation of the one- and two-electron integrals in (3.17)-(3.19) which contain atomic functions on different sites, while the second one describes the tendency of the moving lattice to drag the electron with it [18]. According to the usual technique of time dependent perturbation theory, we expand the solutions of the complete equation of motion (3.1o) in the zeroth-order solutions (3.13):

$$d_{tm}c_{im} = \sum_{k',\{n'\}} a^{k',\{n'\}}(t)\,d_{tm}^{k',\{n'\}}c_{im}(k')\exp\{-(it/\hbar)[E_m(k')+E^{\{n'\}}]\} \tag{3.20}$$

Substituting this expession into Eq. (3.11), multiplying from the left by $d_{tm}^{k,\{n\}*} c_{im}^{*}(k)$, summing over indices t,i and integrating over the vibrational coordinates, we get:

$$i\hbar \frac{\partial a^{k,\{n\}}}{\partial t} = \sum_{k',\{n'\}} a^{k',\{n'\}} \langle k,\{n\}|V|k',\{n'\}\rangle \, e^{-(it/\hbar)\left\{\left[E_m(k')+E^{\{n'\}}\right]-\left[E_m(k)+E^{\{n\}}\right]\right\}}, \tag{3.21}$$

where the matrix element is given by

$$\langle k,\{n\}|V|k',\{n'\}\rangle = G^{-1} \sum_s \sum_l \sum_t \sum_i e^{i(k'R_s^o-kR_t^o)} c_{im}^{*}(k) c_{lm}(k') \times \tag{3.22}$$

$$\times \left[(\nabla H_{il}^{ts})_o \langle \chi^{\{n\}}|x_s-x_t|\chi^{\{n'\}}\rangle - \frac{\hbar^2}{M}\langle \chi^{\{n\}}|\nabla_{x_s}|\chi^{\{n'\}}\rangle\langle \phi_i^t|\nabla_{x_s}|\phi_l^s\rangle\right].$$

Using now the expansion (3.2) of the lattice displacements, and its analogue [5]

$$\nabla_{x_s} = \sum_q \left(\frac{M\omega(q)}{2\hbar G}\right)^{1/2} (a_q - a_{-q}^{+}) e^{iqR_s} \tag{3.23}$$

for the lattice velocities, and introducing the interaction constants

$$V_m^u(k,k') = \sum_i \sum_l c_{im}^{*}(k)\; c_{lm}(k')(\nabla H_{il}^{ou})_o \tag{3.24}$$

$$W_m^u(k,k') = \sum_i \sum_l c_{im}^{*}(k)\; c_{lm}(k')\langle \phi_i^o|\nabla_{x_u}|\phi_l^u\rangle, \tag{3.25}$$

we can write down the final expressions of the scattering matrix elements in the compact form

$$\langle k,\{n\}|V|k',\{n'\}\rangle = \sum_{\pm} \sum_u \sum_q \Big[\left(\frac{\hbar}{2MG\omega(q)}\right)^{1/2} (1-e^{\mp iqR_u^o}) V_m^u(k,k') \mp$$

$$\mp \left(\frac{\hbar^3\omega(q)}{2MG}\right)^{1/2} W_m^u(k,k')\Big] e^{ikR_u^o} \delta_{k,k'\pm q+K}\, \delta_{n_q,n_q'\mp 1} (n_q'+1/2\pm 1/2) \tag{3.26}$$

The plus sign in this expression refers to scattering processes in which the electron emits one phonon of wavenumber q, while the minus sign indicates the absorption of phonon q.Furthermore, K stands for an arbitrary lattice vector in the reciproc space.

3.3 The Interaction Between Charge Carriers and Optical Phonon Modes

Energy band structure calculations on polymers [1] have shown that in many cases the electronic bandwidths are only of some tenths of an eV. Consequently, the 'localization time' $\tau_{loc} \sim (\hbar/\Delta E)$ of the charge carriers will be comparable or even longer than $\tau_{opt}=2\pi/\omega_{opt}$, the period of the intramolecular (mostly bond-stretching type) vibrations. In this case the electron-phonon interaction is expected to be very strong, since these high frequency vibrations can accomodate to the presence of the particle by changing their frequencies and equilibria corresponding to the molecular ionic state. Physically, this means that the moving particles in the lattice are accompanied by a cloud of the created virtual optical phonons (infrared polarization effect) i.e. the new quasi particles are polarons consisting of a 'bare' particle plus the 'dressing' phonon field. Due to its strength this electron-vibration interaction cannot be treated as a small perturbation, instead, it has to be built into the zeroth-order Hamiltonian. Fortunately, the mathematical handling of this interaction is not very difficult and it can be shown that the new quasi-particles will move in polaron-bands [22] which arise from the original Bloch bands by a shift with the polaron binding energy and by a multiplication of the original dispersion with a vibrational overlap factor.

Since the predominant mechanism for this electron-molecular vibration interaction is the modulation of the molecular orbital energies by the motion of the nuclei [23], it is suitable to expand the molecular energy levels E_γ in terms of the vibrational normal coordinates

$$E_\gamma(\ldots\xi_i\ldots)=E_\gamma(o)+\sum_i(\partial E_\gamma/\partial\xi_i)_o\cdot\xi_i+\ldots \tag{3.27}$$

Keeping only the first order term of Eq. (3.27) we get immediately the linear electron-vibration interaction Hamiltonian, which can be written analogously to Eq. (3.3) in the form

$$H_{el,ph}=\sum_{i,\gamma}\Gamma_{\gamma i}\hbar\omega_i(b_i^+ + b_i)c_\gamma^+ c_\gamma\ , \tag{3.28}$$

where $b_i^\dagger$ creates a vibrational quantum $\hbar\omega_i$, and the electron creation operator $c_\gamma^\dagger$ is associated with the molecular orbital γ. Furthermore, the dimensionless coupling constants $\Gamma_{\gamma i}$ are defined by [23]

$$\Gamma_{\gamma i}=(\sqrt{2}\,\hbar\omega_i)^{-1}(\partial E_\gamma/\partial\xi_i)_o\ . \tag{3.29}$$

With the aid of the coupling constants one can calculate the modification of the particle self-energy due to the formation of the polaron state from the expression [22]

$$E_b^{(\gamma)}=\sum_i\Gamma_{\gamma i}^2\hbar\omega_i \tag{3.30}$$

This quantity is often referred to as 'polaron binding energy', though it leads to actual binding only in the case of a particle in the conduction band. Due to the change of the lattice wavefunction in the course of the polaron formation the original bandwidths will be reduced by the vibrational overlap (Franck-Condon) factors which have the form (assuming transitions only from the undistorted ground state into the distorted ground state):

$$f^{(\gamma)}=\exp\{-\sum_i\Gamma_{\gamma i}^2\}. \tag{3.31}$$

The most troublesome part of these calculations is of course the determination of the coupling constants $\Gamma_{\gamma i}$. One has to determine first the vibrational normal modes which can interact with a molecular level of given symmetry. Calculating then the electronic structure of the molecule for different atomic positions corresponding to a given normal mode one can extract the coupling constants from the slopes of the orbital energies versus the magnitude of the normal mode displacements. Such calculations have been performed by Lipari and coworkers for some highly symmetrical molecules [23,24] and we will apply in the next Section the coupling constants obtained by them for the TCNQ molecule. In the case of molecules with low symmetry, like for instance the nucleotide bases, this normal mode analysis would be much more tedious and one is left to use cruder approximations to guess the order of magnitude of the polaron binding energy.

4. TRANSPORT CALCULATIONS IN POLYMERS

In this Section we will apply the results of the foregoing discussions to actual polymeric systems. First, we will show through some examples how the relaxation time method works in practice since it is in reality the most widely applied procedure for real solids. We will investigate in this relation two types of conductors, metals and semiconductors, first with $\vec{B}=0$, then with $\vec{B}\neq 0$, respectively. Afterwards, we will present some calculations with the variational method in the case of polymers for which no meaningful relaxation time can be defined. Finally, we will discuss shortly an alternative formulation of the transport problem when the Bloch-type band model breaks down.

4.1 Applications of the Relaxation Time

4.1.a. Metals, $\vec{B}=0$:

Combining Eqs. (2.25a) and (2.21a) the <u>electrical conductivity</u> tensor of an isotropic material becomes a constant σ times the unit tensor and we get as zeroth order solution ($\vec{B}=0$):

$$\sigma = \frac{-e^2}{12\pi^3} \iint \tau v^2 \frac{df^o}{dE} \cdot \frac{dS_E \, dE}{|\nabla_{\vec{k}} E|} \quad , \tag{4.1}$$

where dS_E is the element of area of the constant energy surface. Defining the energy dependent mean free path between two collisions by $\lambda_E = (1/S_E)\int \tau v dS_E$, the conductivity reduces to

$$\sigma = - \int \sigma_E \, (df_o/dE) \cdot dE, \tag{4.2}$$

where $\sigma_E=(e^2/12\pi^3\hbar)S_E \cdot \lambda_E$. Eq. (4.2) is valid both in metals and in semiconductors, but in a highly degenerate electron gas like the one in metals we have $(df^o/dE) \sim \delta(E-\eta)$, and therefore

$$\sigma = \sigma(\eta) = (e^2/12\pi^3\hbar)\, S_\eta \cdot \lambda_\eta \,. \tag{4.3}$$

Eq. (4.3) expresses the fact that in a metal only the electrons situated within the thermal layer on the Fermi surface can take actively part in the transport. In the case of a spherical Fermi surface $S_\eta = 4\pi k_F^2$, $\lambda_\eta = v_F \cdot \tau = \hbar k_F\, \tau/|m^*|$, and introducing the particle density in $\vec{r}$ space by $n=(1/4\pi^3)\cdot 4\pi k_F^3/3$ we get from Eq. (4.3) the well known relation $\sigma = ne^2\tau/|m^*|$.

In the expression (2.25b) of the thermal conductivity the second term is negligible in well conducting metals thus applying the same way of reasoning as above and using the Sommerfeld formula [5]

$$\int \phi(E)\, \frac{df^o}{dE} \cdot dE = -\phi(\eta) - \frac{\pi^2}{6} (k_BT)^2 \left[\frac{\partial^2\phi}{\partial E^2}\right]_{E=\eta} - \ldots \tag{4.4}$$

we get the Wiedemann-Franz law:

$$\mathcal{K} = -\frac{1}{e^2T}\int \sigma(E)\, (E-\eta)^2\, \frac{df^o}{dE} \cdot dE = \frac{\pi^2 k_B^2 T}{3e^2}\, \sigma(\eta) \tag{4.5}$$

Similarly, from Eqs. (2.25.c) and (2.21.b) we obtain for the thermoelectric power:

$$Q = \frac{1}{eT\sigma}\int \sigma(E)\, (E-\eta)\, \frac{df^o}{dE} \cdot dE \,. \tag{4.6}$$

In the case of metals we can write $\sigma(E) = \sigma(\eta) + \sigma'(\eta)(E-\eta)$ and using again Eq. (4.4) we obtain

$$Q = -\frac{\pi^2 k_B^2 T}{3e}\, \frac{\sigma'(\eta)}{\sigma(\eta)} \,. \tag{4.7}$$

We note that for quasi-free electrons in metals, if we neglect the energy dependence of λ, $\sigma'(\eta)/\sigma(\eta) \sim \eta^{-1}$, thus Q is negative with a characteristic magnitude of the order of some μV/degK at room temperature.

4.1.b. Semiconductors, $\vec{B}=0$:

For simplicity let us consider a n-type material with spherical bands. The electrical conductivity can be written from (2.21.a) using the relation (2.11) and keeping in mind that for a non-degenerate electron gas $f^o(\vec{k}) \ll 1$, as follows:

$$\sigma = \frac{e^2}{12\pi^3 k_BT}\int \tau v^2(\vec{k}) f^o(\vec{k}) \left[1-f^o(\vec{k})\right] d^3\vec{k} \approx \frac{e^2}{12\pi^3 k_BT}\int \tau v^2(\vec{k}) f^o(\vec{k})\, d^3\vec{k}. \tag{4.8}$$

Remembering now, that $(3/2)nk_BT = (1/4\pi^3)\int E(\vec{k}) f^o(\vec{k})\, d^3\vec{k}$, we get

$$\sigma = n|e|\mu, \text{ with } \mu = |e|\langle\frac{\tau}{m^*}\rangle \text{ and } \langle\frac{\tau}{m^*}\rangle = \frac{\int_0^\infty E^{3/2}\tau(E)m^*(E)^{-1}\exp\{-E/k_BT\}dE}{\int_0^\infty E^{3/2}\exp\{-E/k_BT\}dE}. \quad (4.9)$$

μ is here the macroscopically defined charge carrier mobility, i.e. the aquired drift velocity per unit electric field. Of course, if we have both electrons and holes, we have to add up the conductivities: $\sigma = |e|(n_e\mu_e + n_h\mu_h)$. Furthermore, it can be shown, that similar expressions describe the conductivity also in semiconductors with ellipsoidal energy surfaces, one has only to introduce different effective masses along the main axes of the ellipsoid and transform back the problem to a spherical one [5].

The thermal conductivity and the thermopower can be also very simply obtained for semiconductors by the obvious generalization of the above procedure to calculate $\langle E^2\tau/m^*\rangle$ and $\langle E\tau/m^*\rangle$, respectively. We note here only, that the thermopower, given by

$$Q = -\frac{1}{eT}\left[\frac{\langle E\tau/m^*\rangle}{\langle\tau/m^*\rangle} - \eta\right], \quad (4.10)$$

is expected to be much larger in this case, than for metals, since η is generally deep in the forbidden gap, so that $|\eta| \gg \langle E\rangle \sim k_BT$.

4.1.c. The Hall Effect

Let us substitute now the second term of the expansion (2.30) into (2.21.a):

$$K_{ij}^{EE} = -(e^3/4\pi^3\hbar c)\int (df^o/dE)v_i\tau^2[\vec{v}\times\vec{B}]\,\nabla_{\vec{k}}v_j\,d^3\vec{k}\;. \quad (4.11)$$

On multiplying this equation by $\mathcal{E}'_j$, summing over j and taking into account again spherical energy surfaces we find for the modification $\sigma^{(1)}$ of the electrical conductivity due to the magnetic field in first order:

$$\vec{J} = (e^3/4\pi^3c)\int (df^o/dE)\vec{v}(\tau^2/m^*)[\vec{\mathcal{E}}'\times\vec{B}]\vec{v}\;d^3\vec{k} = \sigma^{(1)}[\vec{\mathcal{E}}'\times\vec{B}]\;. \quad (4.12)$$

Combining Eqs. (4.1) and (4.12) we can see that, in the presence of a magnetic field, the electric field needed to create the current $\vec{J}$ has two components:

$$\vec{\mathcal{E}}' = \rho\vec{J} + \sigma^{(1)}\rho^2\,[\vec{B}\times\vec{J}]\;. \quad (4.13)$$

Thus, along $\vec{J}$ we have $E_{\parallel} = \rho J$, showing that the apparent resistance $\rho = \sigma^{-1}$ is unaltered by the magnetic field, i.e. there is no magnetoresistance in first order. Due to the presence of $\vec{B}$, however, we have also a transverse field, proportional to the Hall coefficient $R_H = \sigma^{(1)}\rho^2$. Exploiting the spherical symmetry again we can write analogously to (4.2):

$$\sigma^{(1)} = \int \sigma_E \, (e\tau/cm^*) \, (df^o/dE) \, dE. \quad (4.14)$$

In the case of metals this integral can be easily evaluated and using the same arguments as after Eq. (4.3) we get:

$$R_H = -(e\tau_\eta/\sigma_\eta \, cm^*_\eta) = -(1/nec)\cdot(|m^*|/m^*). \quad (4.15)$$

In a similar manner one can calculate also higher order effects with respect to the magnetic field by taking into account further terms in the expansion (2.32) of the distribution function. An excellent treatment of the various galvanomagnetic phenomena can be found for instance in the review article of Beer [25].

4.1.d. Deformation Potential Calculations in Polymers

In a large class of polymers the valence- and conduction band widths range from some tenths of an eV to ~ 2 eV, i.e. they are definitely larger than the room temperature thermal energy (0.026 eV at 300°K). In these systems we may expect that the electron-phonon scattering can be reasonably well described by the use of the deformation potential approximation. Upon substitution of the corresponding expression (3.6) of the scattering probability into the defining Eq.(2.36) of the relaxation time, we get:

$$\tau(E) = \frac{\pi^2 M v_s^2 \hbar^4 E^{-1/2}}{4\sqrt{2}(m^*)^{3/2} D^2 k_B T}. \quad (4.16)$$

Here we have used the simple dispersion $\omega(q) = v_s q$ for the acoustic phonons and the sound velocity v_s is determined by the relation $v_s = (c_l/\rho)^{1/2}$, where c_l is the longitudinal elastic constant and ρ is the mass density. From Eq. (4.9) we can calculate the charge carrier mobilities and the free paths (defined in this case by $\lambda = \langle \tau v \rangle$):

$$\mu = \frac{2\sqrt{2}}{3} \frac{|e| c_l \hbar^4}{D^2 (m^*)^{5/2} (k_B T)^{3/2}}, \qquad \lambda = \frac{\pi \hbar^4 c_l}{(m^*)^2 D^2 k_B T}. \quad (4.17)$$

Similarly, for the preexponential factor of the electrical conductivity, defined by $\sigma = \sigma_o \exp\{-\Delta E/2k_B T\}$, where ΔE is the band gap, we get from (4.9):

$$\sigma_o = \frac{2e^2 \hbar c_l}{3\pi m^* D^2} \quad (4.18)$$

In Table I we show the results of some calculations performed on four different polymeric systems. The first one, the homopolynucleotide polyadenine was investigated as a simple periodic model

Table I.

Transport properties of some wide band polymers calculated with the deformation potential method at T=300°K.

Polymeric system	polyadenine[a]		polyglycine (main chain)[b]		polyglycine (H-bonds)[b]		K^+TCNQ^-[c]	
	electrons	holes	electrons	holes	electrons	holes	electrons	holes
Bandwidth (eV)	0.246	0.320	1.376	2.098	0.141	0.291	1.163	0.979
Effective mass (in m_{el})	5.488	4.221	1.049	0.688	4.834	2.342	0.665	0.778
Deformation potentials (in eV)	0.448	0.352	2.487	1.924	4.764	4.102	1.846	1.376
Elastic constant ($dyn\cdot cm^{-2}$)	3.6×10^{11}	3.6×10^{11}	1.05×10^{12}	1.05×10^{12}	1.97×10^{11}	1.97×10^{11}	2.077×10^{11}	2.077×10^{11}
Drift mobility (in $cm^2V^{-1}sec^{-1}$)	156	485	878	4210	1.03	8.50	1034	1258
Free path (in Å)	265	725	987	2650	1.64	9.43	608	801
Preexponential factor of the conductivity (in $\Omega^{-1}cm^{-1}$)	8.018×10^3	1.688×10^4	3.97×10^3	1.01×10^4	44.1	123	2.25×10^3	3.46×10^3

a) Ref. 26
b) Ref. 29
c) Ref. 30

of DNA [26]. It was built from adenine molecules in the same geometrical position as the different nucleotide bases are situated in the DNA macromolecule. The values of D were calculated by a perturbative method [26], and according to the results the mobility and free path values of both electrons and holes in this polymer fall into the region where the Bloch-type delocalized description of the transport process is reasonable [27]. A further application of the same method [28] to other periodic DNA models with much narrower bands (0.003-0.060 eV) has shown, however, that for such systems the deformation potential approximation breaks down, in accordance with our earlier considerations.

The second and third systems in Table I are periodic protein models built first from glycine molecules according to the geometrical structure of the polypeptide backbones (polyglycine main chain) and in a hydrogen bonded configuration afterwards [31]. The energy band structures of both systems have been calculated by the *ab initio* SCF LCAO crystal orbital method [32]. To obtain the deformation potential constants we repeated these calculations by distorting the lattice in both directions and calculated the derivatives of the appropriate band edges numerically. The results show that the electrical conduction in proteins can be expected to be very anisotropic since the main chain direction shows transport properties similar to elemental semiconductors while along the hydrogen bonds the mean charge carrier free paths are comparable with the lattice constants, thus the existence of a delocalized (band type) transport becomes questionable. We note, that the measured values of σ_0 for different polypeptides scatter in the range of 10^4-$10^6 \Omega^{-1} cm^{-1}$ [33], supporting thus a conduction mechanism with charge carrier delocalization along the main chains.

The fourth system investigated by the deformation potential method is a member of the large family of charge transfer molecular crystals containing as anion the tetracyanoquinodimethan (TCNQ) molecule. The K^+TCNQ^- salt as well as the other TCNQ crystals consists of columns of face-to-face stacked TCNQ molecules co-ordinated by similar columns of the cations [34]. The energy band structures of this crystal have been calculated earlier with the help of the Pariser-Parr-Pople parametrization of the unrestricted Hartree-Fock crystal orbital method [35]. From the last two columns of Table I we can see that this material behaves again as a normal semiconductor with relatively large mobilities and free paths. The experiments confirm this prediction, K^+TCNQ^- has been shown to be a semiconductor in the whole temperature region investigated. Its electrical conductivity follows the $\sigma = \sigma_0 \exp\{-\Delta E/2k_BT\}$ law, and the close agreement between the room temperature microwave conductivity ($3\times10^{-4} \Omega^{-1} cm^{-1}$ [36]) and the corresponding d.c. value ($1\times10^{-4} \Omega^{-1} cm^{-1}$ [37]) makes it probable that the measured single particle gap may be considered to be the intrinsic value and not due to other, disorder limited transport processes. With the calculated gap of ΔE=0.829 eV we obtain from our values in Table I for the room temperature conductivity $\sigma_{RT}=7.29\times10^{-4}\Omega^{-1}cm^{-1}$.

4.1e. Application of the Variational Method of Transport Theory to Polymers

For the majority of condensed organic systems the valence- and conduction band widths lie in the region of some tenth or hundredths of an eV. In these cases the phonon energies are by no means negligible besides the electronic ones and instead of the relaxation time approximation we have to apply the variational method of transport theory. In this Point we will show two representative applications of this procedure. The first one is the neutral TCNQ crystal whose energy bands have been calculated earlier in the PPP approximation [38]. The valence- and conduction bands obtained in the Hartree-Fock scheme (first column of Table II.) have been corrected for electron correlation effects by the electron-polaron method [39]. Furthermore, the analysis of TCNQ [23] has shown, that there are ten vibrational normal modes of the a_g type which can couple linearly to any electronic orbital in this molecule having D_2 symmetry. From the calculated frequencies of these normal modes [24] and from the correlation corrected band widths (second column in Table II.) one expects a strong electron-optical phonon interaction in TCNQ leading to the formation of phonon polarons with binding energies of 0.103 eV and 0.153 eV, in the case of the valence- and conduction bands, respectively. The resulting polaron bandwidths are shown in the third column of Table II.

Table II.

The transport properties of the TCNQ crystal

Transport property		HF	HF+el.pol.	HF+el.pol.+phon.+pol.
Conduction band width (in eV)		0.450	0.352	0.090
Valence band width (in eV)		0.081	0.063	0.034
Electron free path (in Å at T=300°K)		272	186	28
Hole free path (in Å at T=300°K)		12	8.5	4.2
Electron drift mobility (in $cm^2V^{-1}sec^{-1}$)	T=180°K	385	294	61
	T=240°K	305	226	46
	T=300°K	178	147	32
Hole drift mobility (in $cm^2V^{-1}sec^{-1}$)	T=180°K	27.3	24.1	16.8
	T=240°K	19.8	17.2	9.4
	T=300°K	12.3	9.7	5.2

In accordance with our earlier considerations we expect the acoustic type vibrations of the TCNQ molecules along the stacked columns to be the major factor limiting the charge carrier transport. For this vibrational mode we have assumed a Debye-type phonon spectrum and calculated the electron-phonon interaction matrix elements with the method outlined in Point 3.2. To calculate the electrical conductivity and drift mobility we assume first the existence of an approximate relaxation time, defined as the half of the relaxation time calculated from Eq. (2.31) only for the scattering processes $k \rightarrow k'$. This relaxation time is equal to the true one in the case of elastic scattering processes and its use leads to the same mobility values as obtained by the variational method for the (hypothetical) systems given in the first and second columns of Table II. [40]. We have two purposes with this approximate relaxation time: (1) It makes possible to calculate the approximate free paths and (2) making use of the functional form of the obtained relaxation time we can write the distribution function, to be used in the calculation of σ from Eq. (2.41), in the form $\phi(k) = \text{const.} \times \tau(k) \times v(k)$ (see Eq. (2.35)), where the constant cancels from Eq. (2.41). The mobilities are calculated again using the expression $\mu = \sigma / n|e|$. From Table II. it can be seen that the electron free paths are at all temperatures definitely larger than the lattice constant ($a \sim 3.5$ Å). This is not true, however, for the holes, for which consequently the band model seems to be less adequate. The calculated drift mobilities compare very badly with the experimental values of $\mu = 0.4\ cm^2 V^{-1} sec^{-1}$ for both electrons and holes, with the hole mobility about 10% lower than the electron mobility [41]. Furtheron, the experimental mobilities were found to be practically independent of temperature between 200°K and 300°K, clearly in contradiction with the band model calculations. If, however, due to the strong scattering a localized (hopping-like) transport mechanism would be active it should lead to a thermally activated mobility. Therefore, the origin of the above discrepancies is not clear. Since it has been established that both the photo- and dark conduction processes are strongly trap controlled in this material, further investigations seem to be necessary in this direction to get a more realistic picture of the conduction in TCNQ.

TABLE III. The characteristic quantities of the charge carrier transport in periodic DNA models.

Periodic DNA model		δE (eV)	τ_{loc}/τ_{opt}	$\tilde{\Lambda}_{appr.}^{300°K}$ (Å)	$\mu^{180°K}$	$\mu^{240°K}$ (cm² V⁻¹·sec⁻¹)	$\mu^{300°K}$
poly T	electron	0.072	0.36	12	189	57	30
	hole	0.274	0.09	42	297	132	82
poly A	electron	0.244	0.10	32	227	93	65
	hole	0.318	0.08	54	352	132	94
poly C	electron	0.142	0.18	15	183	77	54
	hole	0.310	0.08	41	270	116	87
poly (A–T)	electron	0.268	0.09	25	142	65	40
	hole	0.254	0.10	32	150	65	43
poly (G–C)	electron	0.142	0.18	18	146	71	42
	hole	0.268	0.09	27	254	105	77
poly (T,C)	hole	0.056	0.45	10	82	30	16
poly (T,A)	hole	0.030	0.85	6	78	18	12
poly (G,A)	hole	0.040	0.70	6.8	64	18	9.5
poly (A–T, G–C)	electron	0.046	0.55	8.5	45	16	7
	hole	0.053	0.48	8	63	18	10
poly (A–T, T–A)	hole	0.037	0.68	5.2	82	18	13
poly (G–C, C–G)	electron	0.027	0.90	5	24	8	4
	hole	0.020	1.40	3.5	28	9	4

As a second application of the variational method we present the results of mobility calculations for some more complicated periodic DNA models in Table III. [21]. In the first column we give the band widths (δE) [42], in the next one τ_{loc}/τ_{opt} stands for the ratio of the charge carrier localization time and of the period f a characteristic intramolecular vibration, $\tilde{\Lambda}_{appr.}$ gives again the free paths calculated at T=300°K with the help of the approximate relaxation time, while the last three columns contain the mobility values calculated by the use of the variational method at three different temperatures. From the results it can be seen that the DNA models can be devided into two different groups from the point of view of their transport properties. The group of the homopolynucleotides and of the poly(A-T), poly(G-C) systems has bandwidths of 0.1-0.3 eV and the corresponding mobility values lie between 30 and 100 $cm^2V^{-1}sec^{-1}$, while the bandwidths of the more complicated poly(A-T, G-C) type systems are much smaller (0.03-0.61 eV) and also their mobilities are smaller by one order of magnitude. All the mobility values follow the T^{-n} law but with n=1.5-2 for the former group and with n=2.-2.5 for the latter one. At the same time in the broad band models the carrier free paths are of 20-50 Å at 300°K, hence the band model seems to be applicable to them . For the second group, however, the $\tilde{\Lambda}_{appr}$ values are of the order of the lattice distance (a=6.72 Å), where the delocalized model becomes inadequate. These problems become even more important if we take into account also the possibility of phonon polaron formation for the very narrow-band DNA models. Such calculations have shown [43] that in this case the band model breaks completely down, the free paths become less than 1 Å and the mobilities are some tenths of $1 cm^2V^{-1}sec^{-1}$.

As an alternative to the Bloch type delocalized transport mechanism in the case of polymers with very narrow bands we have to assume that the zeroth order states are localized molecular states and the transport takes place through an uncorrelated 'hopping'-type electron motion between these states. This means that in the course of the application of perturbation theory the zeroth order Hamiltonian contains the molecular potential plus the electron-vibration interaction and we have to treat as perturbation the intermolecular (crystal field) potential, which gave rise earlier to the formation of Bloch functions. The transport becomes thus a diffusion process and the mobility is determined by the Einstein-relation $\mu=ed/k_BT$, where the diffusion constant d is given in the one-dimensional case by $d=Wa^2$. The transition probability per unit time W between two neighbouring localized states can be determined again by the Golden Rule. Here we summarize only the main results of such a calculation performed for the above mentioned very narrow band DNA models [43]:

(1) The polaron binding energies lie between 0.180 and 0.260 eV, the resulting band narrowing factor is about 0.4-0.2.

(2) The hopping mobilities are of 10^{-1}-$10^{-2}cm^2V^{-1}sec^{-1}$ thus by nearly two orders of magnitude smaller than in the band case.

(3) The temperature dependence of the mobility depends on the strength of the electron-phonon interaction. For the obtained polaron binding energies it is proportional to T^{-1}, the temperature activated nature would be observable actually only for much stronger coupling.

5. CONCLUSIONS

We hope that the preceding short review has shown at least the special problems one is confronted in studying the transport properties of condensed organic systems. It is evident that theory made in the past years only the first steps to become a helpful partner of experiment in explaining and predicting the fascinating new physical properties of these materials. The main difficulty was until now that due to the very complicated molecular- and crystal structure of organic systems their electronic structure could be described only with the help of very simple approximate Hamiltonians in which all finer details of the intra- and inter-molecular interactions were completely neglected in consequence of the phenomenological character. The combination of different molecular- and solid state physical calculations methods developed in the past decade [1] will provide in our opinion a more promising basis for these investigations. There seem to be at the moment three main fields whose more intensive theoretical

exploration could efficently contribute to the solution of the above problems:
(1). Since the transport properties of the majority of polymers and molecular crystals seem to be at least not in contradiction with the general features one expects on the basis of band theory further efforts will be necessary to get even more realiable band structures. For the quasi-1D systems they can be calculated at the *ab initio* Hartree-Fock level (such a calculation has been performed for polycytosine and is in progress in our laboratory for the TCNQ crystal). In charge transfer crystals containing large organic molecules as cations, the one-dimensional approach is unsatisfactory since the interchain polarization effects fundamentally alter the electronic structure [45]. The correct 3D *ab initio* treatment of these systems even at a minimal atomic basis level does not seem to be feasible in the next years. Therefore, semiempirical schemes have to be developed and tested with the 1D *ab initio* calculations.
(2). The vibrational properties of organic crystals have to be calculated with better approximations. The method of Lipari et. al. [24] proved to be promising for the calculation of the intramolecular modes. Other methods seem to be applicable for the more correct treatment of the intermolecular phonon modes [46]. Having more reliable phonon spectra, also the electron-phonon interaction matrix elements should be calculated on the basis of the *ab initio* electronic wave functions. The *ab initio* calculation of the matrix elements given by Eq. (3.26) could be performed essentially by the existing crystal orbital programs, and would require approximately the same computational effort as the band structure calculation itself if one uses Gaussian atomic basis (Slater-type functions would be more advantageous also in this respect). Since electronic correlation effects proved to be essential in many organic systems, the band structures had to be, at least, corrected for these effects.
(3.) In many cases the transport theory based on the band model breaks completely down. For such systems more sophisticated computational schemes will have to be applied. Furthermore, in some organic charge transfer crystals (for instance in TTF-TCNQ) experiments suggest the influence of many-body effects [47] strongly enhancing the conductivity in some temperature regions. The explanation of these phenomena will require the investigation of different collective states in these crystals.

ACKNOWLEDGMENTS

The author should like to express his sincere gratitude to Professor J. Ladik, whose interest and support gave him the opportunity to continue his research and whose suggestions basically contributed to the development of the ideas presented in this lecture.

He is further very much indebted to Miss S. Patzak for the

demanding and careful typing of the manuscript.

REFERENCES

1. For a review see: J. Ladik in "Electronic Structure of Polymers and Molecular Crystals", J.-M. André, J. Delhalle and J. Ladik, Eds.; D. Reidel Publ. Co., Dordrecht-Boston (1978).
2. For a discussion of the limitations of the Boltzmann transport theory see for instance: R. Peierls in "Transport Phenomena", G. Kirczenow and J. Marro , Eds., Springer Verlag (1974).
3. R. Kubo, J. Phys. Soc. Japan. 12,570 (1957).
4. A.H. Wilson, The Theory of Metals, Cambridge University Press (1953).
5. J.M. Ziman, Electrons and Phonons, Oxford University Press (1960).
6. A.C. Smith, J.F. Janak, R.B. Adler, Electronic Conduction in Solids, McGraw Hill (1967).
7. H. Jones and C. Zener, Proc. Roy. Soc. (London) A144, 101 (1934).
8. M. Kohler, J. Phys. 124, 772 (1948); ibid 125, 679 (1949).
9. E.H. Sondheimer, Proc. Roy. Soc. (London) A203, 75 (1950).
10. F. Garcia-Moliner and S. Simons, Proc. Cambr. Phil. Soc. 53, 848 (1957).
11. M. Born and K. Huang, Dynamical Theory of Crystal Lattices, Oxford University Press (1954).
12. L.J. Sham and J. M. Ziman in Solid State Physics, F. Seitz and D. Turnbull, Eds., Academic Press, New York (1963) Vol. 15, p. 221.
13. J. Bardeen and W. Shockley, Phys. Rev. 80, 72 (1950).
14. H. Brooks, Adv. in Electronics 7, 85 (1955); W.P. Dumke, Phys. Rev. 101, 531 (1956); C. Herring and E. Vogt, Phys. Rev. 101, 944 (1956).
15. W.L. McCubbin and F.A. Teemall, Phys. Rev. A6, 2478 (1972).
16. L.I. Schiff, Quantum Mechanics, McGraw Hill, Inc., New York (1955).
17. C. Herring, Proc. Inter. Conf. Semicond. Phys. Prague (1960), p. 60 (1961).
18. G.D. Whitefield, Phys. Rev. 121, 720 (1961).
19. L. Friedman, Phys. Rev. 140, 1649 (1965).
20. G. Del Re, J. Ladik and G. Biczó, Phys. Rev. 155, 997 (1967).
21. S. Suhai, J. Chem. Phys. 57, 5599 (1972).
22. T. Holstein, Anuals of Phys. 8, 329, 343 (1961); W. Siebrand, J. Chem. Phys. 40, 2223, 2231 (1964).
23. N.O. Lipari, C.B. Duke and L. Pietronero, J. Chem. Phys. 65 1165 (1976).
24. N.O. Lipari, C.B. Duke, R. Bozio, A. Girlando, C. Pecile and A. Padva, Chem. Phys. Lett. 44, 236 (1976).
25. A.C. Beer, in Solid State Physics, H. Ehrenreich, F. Seitz and D. Turnbull, Eds., Academic Press, New York (1963) Suppl. 4.
26. F. Beleznay, G. Biczó and J. Ladik, Acta Phys. Hung. 18, 213 (1965).
27. S.H. Glarum, J. Phys. Chem. Solids 24, 1577 (1963).

28. J. Ladik, G. Biczo and G. Elek, J. Chem. Phys. 44, 483 (1966).
29. S. Suhai, Biopolymers, to be published.
30. S. Suhai, Solid State Comm., submitted.
31. S. Suhai, Theor. Chim. Acta 34, 157 (1974).
32. S. Suhai and J. Ladik, Theor. Chim. Acta (Berl.), to be published.
33. D.D. Eley and D.I. Spivey, Trans. Faraday Soc. 56, 1432 (1961).
34. R.G. Anderson and C.J. Fritchie, 2nd Natl. Meeting of the Soc. of Applied Spectroscopy, Paper 111, San Diego (1963).
35. S. Suhai, Solid State Comm. 21, 117 (1977).
36. S.K. Khanna, A.A. Bright, A.F. Garito and A.J. Heeger, Phys. Rev. B10, 2139 (1974).
37. R.G. Kepler, J. Chem. Phys. 39, 3528 (1963).
38. S. Suhai, Physics Letters A, in the press.
39. Y. Toyozawa, Progr. Theor. Phys. (Kyoto) 12, 421 (1954); M. Inohue, C.K. Mahutte and S. Wong, Phys. Rev. C2, 539 (1970; A.B. Kunz, Phys. Rev. B6, 606 (1972).
40. S. Suhai, unpublished results.
41. A.A. Bright, P.M. Chaikin and A.R. McGhie, Phys. Rev. B10, 3560 (1974).
42. J. Ladik and G. Biczó, J. Chem. Phys. 42, 1658 (1965).
43. S. Suhai, unpublished results.
44. S. Suhai, Ch. Merkel and J. Ladik, Phys. Letters 61A, 487 (1977).
45. S. Suhai and G. Biczó, Physics Letters A, in the press. S. Suhai, Solid State Comm., submitted.
46. L. Sham, Phys. Rev. B6, 3584 (1972), P.E. Van Camp, V.E. Van Doren and J. T. Devreese, J. Chem. Phys. in the press.
47. M.J. Cohen, L.B. Coleman, A.F. Garito and A.J. Heeger, Phys. Rev. B13, 5111 (1976).

INDEX OF AUTHORS